Authenticity of Foods of Plant Origin

Books Published in *Food Biology* series

1. **Microorganisms and Fermentation of Traditional Foods**
   Ramesh C. Ray, Montet Didier (eds.), 2014
2. **Bread and Its Fortification: Nutrition and Health Benefits**
   Cristina M. Rosell, Joanna Bajerska, Aly F. El Sheikha (eds.), 2015
3. **Authenticity of Foods of Animal Origin**
   Ioannis Sotirios Arvanitoyannis (ed.), 2015
4. **Fermented Foods, Part I: Biochemistry and Biotechnology**
   Didier Montet, Ramesh C. Ray (eds.), 2015
5. **Foodborne Pathogens and Food Safety**
   Md. Latiful Bari, Dike O. Ukuku (eds.), 2015
6. **Fermented Meat Products: Health Aspects**
   Nevijo Zdolec (ed.), 2016
7. **Fermented Foods of Latin America: From Traditional Knowledge to Innovative Applications**
   Ana Lucia Barretto Penna, Luis A. Nero, Svetoslav D. Todorov (eds.), 2016
8. **Lactic Acid Fermentation of Fruits and Vegetables**
   Spiros Paramithiotis (ed.), 2016
9. **Microbial Enzyme Technology in Food Applications**
   Ramesh C. Ray, Cristina M. Rosell (eds.), 2016
10. **Acetic Acid Bacteria**
    Ilkin Yucel Sengun (ed.), 2017
11. **Fermented Foods, Part II: Technological Interventions**
    Ramesh C. Ray, Montet Didier (eds.), 2017

Food Biology Series

Authenticity of Foods of Plant Origin

Editor
Konstantinos Kotsanopoulos
University of Thessaly, School of Agricultural Sciences
Department of Ichthyology and Aquatic Environment
Hellas, Greece

CRC Press
Taylor & Francis Group
Boca Raton London New York

CRC Press is an imprint of the
Taylor & Francis Group, an **informa** business
A SCIENCE PUBLISHERS BOOK

First edition published 2022
by CRC Press
6000 Broken Sound Parkway NW, Suite 300, Boca Raton, FL 33487-2742

and by CRC Press
2 Park Square, Milton Park, Abingdon, Oxon, OX14 4RN

CRC Press is an imprint of Taylor & Francis Group, LLC

Library of Congress Cataloging-in-Publication Data (applied for)

ISBN: 978-0-367-14653-5 (hbk)
ISBN: 978-1-032-16956-9 (pbk)
ISBN: 978-0-429-05294-1 (ebk)

DOI: 10.1201/9780429052941

Typeset in Palatino
by Radiant Productions

Preface to the Series

Food is the essential source of nutrients (such as carbohydrates, proteins, fats, vitamins, and minerals) for all living organisms to sustain life. A large part of daily human efforts is concentrated on food production, processing, packaging and marketing, product development, preservation, storage, and ensuring food safety and quality. It is obvious therefore, our food supply chain can contain microorganisms that interact with the food, thereby interfering in the ecology of food substrates. The microbe-food interaction can be mostly beneficial (as in the case of many fermented foods such as cheese, butter, sausage, etc.) or in some cases, it is detrimental (spoilage of food, mycotoxin, etc.). The *Food Biology* series aims at bringing all these aspects of microbe-food interactions in form of topical volumes, covering food microbiology, food mycology, biochemistry, microbial ecology, food biotechnology and bio-processing, new food product developments with microbial interventions, food nutrification with nutraceuticals, food authenticity, food origin traceability, and food science and technology. Special emphasis is laid on new molecular techniques relevant to food biology research or to monitoring and assessing food safety and quality, multiple hurdle food preservation techniques, as well as new interventions in biotechnological applications in food processing and development.

The series is broadly broken up into food fermentation, food safety and hygiene, food authenticity and traceability, microbial interventions in food bio-processing and food additive development, sensory science, molecular diagnostic methods in detecting food borne pathogens and food policy, etc. Leading international authorities with background in academia, research, industry and government have been drawn into the series either as authors or as editors. The series will be a useful reference resource base in food microbiology, biochemistry, biotechnology, food science and technology for researchers, teachers, students and food science and technology practitioners.

Ramesh C. Ray
Series Editor

Preface

This book was written as an effort to summarise the latest developments around authenticity of foods of plant origin, the detection of adulteration and methods and techniques commonly employed to fulfil these purposes. As the advances in chemistry, chemical engineering biotechnology and traceability are rapid, there is always a need for updated books on this subject. This book does not intend to cover all existing literature or go in deep depth into individual techniques. It is rather a comprehensive summary of the most important elements and information categorised into clearly defined chapters that the reader can use to quickly refer to and get briefed on the developments around a particular category of food of plant origin or relevant subject. The book outlines the reasons around the need for the development of new and improvements on existing techniques that can verify the authenticity of foods of plant origin or detect adulteration, as well as the ever-increasing need for advanced traceability methods. Focus is then given on the most commercially and socially important fruits, vegetables and crops for the world's population and various techniques that have been employed to test them, ranging from chromatographic, spectroscopic, molecular, biochemical techniques and others, are briefly discussed. An effort is also made to briefly review the legislation around these matters and offer an insight into potential future trends. The book intends to reach food industrialists, food chemists, food quality control scientists and technologists, academicians, food microbiologists, analytical chemists, food physical chemists and everyone with an interest on the authenticity of food.

Contents

1

Traceability

Konstantinos V Kotsanopoulos[1,*] and
Parthena P Vartholomidou[2]

1. Introduction

The establishment of traceability is essential for providing safer food supplies and connecting producers and consumers. Foodborne diseases, such as bovine spongiform encephalopathy (BSE), and the availability of products that require specific labelling [e.g., genetically modified organism (GMO)] render the use of traceability mandatory, in order to control all links in the food chain (Regattieri et al. 2007). In recent years, traceability's importance has been widely recognized and the tool has been established as one of most important measures for guaranteeing food safety and quality. The development of traceability systems though can only be successfully performed via thorough understanding and re-organization of food supply chains (Dabbene et al. 2014). A traceability system of foods of plant origin across the farming and food supply chains can be defined as a documented identification of all processes involved in the production and sale of these foods. Its objective is the identification of all steps of the supply chain and the monitoring of the product as it moves through this chain. The system must be of high efficiency and effectiveness, accurately storing relevant information and making this information available as needed. Tools such as "Failure Mode Effect and

[1] School of Agricultural Sciences, Department of Ichthyology and Aquatic Environment, University of Thessaly, Fytoko St, 38446 Nea Ionia Magnesias, Volos, Hellas, Greece.
[2] School of Science & Technology, Hellenic Open University, Parodos Aristotelous 18, 26 335, Patra, Greece.
* Corresponding author: kostaskot@yahoo.gr

Criticality Analysis" (FMECA) can be used to detect possible critical points of a traceability system and propose improvements (Bertolini et al. 2006). Traceability in food manufacturing varies from in-house traceability within manufacturing sites to traceability across supply chains, from raw material cultivation to retailers. Carefully developed traceability systems are very important in order to achieve optimal benefits from quality and production control as well as to meet consumer demands (Moe 1998).

Borit and Santos (2015) stated that traceability is a tool that regulators use for managing risk in multiple supply chains, including, e.g., supply chains of GMOs. As a result, traceability can be used for supporting various claims. Schwägele (2005) noted that traceability systems are needed to provide information on the origin, processing, retailing and final destination of foods. The use of these systems can therefore increase consumers' confidence in the food market and allow the effective detection and withdrawal of hazardous food. Animal feed production as an important step in the supply chains should also be covered by a harmonized traceability system facilitating the detection of fraud and hazardous substances by authorities.

Olsen and Borit (2018) reported that food traceability has been extensively covered by both regional and national legislation, research and projects as well as a very high number of scientific articles. Their study provided a structural approach for the description and analysis of a traceability system focusing particularly on differentiating between the system mechanisms and the characteristics of the units to be traced. The differentiation was made based on practical experience and engagement in international standardization processes in relation to traceability. For the purposes of their study, a traceability system encompasses the principles, practices, and standards that are needed to establish traceability in the food industry, regardless of how these are put in place. As these researchers mentioned, practically, computerized systems are principally employed to maintain traceability and determine the units under consideration, maintain records of the fragmentation or joining of these units and record unit characteristics. The distinction between the different components is of particular importance for the description and comparison of traceability systems, as well as for introducing improvements.

The European Union has issued Directive 2000/29/EC establishing rules for the control of the movement and trade within its territory of plants and products thereof and materials that can potentially be carriers of organisms harmful to plants. These plants and plant products are listed in Part A, Annex V of this Directive. Within the European Union, the rules dictate the conduction of controls and inspections as these plants and materials are produced during the growing seasons and post-harvest, issuing official producer registrations and plant passports, that are generated once controls are finalized (https://ec.europa.eu/food/

plant/plant_health_biosecurity/trade_eu_en). For certain plants, plant products and other objects (listed in Part B, Annex V—Directive 2000/29/EC) that are imported into the European Union, phytosanitary certificates are required to certify that they have been subjected to inspections, do not carry harmful organisms and meet the plant health regulations of the member state they are imported into (https://ec.europa.eu/food/plant/plant_health_biosecurity/non_eu_trade_en). These requirements can only be met if robust traceability systems are in place to monitor the movement of these products.

An examination of traceability in the food supply chain was conducted by the Institute of Food Technologists in the United States. Data on traceability were provided by industry representatives from food and non-food companies. It was shown that generally all food companies that participated in this assessment recognized that it is very important that a robust and efficient traceability system needs to be implemented within their supply chains. Most of the companies had systems in place to keep records, although these systems ranged from manual to sophisticated electronic-based systems. Various tracing practices have been seen in the industry, possibly as a result of the complexity of the food chains. However, similarities between the companies have been found in relation to the types of data collected, the ways data were captured and how data were shared within a site and among trading partners. It is important to mention that in comparison to non-food companies, the food supply chains were significantly more complex with a very high number of contributors around the globe, handling various products, which could be combined or processed (McEntire et al. 2010).

The costs but also the advantages of traceability systems when used to trace contaminated foodstuffs were discussed by Mejia et al. (2010). Among the costs considered were those of product-tracing systems and technologies, capital equipment, personnel, supplies and materials, social costs and value to society. Their article concluded that companies that have developed and implemented robust tracing systems benefit considerably via improved supply chain management, better inventory control, better access to contracts and markets, by being able to deliver more reliable product assurances, and effective and less expensive recalls. Product tracing systems facilitate the compartmentalization and reduction of the type and amount of product that may need to be recalled. Other benefits include the protection of brand names, maintaining consumer confidence and limiting liability claims. Although the benefits are considerable, the relevant costs of implementing robust traceability systems can be high. This is more noticeable in companies processing big amounts of materials that need to be traced into final products or when paper-based systems are used.

Food safety and quality are at the heart of consumer and industry interests. Incidents of food alteration, particularly when covered by the media, have a major effect on public opinion. The demands of improving quality controls are increasing thus great efforts are made to develop accurate and reliable molecular techniques for food analysis. DNA barcoding is a molecular-based method commonly used nowadays for the identification of biological specimens and can be effective for both raw and processed foods. An assessment of DNA barcoding as a traceability tool in the food industry showed that the method is widely used. This is a result of the continuously dropping cost of molecular analyses, the increasing availability of laboratories with the right equipment and trained personnel, the increasing availability of web-based resources and the increasing number of consumers requiring food of better quality (Galimberti et al. 2013).

2. Food Labelling

Labels are used to provide certain information about the ingredients and properties of foods, and they are used to allow consumers to take better-informed decisions with regards to the products they purchase and consume (https://www.bmel.de/EN/Food/Food-Labelling/food-labelling_node.html). Within the EU, the Regulation (EU) No 1169/2011 of the European Parliament and of the Council of 25 October 2011 on the provision of food information to consumers, lays down the requirements for the information that needs to be provided on labels of pre-packed products. Article 7 of this Regulation states that the information provided should not mislead the consumers as to its origin, while Article 26 renders the labeling of the country of origin or place of provenance mandatory when its omittance could be misleading, for example when the presentation of the product (its name, pictures, etc.) may imply it originates from a different country or area (Regulation (EU) No 1169/2011). It is therefore clear that traceability is required to meet this requirement. Legislation in the United States also enforces the provision of information on the country of origin at the point of purchase for various foods of plant origin (Newman et al. 2014). Specifically, United States requires that all retailers provide information to their customers in relation to the source of certain foods of plant origin, including fresh and frozen fruits and vegetables, peanuts, pecans, and macadamia nuts and ginseng (https://www.ams.usda.gov/rules-regulations/cool).

How the mandatory and voluntary labeling requirements will affect consumers' choices is still not clear. Therefore, as labelling on the origin of products becomes more and more common around the globe, the need for additional theory-driven research increases, particularly from a macromarketing scope, to produce more generalizable knowledge on how this type of labelling can aggregate food marketing systems (Newman et al. 2014).

2.1 Identification systems (EAN, SSCC, GTN, and GLN)

2.1.1 EAN

EAN/UPC barcodes are practically used in the vast majority of consumer products. They are one of the first codes established and the most widely applied of all GS1 barcodes (https://www.gs1.org/standards/barcodes/ean-upc). The EAN/UPC group of symbols involves the GS1 barcodes. EAN/UPC barcodes can be very effectively used at a retail point-of-sale due to their design that renders them suitable for uses in high volume scanning environments. They can be used in GTIN, Restricted Circulation Numbers, coupons, and in-store codes as well as general distribution and logistics. EAN/UPC minimal sizes are increased in cases that they need to be scanned at both retail point of sales and in general distribution (https://www.gs1.org/docs/barcodes/GS1_Barcodes_Fact_Sheet-GS_%20EAN_UPC_family.pdf). The EAN barcode is very common in products sold in convenience stores and, due to its compatibility with UPC of U.S.A. and Canada and JAN of Japan, it is considered to be a universal code across the globe (https://www.keyence.com/ss/products/auto_id/barcode_lecture/basic/jan/).

2.1.2 SSCC

The SSCC is the GS1 Identification Key that is used for the identification of logistic units and its use is mandatory only in the case of logistic labels. By scanning a logistic unit's SSCC barcode, its physical movement is matched with the electronic business messages that are linked to it. This identification key consists of an Extension Digit, a GS1 Company Prefix, a Serial Reference, and a Check Digit. The SSCC can be employed for all logistic units that need to be individually identified, including container loads. It can also though be used for the identification of pallets or units when moved within a warehouse or business (https://www.nicelabel.com/resources/files/doc/resources/local/en_US/3-STILL-SSCC-labeling-logistic-units.pdf).

Therefore, the logistic units that can be identified using this code can be any type of items packed together to be stored or transported (cases, pallets, parcels, containers). The SSCC is therefore very important to maintain traceability, as it allows the unique identification of each logistic unit and its content. It can also be encoded in a barcode or EPC/RFID tag, facilitating the identification of the units as they move within the supply chain worldwide. When SSCC data is shared via EDI or EPCIS, companies can exchange information about the status of the logistic units at any time.

The use of SSCC allows companies to access additional information about the logistic unit, which can be shared via a Despatch Advice or Advanced Shipping Notice before the logistic units arrive. Upon receipt, scanning of the SSCC provides data that can accelerate the receipt of goods and any invoicing processes that follow. The SSCC is in full compliance with ISO/IEC 15459–part 1: unique identifiers for transport units, which is required for tracking and tracing logistic units in several international supply chains (https://www.gs1.org/docs/idkeys/GS1_SSCC_Executive_Summary.pdf).

2.1.3 GTIN

Companies can use the Global Trade Item Number (GTIN) for the unique identification of trade items. The GTIN can be employed for identifying types of products regardless of packaging (e.g., consumer unit, inner pack, bag, pallets). Identification of groups of trade items produced via similar processes and sharing common characteristics can be achieved using the batch/lot numbers, expiry dates, and other data. The identification of individual items can be achieved with the use of a GTIN and a serial number. The assignment of a GTIN to a trade item allows all entities of a company worldwide to uniquely identify the item and facilitates the communication of data about the item. A barcode or an EPC/RFID tag can be used to encode the GTIN. Scanning the barcode or EPC/RFID tag allows the efficient and accurate processing of products and relevant information, e.g., when products are scanned in a store, upon receipt of goods in a warehouse or site. GTINs can also be used for the unambiguous identification of products online, for example in catalogues, purchase orders or invoices (https://www.gs1.org/docs/idkeys/GS1_GTIN_Executive_Summary.pdf).

2.1.4 GLN

The Global Location Number (GLN) can be employed for the identification of the location of companies, allowing them to be fully flexible. These physical locations can include a store, a warehouse, or a berth in a port. The level of identification can be anything from a corporate entity to a company's legal and functional entities that may engage as parties in a business transaction (e.g., buyer, seller, transporter, etc.). The GLN can be incorporated in either a barcode or EPC/RFID tag to identify locations including, for example, the destination or origin of a product. It can be easily used in electronic messages or databases to provide information to traders about organisations, their GLNs and related GLN data. The GLN extension component can be incorporated to a GLN to allow data to be more accurately shared and recorded. For example, an extension component could be used for the identification of sub-locations such as storage bins or scan and read points. In the public sector, GLN is used by governments for the identification of their agencies or companies that need access to government databases. GLN is fully compatible with the ISO/IEC 6523 standard and its international code designator is: "0088" (https://www.gs1.org/docs/idkeys/GS1_GLN_Executive_Summary.pdf).

2.1.5 Batch numbers/codes

The assignment of batch or lot numbers is of paramount importance for traceability. The idea is that all items have the same characteristics in all respects, including composition, properties, etc., and are covered by the same manufacturing practices and risk assessments. That means that ingredients sourced from different suppliers should have different batch numbers even if all other characteristics are the same. This allows companies to effectively trace a product in cases of quality or safety concerns associated with it. Therefore, in these cases, products with the same batch code can be recalled or withdrawn if required. As a result, it is obvious that the smaller the batch, the less expensive the recall for a manufacturer (https://blog.matthews.com.au/speaking-code-5-codes-every-manufacturer-must-know/). A batch number or lot number is therefore an identification code usually generated by a manufacturer, to allow the identification of small quantities of a product. It can be linked to the date of production, the source of raw materials or the machinery used for the production of a product. The batch number may also be related to a production plan which is commonly controlled by an MRP (material requirements planning) system. The effective tracking of lot numbers is now an absolute requirement of retailers when food for human consumption is supplied (https://www.claruswms.co.uk/lot-batch-number-warehouse-tracking/). A batch, according to the US FDA, is a specified quantity of a drug or other material that, within certain determined limits, has uniform qualitative characteristics and properties, and has been manufactured in a single manufacturing cycle. The homogeneity of the product should be possible to be demonstrated (https://www.lfatabletpresses.com/articles/why-batch-produce-use-batch-numbers).

2.2 Radio frequency identification (RFID): Theory, functioning principles and case studies

In 2001, Sarma et al. (2001) had noted that remotely scannable, low cost integrated circuit chips capable of accessing large amounts of constantly updated data were expected to be commercially developed. These researchers noticed that the development of new technology and the reduction of the price of chips would allow the widespread use of automatic data and identification systems using the Internet. A few years later, in 2004, Weis et al. (2004) reported that similarly to several other technologies, low-cost Radio Frequency Identification (RFID) systems would become very common and would be used in everyday consumer items in the form of "smart labels". They did however also mention that although this technology would lead to increased productivity and efficiency, it may also create new security and privacy threats for individuals and organizations.

Although the technology was first conceived in 1948, it was developed throughout the years to the point where it is affordable enough and reliable for widespread use. Its use though has indeed raised concerns over privacy and security (Roberts 2006). Nowadays, RFID is a well-known wireless application used for traceability, logistics, and access-control purposes. It is now widely used in the industry and as a standardized technology. It offers a number of advantages, such as unitary identification, wireless communication, and low cost of tags (Duroc and Tedjini 2018). In the study of Adhiarna et al. (2013), a comprehensive framework was presented that aimed at identifying dynamic RFID adoption and diffusion from three points of view: adoption levels, degree of analysis, and domain of issues. The study focused mainly on adoption levels, covering three phases linked to the quality of the RFID project and how sophisticated the relevant applications and the RFID technology were. Different levels of analysis were used ranging from an organization level to an industry and country levels. The dimensions of RFID issues were analyzed following an approach based on strategy, technology, organization, people and environment (STOPE). All available literature was reviewed to identify a number of RFID success factors and the Delphi method was employed to find positions of these factors within the framework. It was revealed that certain factors belonged to early adoption levels, while others belonged to later stages of adoption. When the country level was considered, factors such as RFID national policy, R&D policy and income per capita were generally acceptable at the preliminary and intermediate stage. Practical implications of the framework were investigated via the conduction of a case study of Indonesia at all levels of analysis. An analytic hierarchy process was employed for the identification of the main factors. At the country level, RFID policy, vision of leadership and RFID potential market were identified as the factors of highest importance, while strategy and environment were the most important domains.

Abad et al. (2009) published a study that aimed at validating an RFID smart tag designed to enable real-time traceability and cold chain monitoring for food products. The system was compiled of a smart tag and a commercial reader/writer. The smart tag is attached on the product to be traced and has integrated light, temperature and humidity sensors, a microcontroller, a memory chip, low power electronics as well as an antenna that allows the RFID data to be communicated as required. These sensor data and the traceability information can be stored in the memory. The reader/writer's role is to read and write data on the smart tag and has a wireless reading capability of 10 cm. The study demonstrated that this system has several advantages over conventional traceability systems or other commonly used temperature loggers such as more memory, reusability, potential automation, no need of tag visibility for reading, the potential of reading many tags simultaneously and greater resistance to humidity and environmental conditions. Therefore, the study once again proved the significance and potential of the technology when used for the traceability of foodstuffs.

3. Protected Geographical Origin and Indication

Europe's food and drink products reflect the cultural diversity and production capabilities of the region. European geographical indication schemes have been developed for the protection of this expertise on certain production techniques, their authenticity and the importance of particular agro-environmental conditions. Geographical indications can refer to protected geographical indications (PGI) or protected designation of origin (PDO). The aim of these schemes is the protection of the name of products that originate from certain regions and are produced following particular traditional production methods. In the case of PDOs, the raw materials used for the production of the products can only be sourced from the region of origin where the whole production has to be carried out. In the case of PGIs one the other hand, at least one production stage needs to take place in that particular region (https://ec.europa.eu/info/food-farming-fisheries/food-safety-and-quality/certification/quality-labels/eu-quality-food-and-drink_en).

3.1 (Protected) geographical origin [(P)GO]

The globalization of the food industry and the modern technology that allows food commodities to be easily transported through and between countries and continents have raised concerns over the origin of food. A lot of research has been conducted in the past few years on the use of natural isotope variation and the abundance and concentrations of certain elements, as geographical indicators for determining the provenance of food. Systematic global variations of stable hydrogen and oxygen isotope ratios in conjunction with levels of certain elements, including heavy isotope variations (e.g., strontium-87) and other biogeochemical indicators have been investigated. Although isotope ratio analysis is very promising in establishing the geographical origin of foodstuffs, other techniques such as ^{1}H NMR should also be considered as they allow additional information to be generated. As databases are developed and the number of samples analyzed increases, the interpretation of data becomes more and more complex as there is considerable overlap in measured parameters. It is therefore of paramount importance to evaluate the repeatability and reproducibility of multi-element and multi-isotopic analyses, as well as the uncertainty of the measurements. A better understanding of the traces that meteorological and geochemical conditions leave on foods, could allow the development of isotopic and multi-element maps that could in turn be used to verify the traceability of foodstuffs (Kelly et al. 2005). It is clear that consumer requirements for quality and cultural identification have led to a growing demand for foods of plant origin that are clearly identified with a particular geographical region. Many of these foodstuffs are currently protected by European Legislation (Loureiro and McCluskey 2000, Zhao et al. 2014).

The study of Profeta et al. (2010) aimed at evaluating the use of the European system of PGOs in accordance with Council Regulation (EC) 510/06 (which is no longer in force) using Germany as an example to demonstrate how it was implemented as well as to evaluate how it could be further developed. The study predicted that the number of products covered by the system would exceed 1,000. By then, the system had not been extensively applied in Germany but an increase in products covered in that country was already expected.

3.2 (Protected) geographical indication [(P)GI]

A geographical indication (GI)—one of the intellectual property rights—is a sign used on foodstuffs that originate from a specific geographical area and have particular qualitative characteristics and a reputation that renders them essentially attributable to that origin (WIPO 2012). Generally, GIs focus on qualitative characteristics, local identity and traditions, as, for example, in the cases of Champagne, armagh bramley apples, Fenland celery, etc. GIs allow for distinct and high value-added products to be provided to local producers in rural areas while they are in high consumer demand (Dogan and Gokovali 2012). According to Tashiro et al. (2018), the internal processes of geographical indications (GIs) by applicants in certain rural regions have not yet been explored, although GI is a form of retaining long-term accumulated knowledge in the manufacture of traditional ethnic foods.

One of the main objectives of GIs is to foster rural development. A very low number of studies, though, have quantitatively examined the extent of the effectiveness of this policy in achieving its purpose. The current literature focuses instead on particular GIs, examined in case studies, which aim at identifying which parameters affect the success or failure of certain initiatives. An effort was therefore made to quantify the impact of such policy on an indicator of rural development: the added agricultural value. The assessment of the impact was made by building an index, measuring the number of GI schemes implemented at the NUTS3 level in the Italian regions. The implementation of a fixed effect estimator was then carried out following a difference-in-difference evaluation strategy based on an explicit theoretical model. The model and its variables were chosen using a directed acyclic graph. It was demonstrated that there was an overall positive effect of GI protection on added agricultural value in Italy, thus indicating that there was a positive impact of the European policy on rural development (Cei et al. 2018).

4. Conclusions

The traceability in the food industry has become an absolute priority. It is of paramount importance that consumers trust that the supply chains are transparent and secure and can support their demands for healthier food of higher quality and improved nutritional value. It is, however, equally important for manufacturers, as it is an important tool for monitoring the quality of raw materials introduced into the supply chains, allowing certification and accreditation of products, facilitating the removal of sub-standard food products from the market, and allowing the implementation of control systems to prevent fraud and unfair competition (Espiñeira and Santaclara 2016). It is therefore clear that the implementation of effective traceability systems in the food industry and supply chains related to foods of plant origin is not only a legal requirement but also a valuable measure of reducing food safety risks and protecting the rights of producers, distributors, retailers and consumers. A series of tools have been developed that facilitate traceability, with barcodes and electronic systems currently being the most commonly used. Traceability is not only a method of facilitating food quality and safety assurance but is also a tool against food fraud and adulteration.

References

Abad, E., F. Palacio, M. Nuin, A. González de Zárate, A. Juarros, J.M. Gómez and S. Marco. 2009. RFID smart tag for traceability and cold chain monitoring of foods: Demonstration in an intercontinental fresh fish logistic chain. Journal of Food Engineering, 394–399.

Adhiarna, N., Y.M. Hwang, M.J. Park and J. Jeung Rho. 2013. An integrated framework for RFID adoption and diffusion with a stage-scale-scope cubicle model: A case of Indonesia. International Journal of Information Managenment, 378–389.

Bertolini, M., M. Bevilacqua and R. Massini. 2006. FMECA approach to product traceability in the food industry. Food Control, 137–145.

Borit, M. and J. Santos. 2015. Getting traceability right, from fish to advanced bio-technological products: a review of legislation. Journal of Cleaner Production, 13–22.

Cei, L., G. Stefani, E. Defrancesco and G.V. Lombardi. 2018. Geographical indications: A first assessment of the impact on rural development in Italian NUTS3 regions. Land Use Policy, 620–630.

Cozzolino, D., W. Cynkar, N. Shah and P. Smith. 2011. Technical solutions for analysis of grape juice, must, and wine: the role of infrared spectroscopy and chemometrics. Analytical and Bioanalytical Chemistry, 1475–1484.

Culbert, J., D. Cozzolino, R. Ristic and K. Wilkinson. 2015. Classification of sparkling wine style and quality by MIR spectroscopy. Molecules, 8341–8356.

Dabbene, F., P. Gaya and C. Tortia. 2014. Traceability issues in food supply chain management: A review. Biosystems Engineering, 65–80.

Dixit, V., J.C. Tewari, B. Cho and J.M.K. Irudayaraj. 2005. Identification and quantification of industrial grade glycerol adulteration in red wine with fourier transform infrared spectroscopy using chemometrics and artificial neural networks. Applied Spectroscopy, 1553–1561.

Dogan, B. and U. Gokovali. 2012. Geographical indications: the aspects of rural development and marketing through the traditional products. Procedia-Social and Behavioral Sciences, 761–765.

Duroc, Y. and S. Tedjini. 2018. RFID: A key technology for Humanity. Comptes Rendus Physique, 64–71.

Espiñeira, M. and F.J. Santaclara. 2016. 1-What is food traceability? Advances in Food Traceability Techniques and Technologies, 3–8.

Galimberti, A., F. De Mattia, A. Losa, I. Bruni, S. Federici, M. Casiraghi, S. Martellos and M. Labra. 2013. DNA barcoding as a new tool for food traceability. Food Research International, 55–63.

https://ec.europa.eu/food/plant/plant_health_biosecurity/non_eu_trade_en Accessed 20 August 2019.

https://ec.europa.eu/food/plant/plant_health_biosecurity/trade_eu_en Accessed 21 July 2019.

https://ec.europa.eu/info/food-farming-fisheries/food-safety-and-quality/certification/quality-labels/eu-quality-food-and-drink_en Accessed 20 August 2019.

https://www.ams.usda.gov/rules-regulations/cool Accessed 05 December 2019.

https://www.bmel.de/EN/topics/food-and-nutrition/food-labelling/food-labelling_node.html.

https://www.gs1.org/docs/barcodes/GS1_Barcodes_Fact_Sheet-GS_%20EAN_UPC_family.pdf Accessed 05 December 2019.

https://www.gs1.org/docs/idkeys/GS1_GLN_Executive_Summary.pdf Accessed 05 December 2019.

https://www.gs1.org/docs/idkeys/GS1_GTIN_Executive_Summary.pdf Accessed 05 December 2019.

https://www.gs1.org/docs/idkeys/GS1_SSCC_Executive_Summary.pdf Accessed 05 December 2019.

https://www.gs1.org/standards/barcodes/ean-upc Accessed 05 December 2019.

https://www.keyence.com/ss/products/auto_id/barcode_lecture/basic/jan/Accessed 05 December 2019.

https://www.nicelabel.com/resources/files/doc/resources/local/en_US/3-STILL-SSCC-labeling-logistic-units.pdf Accessed 23 July 2019.

Kelly, S., K.I. Heaton and J. Hoogewerff. 2005. Tracing the geographical origin of food: The application of multi-element and multi-isotope analysis. Trends in Food Science & Technology, 555–567.

Kumšta, M., P. Pavloušek and P. Kárník. 2014. Use of anthocyanin profiles when differentiating individual varietal wines and terroirs. Food Technol. Biotechnol., 383–390.

Loureiro, M.L. and J.J. McCluskey. 2000. Assessing consumer response to protected geographical identification labeling. Agribusiness. An International Journal, 309–320.
McEntire, J.C., S. Arens, M. Bernstein, B. Bugusu, F.F. Busta, M. Cole, A. Davis, W. Fisher, S. Geisert, H. Jensen, B. Kenah, B. Lloyd, C. Mejia, B. Miller, R. Mills, R. Newsome, K. Osho, G. Prince, S. Scholl, D. Sutton, B. Welt and S. Ohlhorst. 2010. Traceability (product tracing) in food systems: an IFT report submitted to the FDA, Volume 1: technical aspects and recommendations. Comprehensive Reviews in Food Science and Food Safety, 92–158.
Mejia, C., J. McEntire, K. Keener, M.K. Muth, W. Nganje, T. Stinson and H. Jensen. 2010. Traceability (Product Tracing) in food systems: An IFT report submitted to the FDA, Volume 2: Cost considerations and implications. Comprehensive Reviews in Food Science and Food Safety, 9: 159–175.
Moe, T. 1998. Perspectives on traceability in food manufacture. Trends in Food Science and Technology, 211–214.
Newman, C.L., A.M. Turri, E. Howlett and A. Stokes. 2014. Twenty years of country-of-origin food labeling research: A review of the literature and implications for food marketing systems. Journal of Macromarketing, 505–519.
Olsen, P. and M. Borit. 2018. The components of a food traceability system. Trends in Food Science & Technology, 143–149.
Profeta, A., R. Balling, V. Schoene and A. Wirsig. 2010. Protected geographical indications and designations of origin: an overview of the status quo and the development of the use of regulation (EC) 510/06 in Europe, with special consideration of the german situation. Journal of International Food & Agribusiness Marketing, 179–198.
Regattieri, A., M. Gamberi and R. Manzini. 2007. Traceability of food products: General framework and experimental evidence. Journal of Food Engineering, 347–356.
REGULATION (EU) No 1169/2011 OF THE EUROPEAN PARLIAMENT AND OF THE COUNCIL of 25 October 2011 on the provision of food information to consumers, amending Regulations (EC) No 1924/2006 and (EC) No 1925/2006 of the European Parliament and of the Council, and repealing Commission Directive 87/250/EEC, Council Directive 90/496/EEC, Commission Directive 1999/10/EC, Directive 2000/13/EC of the European Parliament and of the Council, Commission Directives 2002/67/EC and 2008/5/EC and Commission Regulation (EC) No 608/2004. OJ L 304, 22.11.2011, p. 18.
Roberts, C.M. 2006. Radio frequency identification (RFID). Computers & Security, 18–26.
Sarma, S., D. Brock and D. Engels. 2001. Radio frequency identification and the electronic product code. IEEE Micro, 50–54.
Schwägele, F. 2005. Traceability from a European perspective. Meat Science, 164–173.
Tashiro, A., Y. Uchiyama and R. Kohsaka. 2018. Internal processes of geographical indication and their effects: an evaluation framework for geographical indication applicants in Japan. Journal of Ethnic Foods, 202–210.
Weis, S.A., S.E. Sarma, R.L. Rivest and D.W. Engels. 2004. Security and privacy aspects of low-cost radio frequency identification systems. Security in Pervasive Computing, 201–12.
Zhao, Y., B. Zhang, G. Chena, A. Chen, S. Yang and Z. Ye. 2014. Recent developments in application of stable isotope analysis on agro-product authenticity and traceability. Food Chemistry, 300–305.

2

Authenticity

Quality Control of Agricultural Produces

Konstantinos V Kotsanopoulos[1,*] and
Parthena P Vartholomidou[2]

1. Introduction

Food fraud should be considered a crime, although the investigations and the types of threats are very different when there are human criminal adversaries rather than, for example, microorganisms, such as *E. coli*, to be considered (Spink 2019). An authentic food product is a product that is what it claims to be. The word authentic is therefore used to describe a product that is not false or copied and it is genuine and real. Authentic food is a food product that exactly matches its description and it also meets a consumer's reasonable assumption of its character (https://www.foodfraudadvisors.com/horizon-scanning-for-food-fraud/). Food authentication testing is the evaluation of the authenticity of a product. Food fraud on the other hand is committed when foodstuffs or packs are intentionally substituted, added, tampered with or misrepresented; or incorrect or misleading statements claims are made, having economic gain as the ultimate target. Food fraud mitigation refers to reducing the impact of fraud after it occurs. Food fraud can cover a wide range of actions—from economic profit-motivated adulteration to counterfeiting,

[1] School of Agricultural Sciences, Department of Ichthyology and Aquatic Environment, University of Thessaly, Fytoko St, 38446 Nea Ionia Magnesias, Volos, Hellas, Greece.
[2] School of Science & Technology, Hellenic Open University, Parodos Aristotelous 18, 26 335, Patra, Greece.
* Corresponding author: kostaskot@yahoo.gr

sabotage, and mislabeling. Food authenticity is a process used to prove that a food or food ingredient is not adulterated. The biggest concerns resulting from food fraud are related to the safety of the food that has been adulterated. A good example of a food incident resulting from food fraud is the melamine scandal related to infant formula. It is therefore clear that the safety of adulterated food cannot be guaranteed (https://www.qualityassurancemag.com/article/food-authenticity/, n.d.). Pustjens et al. (2016) reported that food fraud is a major issue, which has grown significantly due to globalization and the difficulty in detecting it and punishing the offenders. Even though economically motivated, it can also pose a risk to the health of consumers. Food adulteration or authenticity need to be verified using analytical methods that can also be used to build upon robust food safety management processes. It is to be expected though that due to the costs associated with analyses, a limited number of samples can be taken and examined and therefore methods for on-line rapid testing are needed to allow routine checks to be implemented.

The origin of food is also a factor that needs to be taken into account when its authenticity is evaluated, especially when claims are made. In the study of Armenta and de la Guardia (2016) state of the art analytical methodologies were employed to authenticate the origin and composition of foodstuffs covered by claims of protected designation of origin for European labels. The purpose of these labels is to guarantee the origin and good production practices of foods. A number of target analytes and analytical methodologies, together with chemometrics allowed the discrimination of foods originating from different geographical areas or produced using different production methods.

Moore et al. (2012) reported that although food fraud and adulteration in their current forms are emerging problems, a comprehensive compilation of information about the most vulnerable ingredients and detection methods did not exist. They therefore made an effort to collect such information from available published studies and general media and organize it into a database. Analysis and review of these data were also performed, aiming at identifying trends. The database that was created included 1305 records, including 1000 records with methods of analysis collected from 677 references. Olive oil, milk, honey, and saffron were the most vulnerable foodstuffs from an adulteration point of view, and potentially harmful issues reported included the use of lead chromate and lead tetraoxide to dilute spices, Chinese star anise substituted with toxic Japanese star anise, and melamine addition in foods with high levels of protein. High performance liquid chromatography (HPLC) and infrared (IR) spectroscopy were the most commonly used analytical techniques employed to identify adulterants, while chemometric analysis was widely used to analyze data.

2. The Importance of Authenticity

During the last few years, the number of fraud and adulteration cases has increased significantly. The potential effects on human health and the impact on the markets have rendered these practices a very important subject of study for scientists. A report on potential biomarkers of food authenticity and safety was therefore published focusing on biomarkers that can be used to specify different botanical or geographical origins, genetic diversity or production systems, molecular evidence of food adulteration or spoilage. This was the first report that presented a combination of results from recent studies in a format that allows an easy overview of metabolites (< 1200 Da) and potentially molecular routes that can be used for food authentication and safety purposes (Medina et al. 2019).

As the detection of food fraud at a global level can be achieved via analytical testing, the analysis of the samples of interest and the comparison of their results to results contained within a food authenticity database is a common approach. It was therefore suggested in the FoodIntegrity EU project that guidance should be provided for the creation of these food authenticity databases. Amongst the most important considerations that should be taken into account when creating such databases are the areas of database scope, analytical methodology, sampling, collection and storage of data, validation and curation (Donarski et al. 2019).

The authenticity of food is very important for consumers, regulators and manufacturers, as fraudulent activities can have a detrimental effect on consumer confidence and render products unsafe, while they can also lead to unfair competition. Next Generation Sequencing (NGS) applications with NGS platforms and sequence databases could potentially be promising to assist in product authenticity. A range of NGS platforms are available and are characterized by different sequence read lengths and outputs to suit different analyses needs. Nevertheless, more and more platforms are being constantly developed and improvements such as reduced error rates could allow their expanded use for food authenticity purposes. The increasing use of DNA sequencing has led to an equally rapid enlargement in sequence databases, and the development of contemporaneous, authenticated databases can be of high importance for the application of NGS to authenticity. This type of application requires robust quality control criteria and testing schemes and therefore a lot of research is currently conducted towards this direction. It therefore seems that NGS is a strong candidate to be used as a supporting tool or even as the main technology to verify regulatory compliance and protect the reputation of brands in potential food fraud situations, especially when complex food matrices are examined (Haynes et al. 2019).

Classification methods are of paramount importance in food authenticity studies. These methods are used to classify unknown food samples to known categories. The development of a classification method can be carried out by estimating the classification rule, using both the labelled and the unlabeled data, in contrast with a number of classical methods that make use of labelled data for estimations. Using this methodology, the data can be modeled as they arise from a Gaussian mixture model with parsimonious covariance structure, similar to the case of model-based clustering. Such methods were used to analyze spectra of food products recorded over the visible (Vis) and near infrared (NIR) wavelength range in food authenticity studies. This classification method yielded very low misclassification rates. The correct classification rate was 15% higher than the correct classification rate for model-based discriminant analysis (Dean et al. 2006).

It has been estimated that food fraud and adulteration cost the global food industry €30 billion per year. Recent food scandals have raised concerns over public health and triggered increasing calls on health service resources. A number of initiatives have been started globally, regionally and at national levels, making it hard to maintain a concrete view. There is also no integrated approach currently to cover analytical, regulatory, traceability (vulnerability), risk assessment and management, defense, and other factors. Such an approach would require a combination of adequate tools for vulnerability assessment and analytical tools that would allow the risk of food fraud to be minimized. Only the combined use of these factors could render a procedure applicable to the needs of both producers and authorities. The impact of such a procedure will be substantial for both public and private sectors. This new initiative should allow science to progress as well as allow the identification of the best integrated practices to ensure safe, genuine products and prevent food adulteration. The development of a consolidated view on the suitable mitigation of food fraud is beneficial not only for the food industry, but also for regulatory agencies and academics (http://ilsi.eu/task-forces/food-safety/authenticity-of-food-potential-new-task-force/). Extra Virgin Olive Oil, for example, has frequently been found not to contain what is stated on its labels. It remains one of the most frequently adulterated foods in the world. It has been estimated that around 6 out of 11 olive oils are mislabeled, while 69% of olive oil products labeled as extra virgin do not meet this quality standard. Fraudsters substitute part of the whole olive oil amount with cheaper oils (https://www.vaneerden.com/fake-food-scandal/).

As Anklama and Battaglia (2001) noted, food safety and quality control systems aim at protecting consumers. Taking into account that risks can never be completely eliminated, controls are needed to allow the accuracy of information provided on labels to be verified and food safety to be safeguarded. Historically, there has always been an interest in food safety and quality, while the risk of adulteration has existed since food was first traded. Spink and Moyer (2011) also reported that food fraud, including economically motivated adulteration, is a risk gaining increasing recognition and a major cause of concern. Regardless of the type of fraud, adulteration of food is the responsibility of both the industry and the governments. Food fraud is an intentional act aiming at economic gain, while a food safety incident is unintentional. Economically motivated adulteration may have purely economic gain motives but the related public health risks can frequently be more severe than traditional food safety risks, as the contaminants can be unconventional. Current intervention systems have not been developed to search for a near infinite number of potential adulterants. The aim of this study was to develop a base reference document that could be used to define food fraud, focusing particularly on public health threats, and allow a more proactive approach to be implemented. This could then provide a framework for future research. Deconstruction of fraud was conducted using criminology and behavioral science (the crime triangle and the chemistry of the crime). The study provided a food risk matrix and allowed the identification of food fraud incident types. But as Ruth (2019) noted, how vulnerable a company is, depends on a high number of factors and therefore supply chain mapping is one of the basic measures against fraud.

3. Representative Examples of Major Authenticity Scandals

Global incidences of food mis-description and adulteration are becoming more frequent and international food trade faces more than ever challenges over food safety and quality requirements. It is therefore necessary that continuous monitoring together with improved detection methodologies and stringent regulations are used to minimize authentication issues in the future (Premanandh 2013). A large number of food-related fraud scandals—related to foods of plant origin—have been revealed in the past few years. Information on some of the most important of them that have impacted the industry and consumer confidence, are described below.

3.1 der Weinskandal

The scandal was revealed in 1985. It was found that a number of wineries in Austria used diethylene glycol to give body to their wines and make them sweeter. The incident is also known as "the antifreeze scandal" due to the fact that diethylene glycol is a common ingredient in antifreeze products. The substance is not fit for human consumption and its addition to wine is illegal. An investigation was conducted following an anonymous tip to authorities in Vienna. Prior to this, no tests for diethylene glycol were being carried out. The authorities had found it highly suspicious that a wine-maker tried to offset an excessively large amount of antifreeze against tax, while only possessing a small tractor. Following the investigation, several wine-makers were arrested and millions of liters of wine were destroyed. The incident had a serious, detrimental long-term effect on the wine industry of Austria, with exports returning to the same levels only after 16 years. It also resulted in Austria implementing some of the strictest wine regulations worldwide, which included the introduction of the red and white quality stamp on every bottle (https://secretvienna.org/a-scandal-made-of-antifreeze-and-wine/). In the short-term following the scandal, Austrian wine exports collapsed. The impact to the industry was such that the number of wine control officers increased drastically while a new wine law was introduced (Reinhard 2010).

3.2 Bihar's school food poisoning scandal

In 2013, 22 children died and many more fell sick after eating lunch at a school in the village of Dharmasati Gandaman, in Bihar (https://www.bbc.co.uk/news/world-asia-india-23337445). Early tests indicated that the food could have been contaminated with pesticides used on rice and wheat locally (https://www.theguardian.com/world/2013/jul/17/school-meals-kill-indian-children). Forensic tests revealed that the meal had been contaminated with monocrotophos, a highly toxic pesticide, highly poisonous if consumed that is banned in many countries, including the EU and USA (https://www.theguardian.com/world/2013/jul/24/indian-food-poisoning-school-headteacher-arrested).

3.3 Salmonella poisoning from peanuts—USA 2008 and 2009

A major outbreak of salmonella poisoning was recorded in the US in 2008 and 2009. Nine people died and over 700 were reported ill. A peanut processing plant in Georgia was identified as the source of the outbreak. Soon after the plant was shuttered and liquidation of the company followed (https://www.nytimes.com/2015/09/26/opinion/joe-nocera-of-peanuts-and-prosecutions.html?rref=collection%2Ftimestopic%2FPeanut%20Corporation%20of%20America&action=click&contentCollection=timestopics&region=stream&module=stream_unit&version=latest&contentPlacement=2&pgtype=collection). It had been reported by the centers for disease control and prevention that as of 9 pm EDT, Monday, April 20, 2009, 714 people across 46 States had been infected with the Typhimurium strain of Salmonella (https://www.cdc.gov/salmonella/2009/peanut-butter-2008-2009.html). While the issue was ongoing, a high number of consumers simply stopped purchasing peanut-containing foodstuffs, as they were unable to identify which products had been affected and which ones could safely be consumed. As a result, a number of companies trading peanut products, including companies that had not been involved in the supply chain of the contaminated materials, suffered severe consequences. The fact that consumers stopped purchasing peanut-containing products was also triggered by an advisory published by the Food and Drug Administration and the Center for Disease Control and Prevention suggesting that consumers should stop eating these products if they had doubts about their safety. According to The New York Times, peanut butter sales decreased by 25% during the period of the crisis (Terndrup 2010).

3.4 Epidemic dropsy

Epidemic dropsy has been described as an acute non-infectious disease resulting from the consumption of oil adulterated with *Argemone mexicana* seed oils. Its occurrence is more common in areas where mustard oil is typically used as a cooking medium (North, North-Western and North-Eastern states of India) (https://www.nhp.gov.in/disease/non-communicable-disease/epidemic-dropsy). Sanguinarine and dehydrosanguinarine are toxic alkaloid compounds of *Argemone* oil, which can lead to capillary dilatation, proliferation and enhanced capillary permeability. Oedema has also been observed as a result of plasma substances with high protein levels leaking into the extracellular compartment. Haemodynamically, this vascular dilatation and permeability causes hypovolemia. The symptoms are initially gastroenteric, while cutaneous erythema and pigmentation, coughing, and difficulty in breathing can follow and can also progress to frank right-sided congestive cardiac failure, anaemia, hypoproteinaemia, renal azotemia, retinal haemorrhages, and glaucoma (Sharma et al. 1999). The epidemic in 1998 at New Delhi was the largest known outbreak (https://www.nhp.gov.in/disease/non-communicable-disease/epidemic-dropsy) with 2552 cases reported and 65 deaths recorded between 5 August and 12 October. It is likely that patients who presented mild symptoms did not report them and therefore the actual numbers of individuals affected may be significantly higher (Sharma et al. 1999).

4. Development of Advanced Techniques

Analytical testing is of paramount importance for the detection of fraud and for food authentication purposes, but it unfortunately cannot be employed for the identification of all food fraud cases. It is only one of the parts of the overall strategy towards the mitigation of fraud risk. Several modern tests rely on the comparison of a pattern of measured values in a test sample to patterns obtained from a database of authentic samples. How the results are interpreted depends heavily on how robust the data of the database are, and whether they cover all potential authentic variables and types of samples. This is also information that a laboratory may not be willing to release. The interpretation of such data is usually not definite, and analytical results are frequently used to suggest whether further analysis is required (including unannounced audits or mass-balance checks) instead of allowing a compliance decision to be made (https://www.ifst.org/resources/information-statements/food-authenticity-testing-part-1-role-analysis). Food Fraud becomes more and more difficult to detect as the sophistication of methods fraudsters use increases. There is therefore an urgent need for the development of advanced analytical techniques such as isotopic analysis, molecular biology and other physicochemical and biological methods to test for food authenticity and detect non-compliant products (https://www.eurofins.co.uk/food-testing/food-authenticity/). The employment of innovative analytical techniques in the food science sector follows the increased consumers' concern about the content and safety of the food they consume. Furthermore, the Nutrition Labeling and Education Act in the USA, European regulations in the EU (e.g., Regulation EC 258/97 or EN 29000 and subsequent issues), as well as the Montreal Protocol have had a significant impact on food laboratories. As a result, more time and cost-effective, but also powerful and cleaner analytical procedures are needed by food scientists, regulatory agencies, and quality control departments to meet these requirements. The development of such analytical techniques is not easy, as a number of problems need to be resolved in each case, while the methods should be adequate for use in processing, quality control and for ensuring compliance with food and trade laws (Ibañez and Cifuentes 2001).

Both conventional and emerging analytical methods are used for the identification of adulteration. The techniques can be based, among others, on sensory, physicochemical, DNA-based, chromatographic and spectroscopic principles and are often used in combination with chemometrics. Despite continuous advances, there is always an urgent need for more sensitive and widely applicable methodologies to cover all possible aspects of food adulteration (Hong et al. 2017). Based on the study of Herrero et al. (2012), research in food science is in a transition period from established methods to sophisticated analytical strategies, where Mass Spectrometric (MS)-based methods are of paramount importance. Danezis (2016) suggested that techniques based on chromatographic and molecular methods are major approaches to the verification of food authentication and are widely implemented. Isotopic, vibrational, Ultraviolet (UV) and fluorescence spectroscopy, elemental techniques and Nuclear Magnetic Resonance (NMR) are also of high importance. Other technologies such as non-chromatographic MS, microbial fingerprinting and sensory analysis, have not been fully exploited yet, but are expected to find a range of applications in the near future. Cross scientific collaboration is therefore one of the strongest tools the industry and governments now have (Danezis 2016). Fourier transform infrared (FTIR) spectroscopy in combination with suitable chemometric techniques are good candidates for authenticating high-value edible oils (Rohman and Man 2012) while DNA barcoding, allows an alternative approach to be followed for rapid and reliable plant species identification. DNA barcodes can be successful in their field if the barcode loci have sufficient information to allow the unambiguous differentiations between closely related plant species and the discovery of new cryptic species. In the field of herbal plant identification, matK, rbcL, trnH-psbA, ITS, trnL-F, 5S-rRNA and 18S-rRNA are DNA barcodes that have been successfully used. Novel techniques based on DNA barcoding coupled with next-generation sequencing and high-resolution melting curve analysis now allow species to be successfully identified in samples of finished herbal products. Moreover, development of multilocus methods have opened new horizons on the DNA barcode-based plant identification (Mishra 2016).

5. Role of Retailers, Authorities and Governments

5.1 Retailers/Manufacturers

Food fraud is the combined result of motivated offenders, existing opportunities, and ineffective controls. The supply chain may be less or more vulnerable to fraud at its different stages, but no insights exist on prime fraud drivers and enablers. Across the supply chain, the wholesale/traders appear to be the most vulnerable, while the retailers and processors follow (van Ruth et al. 2018). It is clear that a higher level of cohesiveness and consistency is needed across the industry in how food fraud vulnerability risks are assessed in order to address both intrinsic and extrinsic risk factors. The differences between food fraud vulnerability assessment and traditional food safety hazard analysis or risk assessments, mainly emerge from the fact that in vulnerability assessment socio-economic factors, the level of knowledge of an organization and the study of criminal behavior also need to be taken into account. What is therefore required is to further develop vulnerability assessment tools to render them as effective as possible in assessing existing vulnerabilities and overcoming knowledge gaps and therefore allow the industry to better understand where and how fraud might take place and identify the situational vulnerabilities of each company (Manning and Soon 2019).

Intelligence gathering to prevent or detect food fraud is carried out at an international level via several government-led initiatives, but industry intelligence gathering is not developed enough yet. Moreover, retailers and major food businesses are themselves finding the extent of food fraud too high, especially when they need to consider the detection methods needed and the range of products that may be affected (NSF International 2014).

5.1.1 VACCP and TACCP

A vulnerability assessment, now widely known as Vulnerability Assessment and Critical Control Points (VACCP) system, is one of the food vulnerability management systems that is commonly used nowadays. It is implemented to allow businesses to evaluate any risks of food fraud that can affect their organizations. As part of this assessment, the organization should consider whether each ingredient might be vulnerable to fraud. Upon completion of the assessment, the business is expected to be able to identify with confidence what preventative actions are needed to allow the mitigation of the likelihood of fraudulent food. Any vulnerabilities identified within the supply chain should then be subjected to controls in order to limit the likelihood of the business using fraudulent food (https://www.highspeedtraining.co.uk/hub/food-vulnerability-assessment-checklist/). To start with, processes must be implemented for the approval of suppliers' production sites, following a risk-based assessment (e.g., raw material risk level, location of food safety control measures, supplier performance). Upon qualification of suppliers following a robust approval process, it is then necessary to maintain a relationship between buyer and supplier in order to support any investigation and exchange of information. A streamlined upstream supply chain can be more transparent and facilitates traceability and the management of safety and quality standards. It also limits any opportunities fraudsters have to penetrate the supply chain (Nestec Ltd 2016).

Threat Assessment and Critical Control Point (TACCP) assists food business operators in identifying vulnerable stages in their supply chain and processing activities that may expose the companies to higher fraud risks. It then helps businesses to limit as far as possible the possibilities of such attacks. The British Retail Consortium covered the requirements of such an assessment with the publication of its BRC Global Standard for Food Safety version 7. This standard requires companies to complete a documented Vulnerability Assessment covering all raw materials they use. BRC certificated businesses are now required to have implemented the assessment, or risk losing their certificates. Other food auditing organizations and major food retailers are also introducing similar requirements. TACCP relies on PAS96:2014, Guide to protecting and defending food and drink from deliberate attacks. It is sponsored by

DEFRA and the FSA and is produced by the British Standards Institution (Leathers 2015).

5.1.2 BRC

As previously mentioned, BRC version 7, but now also BRC Issue 8, include a requirement for horizon scanning for the prevention of food fraud, despite the fact that the words 'horizon scanning' do not appear in the standards as such. The BRC considers horizon scanning to be essential for the understanding of risks to food authenticity. Horizon scanning considers developing threats, or in other words, threats that might occur in the future, rather than known incidents that have previously occurred (https://www.foodfraudadvisors.com/horizon-scanning-for-food-fraud/). Clause 4.2.1 of BRC version 8 specifically states that to meet the requirements of the standard, a company is required to perform a documented threat assessment of the potential risks to products that could result from deliberate attempts to inflict contamination or damage, covering both internal and external threats. The assessment and its outcome should be clearly documented in a threat assessment plan. Regular reviews (annually as a minimum) of the plan should be carried out to identify changes in the markets and general circumstances. The plan should also be reviewed every time a new risk is identified, or an incident occurs that has the potential to affect the product's security or the system's food defence. Clause 4.2.2 of the same standard states that in cases where raw materials or products are identified as being at significant risk, the threat assessment plan should be adjusted to implement controls for the mitigation of these risks. However, when the risks cannot be prevented or a risk cannot be eliminated, the organization should have systems for identifying any tampering. Finally, Clause 4.2.3 of the standard states that areas that are identified as being at particular risk shall be clearly defined, monitored and closely controlled. Such areas shall cover external storage facilities and the intake products, raw materials and packaging. Policies and systems should be implemented to ensure that authorized personnel are only allowed in production and storage areas and all access is adequately controlled. A reporting system for visitors should also be in place while personnel should receive training on site security measures and food defence (BRC Global Standards 2018).

5.2 Authorities and governments

In the United States, a number of laws address the illegal nature of food adulteration and misbranding. Both federal and state laws prohibit adulterated foods. According to US legislation, when foods of vegetable origin are considered, adulterated food is defined as:

- Food containing poisonous or deleterious substances.
- Food containing a pesticide chemical residue that renders the food not safe for consumption or food containing any food additive that is not safe if consumed.
- Food consisting in whole or in part of any filthy, putrid, or decomposed material, or is in any other way not suitable for consumption.
- Food prepared, packed, or maintained under unhygienic conditions whereby it may have become contaminated with filth, or been rendered injurious to human health.
- Food packed in a container composed of a poisonous or deleterious substance that can render the contents unsafe for consumption.
- Food that has been irradiated, except if the use of radiation conformed with a regulation or was exempted under the law.
- Food wherein any valuable ingredient has been in total or partly omitted, substituted, damage concealed, or a substance has been added to increase the food's bulk or weight.
- Food containing a color additive that is not safe for consumption.
- Confectionery containing alcohol or nonnutritive substance.
- Dietary supplement presenting a significant or unreasonable risk of causing illness or injury or that is prepared, packed or maintained under conditions that are not in accordance with current good manufacturing practice regulations.
- Additives, microbes, and conditions that render the food contaminated, when necessary measures have not been taken to eliminate these conditions and
- Food offered to be imported while it has previously been refused admission, except if the person reoffering it affirmatively establishes that the food is in compliance with any applicable requirements.

The prohibitions around adulterated and misbranded foods under U.S. federal law apply principally to the processing sector of the industry, but also in the retail and food service sectors (https://www.ag.ndsu.edu/foodlaw/overview/adulterated-etc.).

In the European Union, the Regulation (EC) No 178/2002 of 28 January 2002 outlines the general principles and requirements of food law, establishes the European Food Safety Authority and lays down general procedures in relation to food safety. The article 8 of this regulation states that the food law shall aim at protecting the interests of consumers and providing a basis for consumers to allow them to make informed choices with regards to the foods they purchase. It shall aim at preventing:

(a) fraud or deceptive practices;

(b) the adulteration of food; and

(c) any practices that can mislead the consumers (Regulation (EC) No 178/2002).

The EU policies have been established for the prevention and detection of food fraud, and are contributing significantly to fulfilling this responsibility. The EU countries and the Commission have reached an agreement on 19 concrete measures that aim at re-enforcing the EU's action against fraud. A presentation of these methods was given in the AGRIFISH Council on 9 October 2017. The measures include a commitment to strengthen the interaction between the Rapid Alert System For Food and Feed (RASFF) and the Administrative Assistance and Cooperation (AAC) system, which practically means the creation of common contact points. The EU re-shaped its agri-food chain official control policies in an effort to enhance its overall efficiency and build on citizen's trust (https://ec.europa.eu/food/safety/food-fraud_en). Close cooperation between officials with expertise in EU food chains, enforcement authorities, customs officials, judges and prosecution officials is essential, at both national and EU level. Since 2013, the EU Food Fraud Network allows the Member States and some other European countries to participate in exchanging information and cooperating on a voluntary basis in matters related to cross-border violations of EU agri-food chain legislation. This allows the European countries that participate to work, following the rules laid down in the Official Controls Regulation, in cases that authorities at a national level detect possible intentional violations of the EU agri-food chain legislation that may affect regions outside their borders. The EU Food Fraud Network therefore acts as a link between bodies designated by each Member State, a number of other European countries (Switzerland, Norway and Iceland), representatives from the European Commission services and representatives from Europol. It allows a more efficient and coordinated communication channel to be established between competent authorities, especially when it comes to transmitting and receiving requests for support. The liaison bodies are expected to exchange data that allow compliance with EU agri-food chain legislation to be verified. Data can also be exchanged with the Commission, where the results of official controls indicate that action in more than one country is needed (https://ec.europa.eu/food/safety/food-fraud/ffn_en).

In China, there is limited trust in the domestic food system, as a result of a number of major food scandals, many of which were fraud-related.

Chinese consumers perceive food fraud to be a major food hazard that threatens the authenticity, quality and reliability of food and increases the risk of making unsafe food available for consumption in the market. Consumers have often put their trust on informal kinship networks to obtain information related to food safety and authenticity (Kendall et al. 2019). Based on the Food Safety Law of the People's Republic of China as translated by the USDA Food Agricultural Service, food or food additives that are rotten, spoiled, dirty, contaminated or contain rancid fat, mold, insects or foreign materials, as well as food that has been adulterated, may not be produced or enter the supply chain. Also, Article 124 of the same piece of legislation states that violations of the law as a result of producing or distributing food that meets the above-mentioned conditions and is therefore illegal, can lead to, among others, confiscation of food or food additives and tools, equipment or raw material used for illegal production or distribution. The organizations involved can also receive fines of RMB 50,000–100,000 if the total value of the commodity is less than RMB 10,000 or fines between 10 and 20 times the total value of the commodity in cases that the total value of the commodity exceeds RMB 10,000. Where serious violations are found, the authorities can also revoke the licenses of these companies (https://gain.fas.usda.gov/Recent%20GAIN%20Publications/Amended%20Food%20Safety%20Law%20of%20China_Beijing_China%20-%20Peoples%20Republic%20of_5-18-2015.pdf).

In the case of India, food legislation was established back in 1860 with particular sections of the Indian Penal Code referencing food adulteration. Nevertheless, food laws in their modern form were established only in the early twentieth century. Before independence (1947), Indian provinces under British rule had their own acts that would allow them, to some extent, to prevent food adulteration (van der Gaur 2016). As van der Meulen (2010) noted, following independence, a number of pieces of legislation were brought forward by the ministry of food and other ministries aiming at preventing adulteration of foods of vegetable origin.

The Directorate General of Health Services, through the Central Committee for Food Standards, under the power of the Ministry of Health and Family Welfare of the Government of India, establishes food standards and has the authority to implement amendments to rules as and when needed. The rules are then employed by the State governments and local bodies. Hundreds of amendments to these rules have been made to meet the current requirements of safety as laid down in Sanitary and Phytosanitary (SPS), Technical Barriers to Trade (TBT) and other agreements of World Health Organisation (WHO) (van der Meulen 2010).

6. Conclusions

It is therefore widely accepted that food fraud can be particularly damaging to economies and can have severe consequences for food safety and quality. Systems and standards such as intelligence gathering, BRC, VACCP, TACCP, supply chain mapping and others are now more and more often developed and applied in the industry to prevent or detect fraudulent activities. A large variety of analytical methods and techniques have also been employed to detect adulteration. Supply chain mapping and the ability to trace the origins of foodstuffs using analytical techniques is of paramount importance. The verification of traceability allows not only corrective actions to be taken when contamination occurs, but also re-enforces food authenticity, assists in preventing or detecting fraudulent practices and deters food adulteration. Any relevant controls that are put in place reinforce the consumer's trust in supply chains, thus reducing barriers and concerns over international trade (https://www.iaea.org/topics/traceability-and-authenticity). The implementation of robust systems to ensure food authenticity are mandated by legislation in all major economies including the countries of the European Union and USA, while the prohibition of adulteration is referenced in legislation worldwide.

Abbreviations

Administrative Assistance and Cooperation	AAC
Fourier transform infrared spectroscopy	FTIR
High performance liquid chromatography	HPLC
Infrared	IR
Mass Spectrometry	MS
Near infrared	NIR
Next Generation Sequencing	NGS
Nuclear Magnetic Resonance	NMR
Rapid Alert System For Food and Feed	RASFF
Sanitary and Phytosanitary	SPS
Technical Barriers to Trade	TBT
Threat Assessment and Critical Control Point	TACCP
Ultra-violet	UV
Visible	Vis
Vulnerability Assessment and Critical Control Points	VACCP
World Health Organisation	WHO

References

Anklama, E. and R. Battaglia. 2001. Food analysis and consumer protection. Trends in Food Science & Technology, 197–202.

Armenta, S. and M. de la Guardia. 2016. 15—Analytical approaches for the evaluation of food protected designation of origin. Advances in Food Traceability Techniques and Technologies, 275–301.

BRC Global Standards. 2018. Global Standard Food Safety. Issue 8. London. UK, pp. 1–114.

Danezis, G.P., A.S. Tsagkaris, F. Camin, V. Brusic and C.A. Georgiou. 2016. Food authentication: Techniques, trends & emerging approaches. TrAC Trends in Analytical Chemistry, 123–132.

Dean, N., T.B. Murphy and G. Downey. 2006. Using unlabelled data to update classification rules with applications in food authenticity studies. Journal of the Royal Statistical Society, 1–14.

Donarski, J., F. Camin, C. Fauhl-Hassek, R. Posey and M. Sudnik. 2019. Sampling guidelines for building and curating food authenticity databases. Trends in Food Science & Technology.

Gaur, P. 2016. Food Adulteration and law. Environment: Traditional & Scientific Research, January–June 2016, 1(1): 57–64.

Haynes, E., E. Jimenez, M.A. Pardo and S.J. Helyar. 2019. The future of NGS (Next Generation Sequencing) analysis in testing food authenticity. Food Control, 134–143.

Herrero, M., C. Simó, V. García-Cañas, E. Ibáñez and A. Cifuentes. 2012. Foodomics: MS-based strategies in modern food science and nutrition. Mass Spectrometry Reviews, 49–69.

Hong, E., S.Y. Lee, J.Y. Jeong, J.M. Park, B.H. Kim, K. Kwon and H.S. Chun. 2017. Modern analytical methods for the detection of food fraud and adulteration by food category. Journal of the Science of Food and Agriculture, 3877–3896.

http://ilsi.eu/task-forces/food-safety/authenticity-of-food-potential-new-task-force/. Accessed 21 July 2019.

https://ec.europa.eu/food/safety/food-fraud_en. Accessed 22 September 2019.

https://ec.europa.eu/food/safety/food-fraud/ffn_en. Accessed 22 September 2019.

https://gain.fas.usda.gov/Recent%20GAIN%20Publications/Amended%20Food%20Safety%20Law%20of%20China_Beijing_China%20-%20Peoples%20Republic%20of_5-18-2015.pdf. Accessed 22 September 2019.

https://secretvienna.org/a-scandal-made-of-antifreeze-and-wine/. Accessed 21 July 2019.

https://www.ag.ndsu.edu/foodlaw/overview/adulterated-etc. Accessed 22 September 2019.

https://www.bbc.co.uk/news/world-asia-india-23337445. Accessed 21 July 2019

https://www.cdc.gov/salmonella/2009/peanut-butter-2008-2009.html. Accessed 18 September 2019

https://www.eurofins.co.uk/food-testing/food-authenticity/. Accessed 19 September 2019.

https://www.foodfraudadvisors.com/horizon-scanning-for-food-fraud/. Accessed 19 September 2019

https://www.highspeedtraining.co.uk/hub/food-vulnerability-assessment-checklist/. Accessed 19 September 2019

https://www.iaea.org/topics/traceability-and-authenticity. Accessed 21 July 2019

https://www.ifst.org/resources/information-statements/food-authenticity-testing-part-1-role-analysis. Accessed 19 September 2019.

https://www.nhp.gov.in/disease/non-communicable-disease/epidemic-dropsy. Accessed 16 November 2019.

https://www.nytimes.com/2015/09/26/opinion/joe-nocera-of-peanuts-and-prosecutions.html?rref=collection%2Ftimestopic%2FPeanut%20Corporation%20of%20America&action=click&contentCollection=timestopics&region=stream&module=stream_unit&version=latest&contentPlacement=2&pgtype=collection. Accessed 21 July 2019.

https://www.qualityassurancemag.com/article/food-authenticity/. Accessed 21 July 2019

https://www.theguardian.com/world/2013/jul/17/school-meals-kill-indian-children. Accessed 21 July 2019.

https://www.theguardian.com/world/2013/jul/24/indian-food-poisoning-school-headteacher-arrested. Accessed 21 July 2019.

https://www.vaneerden.com/fake-food-scandal/. Accessed 21 July 2019.

Ibañez, E. and A. Cifuentes. 2001. New analytical techniques in food science. Critical Reviews in Food Science and Nutrition, 413–450.

Kendall, H., B. Clark, C. Rhymer, S. Kuznesof, J. Hajslova, M. Tomaniova, P. Brereton, L. Frewer. 2019. A systematic review of consumer perceptions of food fraud and authenticity: A European perspective. Trends in Food Science & Technology, 94: 79–90.

Leathers, R. 2015. The rise of TACCP. Campden BRC. Available at https://www.campdenbri.co.uk/blogs/the-rise-of-taccp.php. Accessed 22 September 2019.

Manning, L. and J.M. Soon. 2019. Food fraud vulnerability assessment: Reliable data sources and effective assessment approaches. Trends in Food Science & Technology, 159–168.

Medina, S., J.A. Pereira, P. Silva, R. Perestrelo and J.S. Câmara. 2019. Food fingerprints—A valuable tool to monitor food authenticity and safety. Food Chemistry, 144–162.

Mishra, P., A. Kumar, A. Nagireddy, D.N. Mani, A.K. Shukla, R. Tiwari and V. Sundaresan. 2016. DNA barcoding: an efficient tool to overcome authentication challenges in the herbal market. Plant Biotechnology Journal, 8–21.

Moore, J.C., J. Spink and M. Lipp. 2012. Development and application of a database of food ingredient fraud and economically motivated adulteration from 1980 to 2010. Journal of Food Science, 118–126.

Nestec Ltd. 2016. Food Fraud Prevention Economically-motivated adulteration. Nestec Ltd., CO—Quality Management, Vevey (Switzerland). Nestec Ltd., Corporate Identity & Design with Christian Stuker, 1097 Riex (Switzerland), pp. 1–13.

NSF International. 2014. FINAL REPORT Risk Modelling of Food Fraud Motivation – "NSF Fraud Protection Model" Intelligent Risk Model Scoping Project FS 246004. NSF Safety and Quality UK Ltd, pp. 1–72.

Premanandh, J. 2013. Horse meat scandal—A wake-up call for regulatory authorities. Food Control, 568–569.

Pustjens, A.M., Y. Weesepoel and S.M. Van Ruth. 2016. 1—Food fraud and authenticity: emerging issues and future trends. Innovation and Future Trends in Food Manufacturing and Supply Chain Technologies, 3–20.

Regulation (EC) No 178/2002. REGULATION (EC) No 178/2002 OF THE EUROPEAN PARLIAMENT AND OF THE COUNCIL of 28 January 2002 laying down the general principles and requirements of food law, establishing the European Food Safety Authority and laying down procedures in matters of food safety. Official Journal of the European Communities. Brussels, Belgium, pp. 1–24.

Reinhard, E. 2010. Lehr- und Forschungszentrum für Wein- und Obstbau (Education and Research Institute for Viticulture and Pomology) A-3400 Klosterneuburg Wiener Strasse 74. ANNEX 5 Report: Production and Control of Wine in Austria. Klosterneuburg, Austria, pp. 1–10.

Rohman, A. and Y.B.C. Man. 2012. Application of fourier transform infrared spectroscopy for authentication of functional food oils. Applied Spectroscopy Reviews, 1–13.

Ruth, S.M. 2019. Food fraud and vulnerability assessments. Reference module in food science. Encyclopedia of Food Chemistry, 663–669.

Sharma, B.D., S. Malhotra, V. Bhatia and M. Rathee. 1999. Epidemic dropsy in India. Postgraduate Medical Journal, 657–661.

Spink, J. and D.C. Moyer. 2011. Defining the public health threat of food fraud. Journal of Food Science, 157–163.

Spink, J. 2019. Food fraud and adulteration: Where we stand today. Reference module in food science. Encyclopedia of Food Chemistry, 657–662.

Terndrup, J.S. 2010. Crisis communication—not just peanuts. An analysis of Kellogg, King Nut and Peanut Corporation of America's press releases during the American salmonella crisis in 2009. Master Thesis, Department of Language and Business Communication Aarhus School of Business, Aarhus University. Aarhus, Dennmark, pp. 1–105.

van der Meulen, B.M.J. 1960. Development of food legislation around the world. pp. 5–70. *In*: Boisrobert, C., A. Stjepanovic, S. Oh and H. Lelieveld [eds.]. Ensuring Global Food Safety. Exploring Global Harmonization. Elsevier Inc., London, UK.

Van Ruth, S.M., P.A. Luning, C.J. Silvis, Y. Yang and W. Huisman. 2018. Differences in fraud vulnerability in various food supply chains and their tiers. Food Control, 375–381.

3

Methods and Techniques for Verifying Authenticity and Detecting Adulteration

*Konstantinos V Kotsanopoulos** and *Athanasios Exadactylos*

1. Introduction to Food Authentication

"Authentic" is a term commonly used to define a reliable or genuine state. In the food industry, the exact determination of a species is essential to verify the authenticity status of a foodstuff and should be carried out based on factors not significantly altered throughout food processing (Lüthy 1999). Food authenticity and the detection of adulteration are subjects of paramount importance in the food industry that are attracting more and more attention (Al-Jowder et al. 1997). Food traceability is also mandatory according to the European Union (EU)regulation. The ability to trace food and feed throughout the food chain is of high importance principally to protect consumers, especially when food and feed are of substandard quality or mislabeled. Traceability is defined by the General Food Law Regulation as the ability to trace and follow food, feed, and their ingredients throughout the production, processing and distribution processes. It allows faulty food/feed to be easily withdrawn from the market, ensures that consumers are provided with accurate information on the products they purchase and ensures all supply chain is covered,

School of Agricultural Sciences, Department of Ichthyology and Aquatic Environment, University of Thessaly, FytokouStr, 38446 Volos, Volos, Hellas.
* Corresponding author: kostaskot@yahoo.gr

including importers who need to be able to identify who was the exporter in the country of origin, while it also enforces businesses to establish measures to ensure they can identify at least their direct supplier of a product and their direct customers (excluding the supply of products from retailers to final consumers) (https://ec.europa.eu/food/safety/general_food_law/general_requirements_en).

Throughout history, efforts have been made to establish controls for the qualitative and safety characteristics of products. More than 2,000 years ago, India had already established regulations to prohibit the adulteration of grains and fats. The laws of Moses contained decrees that have similarities with particular aspects of modern food laws. An example of a very early control is the weights and measures in foods and other commodities. Other ancient civilizations (Chinese, Hindu, Greek, and Romans) also had certain food regulations (Edith and Ochubiojo 2012). Nowadays, advanced analytical methods and techniques have been developed to detect food adulteration and ensure the authenticity of food.

Fraud cases have become more frequent in the last decades, mainly as a result of the opening of international markets and the competition between companies at a global level. Profit is the principal reason for conducting such illegal activities. It is therefore essential to establish accurate controls to ensure foodstuffs are authentic and safe for consumption (Cotte et al. 2003). A processor that deals with the production or packaging of food is frequently required to determine the relative importance of authenticity testing in comparison to alternative types of analysis. The overall quality of the commodity is the most important factor in these cases. It is meaningless to produce an authentic product that is of such poor quality that it cannot be accepted or it can even be unsafe. Commonly, raw materials (commodities) testing at the pre-package stage is approached based on: (1) 'Value for money' testing on the intrinsic characteristics of a product, aiming at ensuring that a buyer has obtained what was purchased, absence of contamination, alignment to specification, etc. (2) Testing of the organoleptic attributes, which ensure that a product tastes, smells, etc., as expected by the consumer. (3) Biological/microbiological testing: mainly aiming at ensuring that a product is free from factors that can lead to microbial spoilage. (4) Any additional testing that could be necessary depending on the product (e.g., heavy metals content such as lead and cadmium, nutritional values, etc.). It is therefore clear that authenticity is indeed part of the value-for-money testing. The manufacturer has then to determine the frequency of testing, but also the most suitable methods of testing and it seems that retailers can influence these decisions (Dennis and Ashurst 1996).

In the UK, the Food Authenticity Steering Group's aim is to assess the need for research and surveys that expose food fraud and mislabeling, as well as the identification of global trends and drivers that can influence food labeling, standards and the possibility of food fraud and misleading actions. It also aims at identifying overarching authenticity research and survey needs that are necessary to reduce deceptive or misleading practices, therefore protecting consumers and encouraging the competitiveness, resilience and sustainability of the food supply chain. The Group also focuses on offering advice with regards to what authenticity research needs to be prioritized and the development of methods to allow that to happen. Finally, it assists in knowledge transfer activities between authorities and stakeholders, but also cooperates with the Food Standards Agency, other organizations, delivery bodies and any trade associations that are concerned (https://www.gov.uk/government/groups/food-authenticity-steering-group#contact-details). In the United States (US), Food and Drug Administration (FDA) is committed to rendering the application of the Intentional Adulteration rule practical and as easy to be applied as possible. The largest businesses will need to be compliant by July 2019 (Gottlieb 2018).

2. Methods and Techniques used for Food Authenticity and Quality Assurance

2.1 Spectroscopic techniques

Spectroscopy is a method that takes advantage of the absorption and emission of electromagnetic radiation as a result of the interaction between matter and energy. This energy depends on a specific wavelength of electromagnetic radiation. Spectroscopy has had great research potential in numerous sectors including chemistry, physics, biology, medicine and ecology. The area of spectroscopy has grown significantly and scientists now explore the use of new techniques that make use of this method (Farrukh 2012). Since the 19th century, scientists have explored many regions of the electromagnetic spectrum, and analytical techniques relying on the energy of each of these regions have been developed. A number of these methods, especially the non-destructive ones are, or could be used nowadays, for the analysis of food attributes and particular components (Scotter 1997). The spectroscopic methods are particularly important in the study of biological systems at a molecular level. Nevertheless, no single spectroscopic method can be used to provide all information required with regards to the structural characteristics and function of molecules in biological systems. The technique chosen depends on the needs and targets of each study. There are techniques that can be carried out both *in vivo* and *in vitro*, while others cannot. Some techniques have timescales of seconds or even picoseconds, while for some the use of a perturbing probe molecule is necessary (Haris 2009). The majority of spectroscopic methods employed for food authenticity testing rely on near infrared (NIR), mid-infrared (MIR), Fourier transform infrared (FTIR), Raman or nuclear magnetic resonance (NMR) spectroscopy or hyperspectral imaging (Esteki et al. 2018).

2.1.1 Infrared spectroscopy

Infrared (IR) spectroscopy has been used traditionally to determine small molecules based on their structure, mainly due to the fact that the technique is very sensitive to the chemical composition and architecture of molecules. Measurements of bond lengths and bond angles can be effectively carried out, while the detection of bond distortions can be achieved with picometer precision. The potential of IR spectroscopy in analytical chemistry was made apparent mainly at the end of the 19th and the beginning of the 20th century (Barth and Haris 2009). IR spectroscopy is based on the interaction between a molecule and the radiation of the IR region of the spectrum (IR region = 4000–400 cm^{-1}). IR radiation leads to the excitation of the vibrations (stretching and bending) of covalent bonds within a molecule. An IR spectrum illustrates the absorptions of energy as the IR region of the spectrum is scanned. Generally, it can be practical to split an IR spectrum into two regions: 4000–1000 cm^{-1}, which is known as the functional group region, and < 1000 cm^{-1} known as the fingerprint region (http://www.chem.ucalgary.ca/courses/351/Carey5th/Ch13/ch13-ir-1.html).

IR spectrometers are used for measuring the wavelength and intensity of the absorption of IR light by a sample. They can be categorized into two major groups: the FTIR spectrometers, which are single-beam instruments that employ interferometers to spread a sample with infrared light and then measure the intensity of the IR light not absorbed by the sample. FTIR instruments convert the resulting time domain digital signal into a frequency domain digital signal, while dispersive IR spectrometers employ visible lasers, a grating, and charged coupled devices (CCDs) for the collection of data. The laser plays the role of a light source and is used for irradiating the sample. The highest amount of radiation scatters elastically at the same energy as the incoming laser radiation, while a very small amount (0.00001%) scatters from the sample at a wavelength, which is shifted slightly from the original wavelength (https://www.globalspec.com/learnmore/labware_scientific_instruments/spectrometers_analytical_photometers/infrared_spectrometers). IR spectrometers have been in the market since the 1940s. Back then the instrumentation was based on prisms that acted as dispersive elements. By the mid-1950s though, the introduction of diffraction gratings into dispersive machines became common. The most important development in IR spectroscopy though occurred when FTIR spectrometers were introduced. These instruments are equipped with an interferometer and make use of a Fourier-transformation mathematical process. FTIR spectroscopy led to significant improvements in the quality of IR spectra and led to minimization of the time needed to obtain data. Additionally, advancements were rendered possible due to the progress in computer science and technology (Stuart 2004).

The fact that the method requires no specific reagents is its biggest advantage. Automation and repetitive analyses could thus be cost-effectively implemented. The effectiveness and the advantages of the method have led to the conduction of further research and the development of IR spectrometers that combine high acquisition speed with advanced spectral sensitivity. Chemometric algorithms of great potential and sophisticated softwares have been developed simultaneously with advanced hardware, leading to new applications of the method. Instead of relying upon reagents for the promotion of color reactions, IR is based on the spectrum of IR colors that is characteristic of the analyte of interest. If a substance produces an IR absorption spectrum and is available at concentrations that allow the spectrum to significantly contribute to the IR absorption profile, it is possible that its quantification can be carried out using IR spectroscopy (Shaw and Mantsch 2006).

2.1.2 Mid-infrared (MIR) spectroscopy

MIR spectroscopy is a vibrational spectroscopy technique used for the identification of substances based on how they interact with electromagnetic radiation in the mid-infrared region (400–4000 cm^{-1}). Molecular absorbance of mid-infrared light leads to rotations and vibrations at a molecular level and their classification is based on chemical functional groups. The MIR spectrum extracted characterizes a specific molecule. Infrared open-path systems produce a beam of infrared light through open air and detect gases of interest. The length of this linear sensor can range from a few meters up to a few hundreds of meters. Point analysis ("benchtop") systems can be employed to analyze condensed phase materials, while some of them can also be used for the analysis of gases in gas cells (https://cbrnetechindex.com/Explosives-Detection/Technology-ED/Molecular-Spectroscopy-ED-T/Mid-infrared-Spectroscopy-ED-MS). The MIR region is the region of wavelengths between 3×10^{-4} and 3×10^{-3} nm. To allow the ease interpretation of numbers used when the technique is employed, IR spectra are commonly reported in μm, although the use of the unit v (nu bar or wavenumber) is also common. A wavenumber can be described as the inverse of the wavelength in cm: $v = 1/\lambda$, where v is in units of cm^{-1}, λ in cm and $E = hcv$. In the case of wavenumbers, the MIR ranges from 4000 to 400 cm^{-1}. Wave numbers increase by increasing the energy input (http://orgchem.colorado.edu/Spectroscopy/irtutor/IRtheory.pdf).

An MIR transmission spectrometer consists of a radiation source, a monochromator, a sample cell and a detector. Glass cannot be used to manufacture the optical components and therefore salts are usually used. The standard spectrometer for studies in the MIR region is a double-beam instrument in which radiation passes alternately through the sample and reference cells (Gordon and Macrae 1987a).

MIR spectroscopy is described as a very powerful, rapid and non-destructive technique of high accuracy that can replace traditionally used analytical chemistry. Several reports have lately been mentioned in the literature proving the usefulness of the technique in analyzing antioxidants in a variety of organic matrices (Cozzolino 2015). Several other studies have been conducted using the technique with the aim of detecting adulteration and fraud in foods of plant origin, including coconut oil (Rohman and Man 2011), extra virgin olive oil (Gurdeniz and Ozen 2009, Sinelli et al. 2008, Gurdeniz et al. 2007), soybean oil (Wang et al. 2006), organic wine (Cozzolino et al. 2009), maple syrup (Paradkar et al. 2003) and others.

2.1.3 NIR spectroscopy

If chemical bonds showed the same behavior as weak springs, quantum mechanics would lead to restriction of their vibration to only two states, which would mean very limited absorptions in the NIR region. The generation of absorptions in the NIR region (780–2500 nm) is due to fundamental vibrations by two processes; overtones and combinations (http://www.impublications.com/content/introduction-near-infrared-nir-spectroscopy).

NIR spectroscopy has its origin in the early 1800s when Fredrick William Herschel reported the discovery of the first nonvisible region of the absorption spectrum. Very little progress took place between 1800s to the 1950s in the area of NIR spectroscopy, while other analytical methods were being examined that could allow more unambiguous results to be delivered, particularly with regard to explaining molecular structures. It was only in the 1950s that Karl Norris and coworkers recognized the significance of the method. Nowadays, NIR spectroscopy is one of the most important quality control methods in the food and feed industry due to its advantages over other analytical techniques (Manley and Baeten 2018).

In NIR spectroscopy, the illumination of the unknown substance is achieved with a broad-spectrum of NIR light, which can be absorbed, transmitted, reflected or scattered by the sample that is examined. Commonly, the illumination is in the wavelength range of 0.8 to 2.5 microns (800 to 2500 nm). Measurements of the intensity of the light as a function of wavelength are taken prior to and after the interaction with the sample, and the diffuse reflectance, a combination of the level of absorbance and scattering, is calculated. The penetration of NIR goes typically much deeper into a sample when compared to FTIR, and in contrast to Raman, fluorescence does have an effect on it. Therefore, although the chemical specificity of the technique is not as high as that of Raman or FTIR, it can still be effectively used in probing bulk material with minimal or no sample preparation (https://www.thermofisher.com/uk/en/home/industrial/spectroscopy-elemental-isotope-analysis/spectroscopy-elemental-isotope-analysis-learning-center/molecular-spectroscopy-information/nir-technology.html). NIR spectroscopy is regularly employed to analyze the composition, function and sensory characteristics of food ingredients, intermediates or final foods. Since, as mentioned, no sample preparation is often needed, the analysis is straightforward and rapid (15–90 s) and is possible to be conducted on-line. Amongst the most important advantages of NIR technology is the fact that it allows more than one constituent to be measured at the same time. Additionally, for every fundamental vibration a corresponding series of overtone and combination bands exists. Each successive overtone band is about an order of magnitude weaker than the preceding one. This allows for a dilution series to be built in, allowing absorptions of different intensity that contain the same chemical information to become available. Last but not least, a relatively less strong absorption, due to water, facilitates the analysis of foods with high moisture levels. It is important to mention though that NIR depends on less-precise reference methods when it comes to food analysis (Osborne 2006). The use of visible and NIR spectroscopy was studied by Cozzolino et al. (2003) who used the techniques for the discrimination of commercial white wines of different Australian varieties, while it was also used in the study of Downey et al. (2002) for detecting and quantifying sunflower oil adulteration in extra virgin olive oils from the eastern Mediterranean. Other studies using the same technique include those of Liu et al. (2006), who used it in combination with multivariate analysis (MVA) for the geographical classification of Spanish and Australian red wines, Zalacain et al. (2005), who used NIR in saffron quality control, Cozzolino et al. (2010) who used it for the discrimination of the geographical origin of yerba mate, and others.

Process analytical technology in the food industry can allow improvements to be made in efficiency and product quality by allowing the manufacturing processes to be better understood and controlled. NIR is one of the most robust e-sensing technologies employed in process analytical technology, mainly due to its potential to fingerprint materials and at the same time analyze food-related phenomena. NIRS has recently been shown to be very effective in real-time monitoring and modeling of various food processes. It is important to mention though that the majority of studies have been performed at a lab scale only (Grassi and Alamprese 2018).

It is not easy to foresee with accuracy how this technology will evolve to further improve its analytical potential. Based on the current developments in NIR, it is possible that future improvements will focus on miniaturization and imaging. However, improvements may also become possible via hyphenation of NIR with a number of other analytical techniques. The merge of analytical information obtained using NIR and other techniques could allow improvements in analytical results for numerous fields of work. For instance, as NIR responds to the organic matter in a sample, information on the whole matrix using other elemental techniques can be used to correct matrix effects (Pasquini 2018).

2.1.4 Fourier transform near-infrared (FT-NIR) spectroscopy

FT-NIR spectroscopy was introduced in an effort to improve the effectiveness of NIR instrumentation. The use of a prism or moving grating allows the separation of the individual frequencies emitted from the NIR source by the dispersive instrument. Measurements of the level of energy at each frequency that passes through the sample are taken by a detector. The time required for the analysis of each sample is essential and can only be achieved by introducing a method that could measure all IR frequencies at the same time rather than individually. The FT-NIR spectrometer generates a unique type of signal known as interferogram that incorporates all IR frequencies into it. Measurements of the signal can be taken rapidly via collecting multiple scans per second. This increase in the speed of measurement is called Fellgett Advantage. A well-known mathematical technique known as Fourier transformation can then be used to transform the interferogram into a spectrum of frequency versus intensity (Thermo Fisher Scientific Inc. 2006). Nowadays, FT-NIR spectroscopy is used in numerous quality control applications across all industries, and finds applications in pharmaceuticals, food, agricultural products and chemicals. The technique is a valuable alternative to wet chemistry and chromatography by mainly shortening the time required for the analyses. It is a non-destructive technique that does not require any sample preparation or use of hazardous chemicals, while it can be used for both quantitative and qualitative analysis to quickly identify substances. It is also a powerful analysis tool that can support multi-component quantitative analysis of high accuracy (https://www.bruker.com/products/infrared-near-infrared-and-raman-spectroscopy/ft-nir.html). FT-NIR spectrometers are commonly employed for the identification and characterization of chemicals and substances in test samples (https://www.azom.com/materials-equipment.aspx?cat=125). The technique can also be employed to detect certain characteristics of food quality. In the case of green tea, for example, it can be used as a rapid and simple method to estimate the product's content in polyphenols (Sinija and Mishra 2009) but also to rapidly discriminate between roast green tea qualities based on their geographical origin (Chen et al. 2009).

2.1.5 Raman spectroscopy

Raman spectroscopy is a vibrational spectroscopic technique that can be employed to give information about molecular vibrations and structural characteristics of crystals. This technique is equipped with a laser light source. The samples are irradiated and an infinitesimal amount of Raman scattered light is generated, while a Raman spectrum using a CCD camera is used for its detection. The characteristic fingerprinting pattern in a Raman spectrum allows the identification of substances and the evaluation of their local crystallinity, orientation and stress (Nanophoton 2016). Impingement of a beam of light upon a molecule leads to absorption of photons by the material and scattering. Most of the photons that are scattered are characterized by the same wavelength as the incident photons and are known as "Rayleigh scatter". The scattering process leads to excitation of electrons of the incident photons into higher energy levels. These electrons will subsequently decay back to a lower level, emitting a scattered photon. In Rayleigh scattering the electron decays back to the starting level and therefore this form of scattering is frequently referred to as a form of elastic scatter. The Raman effect takes place when a photon is incident on a molecule, interacting with the electric dipole of the molecule (https://kosi.com/about-us/kaiser-production-competencies/raman-spectroscopy-a-tutorial/).

The main advantages of Raman spectroscopy are the facts that it allows for non-destructive analysis, high spatial resolution (even to sub-micron scale), in-depth analysis of transparent samples (with the use of a confocal optical system) and the elimination of the need for sample preparation. It can also be used for the analysis of both organic and inorganic components and for gases, liquids, solutions, solids, crystals and emulsions (Nanophoton 2016).

Raman spectrometers are equipped with one or multiple single colored light emitters (lasers), lenses (for focusing the light onto the sample and collecting the scattered light), filters (for purifying the reflected and scattered light in order to collect the Raman light only), a means of separating the light into its constituent colors (usually a diffraction grating or prism), a detector of very high sensitivity for the detection of weak light and a device (e.g., a computer) that controls the whole spectrometer, displays the spectrum and allows the analysis of the extracted information (http://www.renishaw.com/en/a-basic-overview-of-raman-spectroscopy--25805). A Raman spectrum acts as a molecular "fingerprint" of single cells, enabling the discrimination between cell types, physiological states, nutritional characteristics, and a variety of phenotypes (Huang et al. 2010). In the case of plants of food origin, the technique has been used mainly for detecting adulteration of oils. Some examples of studies are those of López-Díez et al. (2003) who used the technique to assess the adulteration of virgin olive oils with hazelnut oils, Baeten and Aparicio (2000) and Duraipandian et al. (2019) who again used the technique to discriminate virgin olive oil from other edible oils and for the detection and quantification of virgin olive oil adulteration, as well the studies of Baeten et al. (2005) (use of FT-Raman spectroscopy for the detection of hazelnut in olive oil) and Baeten and Meurens (1996) who used FT-Raman spectroscopy to detect adulteration of virgin olive oil with soybean, corn and olive pomace oils.

2.1.6 Ultraviolet-visible (UV-Vis) and fluorescence spectroscopy

The visible region of the spectrum (400–700 nm) corresponds to frequencies of 7 × 1014–4 × 1014 Hz. The human eye perceives radiation at the low-energy high-wavelength end of the spectrum as red, while radiation at the high-energy part of the spectrum is perceived as blue. The spectral region of 180 to 400 nm is the ultraviolet UV region. UV and visible absorption is mainly employed for quantitative analysis. The Beer-Lambert law, log (I0/I) = εcl governs the absorption of light in this region of the spectrum (Gordon and Macrae 1987c).

According to the Royal Society of Chemistry (2009), UV-Vis spectrometers can be employed to take measurements of the absorbance of UV or Vis light by a sample. The method can be employed for both quantitative and qualitative measurements. The light source (consisting of tungsten/halogen and deuterium lamps) penetrates the Vis and near UV radiation. The light emitted is focused onto the diffraction grating which very efficiently splits the light into its component colors. UV-Vis spectroscopy is a technique that can be used for determining the concentrations of substances, thus allowing scientists to examine the rates of reactions determining rate equations for reactions. As a result, UV spectroscopy is commonly used for teaching and research purposes as well as in laboratories to quantitatively analyze all molecules that absorb UV and Vis radiation (Royal Society of Chemistry 2009).

Fluorescence spectrometry is a rapid, inexpensive method that can be used to determine the concentration of an analyte in a solution. The method can be employed for analyses that are considered relatively easy in cases, for example, that the substance of interest (the analyte) is known and quantitative analysis is needed for determining the concentration. Fluorescence is mostly employed to measure substances in solutions. In fluorescence spectroscopy, a beam having a wavelength that may vary from 180 to 800 nm passes through a solution in a cuvette. Measurements of the light emitted by the sample are then taken from an analyte. In fluorescence spectrometry measurements of both an excitation spectrum and/or an emission spectrum can be taken. As the level of the analyte increases, the intensity of the emission increases proportionally.

There are numerous factors that affect the intensity level and shape of the spectra. The intensity, in particular, depends on the:

1. Excitation wavelength
2. Concentration of the analyte solvent
3. Cuvette's path length
4. Sample's self-absorption (Chromedia 2018).

Changes in fluorescence intensity can be used as indicators of changes in the structure or binding of two molecules. In the case of the wavelength of tryptophan fluorescence, for example, a longer wavelength could indicate that tryptophan is present in an aqueous environment, while longer wavelength could indicate that it is buried deep within the protein. Fluorescence polarization anisotropy can be used to examine the mobility of fluorophores. In this case, samples must generally have an absorbance lower than 0.1 at the excitation wavelength to minimize any inner filter effects (Warwick Scientific Services 2017). Some examples of studies using these techniques to detect adulteration in foods of plant origin include those of Giumet et al. (2004) and Kunz et al. (2011) who studied the application of the fluorescence spectroscopy for the detection of adulteration or classification of olive oil or the studies of Martelo-Vidal et al. (2012) and Contreras et al. (2010) who used UV and Vis spectroscopy and chemometrics for the discrimination of wines and the detection of adulteration on tequilas, respectively.

2.1.7 Stable isotope ratio mass spectroscopy (IRMS)

Isotopes are atoms that have the same number of protons, but differ in the number of neutrons, and therefore in mass. Isotopes can be found in both stable and unstable forms and have a similar behavior. The unstable atoms are radioactive, while the stable ones that find applications in biological sciences are light elements, e.g., H, C, N, O and S. These elements are characterized by a different proportion of a minimum of two isotopes. "Abundancy" is the term used to describe the prevalence of isotopes of an element. It is known that some present variations in their abundance through time, at different geographical locations and organisms. C and N isotopes are frequently added to a system in enriched status and are very useful tools for tracing relationships in a food chain and assessing physiological metabolic processes. The measurement of stable isotopes can be conducted by a coupling of an elemental analyzer through an interface to a mass-spectrometer (MS) and an Isotope Ratio Mass Spectrometer (IRMS) (Radboud University 2018). The variability in isotopes as a result of natural processes can be used to study various natural phenomena of high complexity from the origin of samples to tracing biological reactions (Kaur and O'Connor 2007).

In this technique, the initial separation of samples is followed by their introduction into the IRMS using a separate technique such as a gas chromatography (GC). In this case, the injected sample is subjected to vaporization and is carried by an inert carrier gas into the column, where the individual components of the sample are separated as a result of various interactions with the carrier gas and the stationary phase of the column. Subsequently, the compounds elute out of the column and are introduced into the MS (e.g., as a gas via combustion). The sample is then combusted into a simple gas that represents the original analyte (e.g., CO_2, N_2, or H_2). Any excess oxygen is then removed in a reduction reactor, often at a slightly lower temperature. The removal of water can be carried out using a substance called Nafion, and the sample then passes into the MS. In the ion source of the MS, molecules in the gas state are subjected to ionization by a beam of electrons in an evacuated chamber. The ions that are formed are then focused and accelerated. The magnetic field determines the trajectory of the ions leading to their separation. Ions characterized by different m/z ratios can be differentiated from one another and their quantity is determined (The Forensics Library 2018).

The technique has found several applications in the food industry. For example, as the adulteration of vegetable oil becomes more and more difficult to detect, it becomes more possible for techniques such as capillary GC to fail to detect it. Stable isotope ratio mass spectrometry (SIRMS), can therefore be employed to use differences in the abundance of elements such as H, N, C, O and S for the detection of adulteration. SIRMS has been applied in the authentication of various food products, such as fruit juices, wines, spirits, honey as well as flavor compounds (Kelly and Rhodes 2002). Some representative studies that used this technique to analyze samples of vegetable oils include those of Seo et al. (2010) who used chromatography and carbon isotope analysis to detect corn oil adulteration in sesame oil, as well as the study of Horacek et al. (2015) who used Stable Isotope analysis and DNA-based markers to control the geographical origin of sesame oil. When fruit juices are concerned, Antolovich et al. (2001) used SIRMS for the detection of adulteration in orange juices and Jamin et al. (1997) used it to detect sugar addition in apple juice and concentrates. Zhang et al. (2012) used the technique to determine the geographical origin of Chinese teas, while Weinert et al. (1999) used GC-IRMS for the analysis of black Ceylon, Assam and Darjeeling teas.

2.1.8 Liquid chromatography mass spectrometry (LC-MS)

LC-MS is nowadays routinely used with electrospray ionisation (ESI) to provide a simple and robust interface. It has found applications in a vast variety of biological molecules and the use of tandem MS and stable isotope internal standards have allowed the development of assays of high sensitivity and accuracy, even though minimization of ion suppression effects is still required. Fast scanning speeds allow the measurement of many compounds in a single run. As the cost of the equipment decreases and their liability increases, LC-MS becomes more and more important in numerous areas of clinical biochemistry and can be used instead of, e.g., conventional LC or immunoassays (Pitt 2009). The main parts of a High Performance Liquid Chromatography (HPLC) instrument are a pump, a mixing unit (solvent degassing system), an injector (manual/auto), a guard column, analytical columns, detectors, a recorder and integrators. Different types of detectors can be used such as UV detectors, fluorescence, electrochemical and mass spectrometric (Parasuraman et al. 2014). LC-MS can find applications in a lot of areas, from the development of pharmaceuticals to environmental analysis. It allows for easier analysis of samples. Although LC-MS is powerful, certain substances cannot be analyzed using this technique. LC-MS can significantly improve the effectiveness of MS, rendering it suitable for more organic compounds. Due to the fact that GC and GC-MS impart limited or no heat to the analyte molecules, the methods based on these techniques can be used for the majority of organic compounds. Sample types can cover substances ranging from small pharmaceutical compounds to large proteins (Hewlett-Packard Company 1998). Important studies that employed the technique include those of Li et al. (2000) who used it to distinguish between Panax ginseng, Asian Ginseng and North American Ginseng, as well as the study of Ehling and Shannon (2011) who employed the technique for the analysis of organic acids in fruit juices. Also, Cabañero et al. (2010) employed the technique to simultaneously analyze stable carbon isotopes of wine glycerol and ethanol, while Hammerstone et al. (1999) used it to identify procyanidins in cocoa (*Theobroma cacao*) and chocolate. Vaclavik et al. (2011) used HPLC–quadrupole time-of-flight MS coupled with advanced data mining and chemometric tools to discriminate and classify red wines in accordance with their variety, while another important study was that of Wang et al. (2003) who used ion trap LC-MS to characterize anthocyanins in grape juices.

2.1.9 Inductively Coupled Plasma-Mass Spectrometry (ICP-MS)

ICP-MS is an analytical technique used at a commercial level since 1983. An ICP-MS uses a combination of a high-temperature ICP source and a MS. The ICP source is used to convert the atoms of the elements of a sample to ions. MS is then used for the separation of the ions and their detection (USGS 2013). The use of plasma when compared with other ionization methods has significant advantages, such as the fact that ionization occurs in a chemically inert environment, thus preventing oxide formation, and ionization is more complete. Furthermore, the temperature profile of the torch is characterized by some uniformity, which reduces the self-absorption effects. Linear calibration curves can be seen over many orders of magnitude for ionization processes (Bazilio and Weinrich 2012). A peristaltic pump is used for the introduction of the sample solution into the device. The solution is nebulized in a spray chamber and the aerosol that is produced is inserted with an injection into an argon-plasma that has a temperature of 6000–8000 K. Inside the plasma torch, the water is removed and atomization and ionization of the sample are performed. Only a small part of the ions generated in the plasma are inserted into the MS part (https://www.ru.nl/science/gi/facilities-activities/elemental-analysis/icp-ms/). The preparation of the samples is of paramount importance to avoid inaccurate results (Mesko et al. 2011). Liquid introduction, and more specifically, flow injection are the most common sample introduction techniques for ICP-MS. Most ICP-MS applications in the area of elemental analysis use liquid sample introduction regardless of whether the original sample is a liquid or not. How easy or difficult a particular analysis is depends on a number of factors, such as the matrix and the chemical characteristics of the analyte, with water being one of the simplest matrices. There are, however, some analysis problems with some analytes even when this matrix is used. Generally, alkali metals and ammonium ions are soluble when the majority of anions are present. It is known though that even when low concentrations of halide and polyatomic anions are present, many other metals hydrolyze and hydroxides or oxides of low solubility in water can form (http://www.spectroscopyonline.com/sample-preparation-problem-solving-inductively-coupled-plasma-mass-spectrometry-liquid-introduction). The rapid developments in the instrumentation during the last decade in the area of elemental MS, as well as the progress that has been achieved in the development of innovative hyphenated techniques, has led to the increased usage of these technologies. The technique has found applications in a vast range of sectors, including environmental analysis of emerging substances to proteomics (e.g., protein phosphorylation and absolute protein quantification) as a tool for effective quantification. This can be attributed to its unique characteristics, such as compound-independent ionization behavior and high sensitivity. Nevertheless, it is important to mention that ICP-MS can only be used for quantification purposes when appropriate standards for calibration exist (Pröfrock and Prange 2012). The number of research articles investigating the potential of the technique has increased in the last years allowing the natural abundance of isotope variation and concentration of elements to be used as geographical "tracers" for the determination of the provenance of food. Such studies exploit the systematic variations of stable hydrogen and oxygen isotope ratios at a global level and in combination with variations in elemental and isotopes' concentrations (e.g., strontium-87) and other biogeochemical indicators (Kelly et al. 2005). The technique has been used to examine a variety of foods of plant origin including tomatoes (Feudo et al. 2010), wheat (Branch et al. 2003), alcoholic beverages (Rodushkin et al. 1999), rice (Kelly et al. 2002) and barley (Husted et al. 2004).

2.1.10 NMR spectrometry

NMR is a technique that has been increasingly improved and developed from its early use in the 1940s. It is most commonly used for structural analysis and to identify organic molecules. Even though it is characterized by lower sensitivity in comparison to techniques, such as UV and fluorescence spectrophotometry, its implementation is much wider and allows specific structural data to be provided that are frequently sufficient to unambiguously identify organic molecules (Gordon and Macrae 1987b).

The analysis of molecular structures by an NMR instrument involves observation and measurements of the interaction of nuclear spins in the presence of a powerful magnetic field. Molecular structure at the atomic level can also be analyzed using electron microscopes and X-ray diffraction instrumentation, but NMR is still used due to its non-destructive process and the less complicated preparation of samples (https://www.jeol.co.jp/en/products/category_nmr.html).

The output of an NMR analysis is a spectrum. The signals in the spectrum reflect nuclei in molecules, which possess a "spin". Not all nuclei have spin, while ^{1}H, ^{13}C, and ^{31}P are the most important when biologically relevant molecules are considered. The area under each signal and the concentration of the nucleus are directly related and this is a key element for the use of the technique (Yabsley et al. 2012). The technique can be used in areas such as foods, and chemistry, as well as new fields such as battery films (https://www.jeol.co.jp/en/products/category_nmr.html).

NMR spectroscopy is applied in food analysis in quality control departments since it is non-destructive and provides qualitative and quantitative data in relation to the composition of the samples, requiring limited or no pre-treatment. NMR analysis can be used to detect a vast range of compounds found in food in a single experiment, thus simplifying the sample preparation process and shortening the time needed for the conduction of the analyses. Nevertheless, due to the high complexity of NMR datasets from food matrices and the fact that the composition of these samples can have many similarities, the use of chemometrics to complement the use of the methods has become necessary (Filho 2016). In contrast to MS spectrometry, NMR spectroscopy is quantitative and no additional steps are required for sample preparation (e.g., separation or derivatization). Even though the sensitivity of the method has increased significantly and improvements are continuously implemented, this is still a disadvantage of NMR when compared to MS. MS-based metabolomics can offer a combined sensitivity and selectivity platform for metabolomics studies. Furthermore, different MS variations such as different ionization techniques and mass analyzers can be employed to ensure more metabolites can be detected (Emwas 2015). Foods of plant origin that have been analyzed with this technique for authenticity purposes include saffron (Petrakis et al. 2015), rice (Monakhova et al. 2014), paprika powder (detection of Sudan dye I) (Hu et al. 2017), coffee (Defernez et al. 2017) and others.

2.2 Chromatographic techniques

2.2.1 HPLC

HPLC is a type of column chromatography. The analysis is carried out by pumping a sample mixture or analyte in a solvent (the mobile phase) at high pressure through a column with a particular material (the stationary phase). A moving gas stream (helium or nitrogen) is used to carry the sample. The separation and identification of compounds present in a sample can be achieved as long as the sample can be dissolved in a liquid. Trace levels as low as parts per trillion can be detected. Due to its versatile nature, HPLC has found applications in a wide range of industrial and scientific areas, such as pharmaceutical, environmental, forensic science, and chemicals. The sample retention time varies and is related to the interaction between the stationary phase, the molecules of interest, and the solvents used. Interactions occur between the sample and the two phases at different rates, mainly due to the differences in the polarities of the analytes. Analytes that interact less with the stationary phase or interact most with the mobile phase will exit the column faster than the rest (http://hiq.linde-gas.com/en/analytical_methods/liquid_chromatography/high_performance_liquid_chromatography.html). In HPLC, a reservoir contains the solvent, which acts as the mobile phase. A high-pressure pump is employed for generating a predetermined flow rate of mobile phase, commonly ml/min. The sample is injected either automatically or manually into the continuously flowing mobile phase stream that transfers it into the HPLC column. The column contains the stationary phase—material that affects the separation. The compound bands are captured by a detector as they elute from the column. Any separated compound bands contained in the mobile phase can be collected for further study (http://www.waters.com/waters/en_GB/How-Does-High-Performance-

Liquid-Chromatography-Work%3F/nav.htm?cid=10049055&locale=en_GB). HPLC is conducted at ambient temperatures and thus agencies such as the U.S. FDA recommends the technique for analyzing thermally labile, nonvolatile, highly polar compounds (Agilent Technologies 2011). The technique has found numerous applications in the food industry including the detection of adulteration and the detection of certain compounds of different foods, such as berries (Mertz et al. 2007, Määttä et al. 2003), plant oils (Lísa et al. 2009, Cserháti et al. 2005, Cunha and Oliveira 2006, Park et al. 2010, Dionisi et al. 1995, Aparicio and Aparicio-Ruíz 2000), apples (Tsao et al. 2003), fruit juices and drinks (Shui and Leong 2002, Gamache and Acworth 1993) and others.

2.2.2 GC

The term GC is used for the description of a group of analytical separation techniques designed to analyze volatile compounds in the gas phase. In gas chromatography, the individual components of the sample of interest are dissolved in a solvent and subjected to vaporization aiming at separating the analytes via the distribution of the sample between the stationary and mobile phases. In this case, the mobile phase is a chemically inert gas that is used to transfer the molecules of the analyte through the column. GC does not use the mobile phase for the interaction with the analyte. The stationary phase can be a solid adsorbent (gas-solid chromatography), or a liquid on an inert support (gas-liquid chromatography) (https://chem.libretexts.org/Textbook_Maps/Analytical_Chemistry/Supplemental_Modules_(Analytical_Chemistry)/Instrumental_Analysis/Chromatography/Gas_Chromatography). GC runs on the basis of partition chromatography to separate components. The gas flows at a constant rate from the cylinder to the liquid layer, while embedded in a solid medium in the column. After its injection into the injection point,the sample is carried by the mobile gas into the column. In the column, separation of the components is carried out due to the differential partition in between the mobile (gas) and stationary phases (https://www.chromacademy.com/lms/sco10/Theory_and_Instrumentation_Of_GC_Introduction.pdf).

The mobile phase or carrier gas is supplied at a pressure of 10–40 psi with a flow rate of 2–50 cm^3 min^{-1}. The pressure is controlled by either using a needle-valve or a mass flow controller. The flow controller allows the flow rate to be kept steady during separation when the temperature increases. Using a needle valve, the flow rate can decrease by increasing the temperature due to the increase in the viscosity level of the carrier gas. The most common carrier gases are nitrogen, helium and hydrogen, while the gas that is used depends on the column type, the relevant cost and the detector (Fifield and Kealey 2000). Depending on whether the compounds are partitioned into gas or liquid, they come out of the column and detected first or later and a computer is used to record the signals as peaks, which can then be used for the determination of the components and their concentrations (https://www.studyread.com/gas-chromatography-principle/). Modern GC is the fastest of all chromatographic techniques. A column can separate hundreds of components within a relatively short time. By taking advantage of the oven temperature and the stationary phase chemistry (polarity), separations of very high complexity can be conducted, such as separations of chiral and other positional isomers (https://www.chromacademy.com/lms/sco10/Theory_and_Instrumentation_Of_GC_Introduction.pdf).

2.2.2.1 GC with Flame-ionization detection (FID)

FID can be used in GC when organic compounds are concerned, to detect the amount of carbon in a sample. The samples are burned in a hydrogen-air flame, a process that produces carbon ions via combustion. Although the process is not very efficient (only 1 in 10^5 carbon ions will produce an ion in the flame) the total amount of ions is directly proportional to the amount of carbon contained in the sample. Measurements of the current from the ions are taken using electrodes. FID is destructive and is not affected by noncombustible gases and water (https://www.jove.com/science-education/10187/gas-chromatography-gc-with-flame-ionization-detection). GC-FID is commonly used in applications in the petrochemical, pharmaceutical and natural gas sectors. An FID commonly uses a hydrogen/air flame. The sample passes through it and organic molecules are oxidized, producing ions. The collection of these ions leads to the production of an electrical signal, which can then be measured. Similarly, to other GC techniques, a carrier gas is needed with low water and oxygen impurities since these can both interact with the stationary phase, causing problems such as high baseline noise and column bleed in the output gas chromatogram, rendering the analyzer less sensitive and decreasing the lifetime of the column. The FID is also characterized by extremely high sensitivity to hydrocarbon impurities in the hydrogen and air used for the flame. These impurities can increase the baseline noise and render the detector less sensitive (https://www.airproducts.com/industries/Analytical-Laboratories/analytical-lab-applications/product-list/gc-with-flame-ionization-detector-gc-fid-analytical-laboratories.aspx?itemId=D6D6641C668A47139A6F1960D9441B93).

The GC-FID can be effectively used to analyze hydrocarbons (e.g., methane, ethane, acetylene, etc.), organic substances that contain hydrocarbons as well as volatile organic compounds (VOCs). The current flow produced by the particles is of higher strength than the signal generated by the pure carrier gas and the fuel gas flame alone. This signal differential is used for providing information about the sample of interest.

In GC-FID a carrier gas is used for transferring the sample from the injector, through the column and into the detector. An inert gas must be used that cannot be adsorbed by the column material. Helium, nitrogen or hydrogen are commonly used as carrier gases. The detector gases, hydrogen and synthetic air, are used as fuel and oxidizing gas, respectively, during the combustion (http://www.aga.is/is/images/AGA%20HIQ%20Flame%20 Ionization%20Detector%20Datasheet%20UK_tcm648-102263.pdf). Studies that have mentioned the use of GC or GC-FID include those of Peng et al. (2015), who dealt with the adulteration of sesame oils with vegetable oils, Andrikopoulos et al. (2001) who used capillary GC to analyze Olive Oil and Seed Oil Triglycerides, aiming at detecting olive oil adulteration and Pellati et al. (2013) who used the technique to assess the authenticity of commercial essential oils.

Some examples of studies in which spectroscopic and chromatographic methods were used for the assessment of the authenticity of foods of plant origin are presented in Table 1.

2.3 DNA-based techniques

2.3.1 Polymerase chain reaction (PCR)

PCR was developed by Kary Mullis in the 1980s. It relies on the use of DNA polymerase for the synthesis of new strands of DNA complementary to a template. Due to the fact that DNA polymerase can add a nucleotide only where a 3′-OH group preexists, a primer is required for the addition of the first nucleotide. This fact allows for a specific region of template sequence to be delineated depending on the amplification that needs to be carried out for each experiment. The PCR reaction leads to the accumulation of billions of copies (amplicons) of the sequence of interest (https://www.ncbi.nlm.nih.gov/probe/docs/techpcr/). The DNA polymerase that was firstly used for PCR was extracted from the bacterium *Escherichia coli*. Even though this enzyme was effectively used for years, allowing an explosion in DNA sequencing methods, it appears to have significant disadvantages when used in PCR. PCR requires that the reaction is performed at an elevated temperature, in order to denature the double-stranded DNA product following each synthesis round. Heating, however, also leads to irreversible inactivation of *E. coli* DNA polymerase, and therefore the addition of fresh aliquots of enzyme is necessary at the start of each cycle. A DNA polymerase that could remain stable during the DNA denaturation stage (95°C) was therefore needed and the bacterium used to yield it was the *Thermophilus aquaticus*. Moreover, since this bacterium is thermophilic, the enzyme can function optimally at temperatures of approximately 72°C, thus allowing the DNA synthesis to be carried out at temperatures higher than the *E. coli* enzyme could tolerate, thus ensuring that copies of

Table 1. Use of spectroscopic and chromatographic methods for the assessment of the authenticity of foods of plant origin.

Food Type/Target	Adulterant/purpose	Method of Analysis	Effectiveness	References
Virgin coconut oil	Cheaper oils (e.g., sunflower, corn)	FT-MIR	Successful classification using DA	Rohman and Man 2011
Extra virgin olive oil	Corn, sunflower, cottonseed, rapeseed oils	MIR	Prediction of adulterants using chemometric analysis—detection limit of adulteration 10%	Gurdeniz and Ozen 2009
Italian virgin olive oils	Characterization - classification of oils		Classification of non-varietal and industrial oils—correct classification in > 95% of cases	Sinelli et al. 2008
Extra virgin olive oils	Differentiation of mixtures of monovarietal oils	FT-IR	Successful classification of cultivars and mixtures—monovarietal differentiation	Gurdeniz et al. 2007
Extra virgin olive oil	Differentiation from cheaper oils	Raman spectroscopy	Effective when used with multivariate data analysis	Duraipandian et al. 2019
Camelia oils	Soybean oil	Attenuated total reflectance MIR	Rapid cost-efficient classification—quantification (with PLS), easy to perform, avoidance of issues related to sample handling and pre-treatment	Wang et al. 2006
		Fibre optic diffuse reflectance NIR		
Spanish and Australian Tempranillo red wine	Geographical classification	Visible and NIR spectroscopy - multivariate analysis	100% and 84.7% correct classification of Australian and Spanish wines, respectively	Liu et al. 2006
Commercial white wines of different varietal origins	Discrimination of varietal origins	Visible and NIR spectroscopy - chemometrics	Correct classification of 96–100% of wines depending on wine	Cozzolino et al. 2003

Table 1 contd. ...

...Table 1 contd.

Food Type/Target	Adulterant/purpose	Method of Analysis	Effectiveness	References
Extra virgin olive oils	Sunflower oil	Visible and NIR spectroscopy	Prediction of adulteration with standard error of 0.8%	Downey et al. 2002
Saffron	Quality control - chemical composition, geographical origin	NIR spectroscopy	Effective, accurate, time effective	Zalacain et al. 2005
Black rice	Authenticity verification			Chen et al. 2018
Lime juice	Determination of purity (natural/synthetic)			Shafiee and Minaei 2018
Brazilian coffee	Authentication of geographical and farming system origins		High classification performance, time and cost effective	Monteiro et al. 2018
Maple syrup	Cane and beet invert syrups, cane and beet sugar solutions	NIR spectroscopy and chemometrics	Can be used but FTIR was more effective	Paradkar et al. 2003
		FTIR and chemometrics	Superior to NIR	
Yerba mate (*Ilex paraguayensis* St. Hil)	Geographical origin discrimination	NIR and multivariate analysis	76–100% overall classification	Cozzolino et al. 2010
Green tea	Measurement of total polyphenols, identification of tea varieties	FT-NIR with multivariate analysis	Successful application, rapid	Sinija and Mishra 2009
Roast green tea	Geographical origin discrimination	FT-NIR and supervised pattern recognition	100% discrimination	Chen et al. 2009
Spanish Olive Oils	Clustering of types	Excitation–Emission Fluorescence Spectroscopy	Clear Distinguishment	Guimet et al. 2004

Wines	Discrimination between geographical subzones	UV/visible/NIR Spectroscopy with chemometrics	Rapid discrimination using the combination of the techniques	Martelo-Vidal et al. 2012
White and rested tequila	Identification of adulteration	UV-visible spectroscopy and chemometrics	Time effective, trustworthy	Contreras et al. 2010
Sesame oil	Corn oil	GC-FID and IRMS	Good tool for adulteration up to 10%	Seo et al. 2010
Sesame Oil	Control of geographical origin	Stable Isotope Analysis and DNA-based Markers	Very powerful, accurate	Horacek et al. 2015
Australian orange juice	Determination of authenticity	Stable carbon isotope analysis	No significant seasonal variations in 13C/12C ratio. No significant variations in combustion temperature.	Antolovich et al. 2001
Apple juice and concentrates	Sugar addition	13C IRMS	Detection limit 5%	Jamin et al. 1997
Chinese teas	Determination of geographical origin	Stable carbon and nitrogen isotope ratios	Discrimination of geographical origin in some cases only	Zhang et al 2012
black Ceylon, Assam and Darjeeling teas	Methyl salicylate	GC-IRMS	Adulterations with methyl salicylate were detectable	Weinert et al. 1999
Asian and North American ginseng	Discrimination	HPLC-Tandem MS	Can be used to unambiguously identify the species	Li et al. 2000
Fruit juices	Analysis of organic acids	LC-MS	Tartaric and quinic acids detected at 1–5 and approximately 1 mg/L. Technique suitable for adulteration testing	Ehling and Shannon 2011

Table 1 contd. ...

...Table 1 contd.

Food Type/Target	Adulterant/purpose	Method of Analysis	Effectiveness	References
Wine	Simultaneous stable carbon isotopic analysis of glycerol and ethanol	LC-MS	No previous isolation needed, simple, rapid	Cabañero et al. 2010
Cocoa and chocolate	Identification of procyanidins	HPLC-MS	Reliable for quantitative analysis	Hammerstone et al. 1999
Red wines	Varietal discrimination and classification	HPLC, advanced data mining and chemometrics	Correctly classified 96% of samples	Vaclavik et al. 2011
Grape juices	Characterization of anthocyanins	Ion trap LC-MS	Successful identification of individual anthocyanins	Wang et al. 2003
Tomato and tomato pastes	Origin determination	ICPMS	The accuracy of the method was acceptable	Feudo et al. 2010
Wheat	Determination of geographical origin	Isotope ratio ICPMS (13C, 15N)	Successful classification using a single analyte, δ13C	Branch et al. 2003
Premium long grain rice	Determination of geographical origin	Isotopic and elemental analysis	Successful estimation of abundance of elements	Kelly et al. 2002
Arabica and robusta ground roast coffees	Differentiation between two varieties	^{1}H NMR	Detection limit of 10–20% w/w	Defernez et al. 2017
Paprika powder	Detection of Sudan I dye	NMR	Rapid and accurate determination	Hu et al 2017
Blackberries (*Rubus glaucus* and *R. adenotrichus*)	Analysis of phenolic compounds	HPLC with Diode Array Detection and Electrospray Ion Trap MS	Very effective for the determination of the compounds	Mertz et al. 2007
Berries (*Ribes* Species)		HPLC with Diode Array Detection and Electrospray Ion MS		Määttä et al. 2003

Plant oils	Statistical evaluation of triacylglycerol composition	HPLC—Atmospheric pressure chemical ionization MS Data	Clear identification of adulteration (from 1% of added sunflower oil as an adulterant)	Lisa et al. 2009
Apples	Investigation of polyphenolic profiles	HPLC	Detection of five major polyphenolic groups (16 compounds in total)	Tsao et al. 2003
Vegetable oils	Discrimination by triacylglycerols profile evaluation	HPLC/evaporative light scattering detector	Adequate for quality control and authenticity assurance	Cunha and Oliveira 2006
Blends of sesame and soybean oil	Discrimination of oils based on triacylglycerol and fatty acid analyses	HPLC—evaporative light scattering detector and GC	Effective discrimination	Park et al. 2010
Olive oil	Assessment of adulteration through detection of tocopherols and tocotrienols	Reversed-phase HPLC—amperometric detection	Can detect variety of compounds, helpful in assessment of authenticity	Dionisi et al. 1995
Fruit juices and drinks	Separation and identification of organic acids and phenolic substances	HPLC	Accuracy: average recovery 85–106%. Successful for numerous organic acids and phenolic compounds.	Shui and Leong 2002
Sesame Oils	Vegetable oils	GC and multivariate data analysis	Detection limit 5%	Peng et al. 2015
Olive oil/seed oil	Analysis of triglycerides to detect adulteration of olive oil	Capillary GC	Adulteration of olive oil with < 5% seed oils can be detected. Method not suitable for distinguishing mixtures of virgin olive oil with refined or residue olive oil	Andrikopoulos et al. 2001
Commercial essential oils of *Rosa damascena* Mill.	Characterization/detection of authenticity	GC/MS, flame ionization detection and elemental analyzer/isotope ratio MS	Effective and reliable	Pellati et al. 2013

the template DNA strand could be generated with higher reliability due to the greater stringency of primer binding (Bartlett and Stirling 2003).

Practically, the amplification of a segment of DNA using PCR requires that the DNA within the sample is denatured or separated into two pieces of single-stranded DNA by heating the sample. Following that, an enzyme called "Taq polymerase" is used for the synthesis of two new strands of DNA, using as templates the original strands. Via this process, the original DNA is duplicated, while each of the new molecules contains one old and one new strand of DNA. Two new copies can then be created by each of these strands and the process is repeated for as many as 30 or 40 times, generating more than one billion copies of the original DNA segment. The automation of the process allows its completion in just a few hours. A machine known as thermocycler is used to automatically change the temperature of the reaction every few minutes,allowing the denaturation and synthesis of DNA (https://www.genome.gov/10000207/polymerase-chain-reaction-pcr-fact-sheet/).A PCR is therefore conducted in three steps:

a. Separation of the two strands of the parent DNA molecule by heating the sample to 95°C for 15 s.
b. Hybridization of primers—the temperature of the sample is lowered to 54°C, allowing the hybridization of the primers.
c. DNA synthesis—performed by heating the sample to 72°C (optimal temperature for Taq DNA polymerase). The polymerase is used to elongate the primers in the direction of the target sequence.

This three-step cycle is then repeated by changing the temperature of the system. Thanks to the fact that the polymerase is thermostable, the PCR can be performed in a closed container, while no reagents are required after the first cycle (Berg et al. 2002).

The benefits of the automation of the method were highlighted in the study of Wilke et al. (1995). Specifically, according to these authors, the PCR automation led to a six fold reduction in the number of experiments, while the overall contamination rate of the whole process was reduced by 68-fold. In this study, contamination cases among experiments conducted manually were related to the lack of personnel experience. In these manual experiments, several of the 81 no-target false positives were strong enough to invalidate 151 samples and positive controls. The 7 no-target control false positives of the automated system were weak bands that were easy to subtract as background and since none of the negative samples had DNA bands, no invalidation of samples was required. It was concluded that the automation of PCR tests is very promising for the reduction of contamination to levels as low as $\leq 0.1\%$).

Many methods relying on the use of PCR have been recommended in the last years for the identification of species in foods, as well as food allergens and genetically modified organisms (GMO), mainly due to them being very specific, sensitive, rapid and cost-effective (Mafra et al. 2008).

2.3.1.1 Multiplex PCR

Multiplex PCR is a technique used widely to amplify multiple targets by conducting one single PCR experiment. In such an assay, amplification of more than one sequence can be achieved by using multiple primer pairs in a reaction mixture. This PCR technique allows the time needed for the analysis to be shortened significantly without resulting in any compromises on the utility of the experiment. Multiplexing reactions can generally be categorized as:

1. Single Template PCR Reaction: a technique that employs a single template and several pairs of forward and reverse primers used for the amplification of specific regions within a template.
2. Multiple Template PCR Reaction: a technique that employs multiple templates and several primer sets within the same reaction tube. This can result in cross hybridization between the primers and the mis-priming with other templates (http://www.premierbiosoft.com/tech_notes/multiplex-pcr.html). EU legislation requires that efficient methods are used for detecting genetically modified ingredients in food. Real-time-PCR is amongst the most effective methods, but the costs of the relevant equipment may be too high for many molecular laboratories (Burns et al. 2003). The technique has been used to authenticate a small range of foods of plant origin and some of the most important studies are reviewed below.

A representative example of the use of the method for the authentication of foods of plant origin is the study of Moon et al. (2016). In this study, a multiplex PCR method using species-specific primers was generated to simultaneously detect members of the Poaceae family, including adlay, barley, maize, rice, and wheat. The method relied on sequence polymorphism in DNA-directed RNA polymerase beta-prime chain genes (rpoC2). Construction of species-specific primer pairs was carried out with a common forward primer and different reverse primers in terms of sequence polymorphisms and gene location. The primer pairs were constructed in order to carry out a PCR aiming at amplifying products of five chloroplast genes of 443 (rice), 346 (barley), 278 (adlay), 221 (wheat), and 96 bp for rice (maize). Multiplex PCR using a series of template DNA concentrations (0.01–10 ng/μL) was employed for optimizing the amplification of fragments from pooled Poaceae. Additionally, species-specific primers were employed for detecting seven commercial flour-mixed products. The advantages of the method (highly sensitive due to the multiplex PCR and specific due to the primers) highlighted its potential for use in the food industry.

A method based on multiplex PCR and quantitative analysis from data obtained using Lab Chip technology was also published. The technique was characterized by high sensitivity, accuracy and precision, while was also faster and required fewer reagents than other techniques that use Lab Chip technology. This technique could therefore be a very good alternative to real time-PCR for detection and quantification of GMOs (Burns et al. 2003). The development of a multiplex PCR-based DNA assay was also reported for the simultaneous detection of multiple target sequences in genetically modified soybean. Internal control target (lectin gene) was included not only for assessing the efficiency of the reactions but also to exclude any false negatives. The method was both sensitive and specific when tested against samples that contained GM soybean and food samples. After the method was compared with a realtime PCR, it was shown that it could be used for semi-quantitative determinations. Similarly, to the study of Burns et al. (2003) it was again indicated in this study that this method can significantly reduce the cost and the time required for analyses, particularly when high numbers of samples need to be examined (Dainese et al. 2004). Also, a molecular screening method was detailed in the study of Forte et al. (2005) that relied on the use of multiplex-PCR involving amplification of certain soya or maize sequences from plant DNA and the amplification of 35S promoter and NOS terminator. The method is aimed for detecting genetically modified soya and maize. The m-PCR assay allowed the quick discrimination of the target GMO product, while it was cost- and time-effective and showed very good reproducibility.

Multiplex-PCR can also be used to detect plant materials such as soya when these are used as adulterants (cheaper substitutes to meat). An example of a study detailing the use of the method for this purpose is that of Tafvizi and Hashemzadegan (2015) who used it to detect soyabean in hamburgers made of beef. Specifically, 10 brands of beef hamburgers from markets in Tehran City in Iran were collected and amplification of 118, 183, and 274 bp fragments was conducted in all samples from soybean lectin, 12 s rRNA, and mitochondrial cytochrome b genes, respectively. It was proved that the samples indeed contained unlabeled soybean.

From a commercial point of view, *Lentinula edodes* is a very important mushroom worldwide, but identification of its strains is traditionally carried out via phenotypic analysis. A multiplex-PCR-based method was therefore used to distinguish strains of *L. edodes* via the development of specific sequence characterized amplified region (SCAR) markers and the establishment of multiplex PCR assays with the SCAR primers. Ten SCAR markers were developed using the randomly amplified polymorphic

DNA (RAPD) and sequence-related amplified polymorphism (SRAP) techniques, from 10 polymorphic RAPD and SRAP bands. The distinctness in SCAR phenotypes between strains allowed the SCAR markers to be used in the characterization of 6 strains. Multiplex-PCR assays were also effectively developed for the combined use of some SCAR markers. In comparison to similar methods, this method can be performed rapidly, while it is not complicated and therefore suitable for the analysis of large numbers of samples (Wu et al. 2010).

Saffron (*Crocus sativus* L.), is another product of high commercial significance. As it is a very expensive product, it is frequently the target of adulteration attempts. Most commonly used adulterants include safflower stamens and corn stigmas. PCR-based markers were therefore developed to allow the accurate detection of these adulterants in saffron. The method was highly reproducible. Six putative RAPD markers developed by random primers, OPA-14, MG-11, MG-12 and AJ-05, were found to be specific to saffron via comparative RAPD analysis of genuine saffron, safflower and corn. Cloning and sequencing of these RAPD markers followed, and six pairs of SCAR primers were generated. The primers allowed the amplification of saffron DNA but not corn and safflower DNA. In the same study, a primer pair was also developed based on ITS sequences for specific amplification of safflower DNA. PCR reactions also amplified 613 bp of ITS region in the safflower genome. The multiplex PCR assays were further designed to combine the use of some SCAR and ITS markers. The new method was proved to be able to produce results rapidly while it was easy to perform and suitable for the analysis of high numbers of samples. It could therefore be widely employed for the identification of adulterants in saffron samples (Babaei et al. 2014).

Anemarrhena asphodeloides Bunge (Jimo) is a very valuable raw material in countries such as China, Korea, and Japan. A number of products such as Sagan, Chog-chag-yag, and Shig-ban-pung are quite similar to Jimo roots and can therefore be used to test methods for the authentication of Jimo using these products as potential adulterants. PCR analysis of the *trn*L-F region has been proved to be a suitable method for identifying species in the *A. asphodeloides* genus. The identification of a single nucleotide polymorphism (SNP) in Jimo, Sagan, Chog-chag-yag, and Shig-ban-pung allowed specific PCR primers to be developed from it and be used for the differentiation of *A. asphodeloides* from Sagan, Chog-chag-yag, and Shig-ban-pung via multiplex PCR. The method was rapid and effective when used for the identification of *A. asphodeloides* (Jigden et al. 2010).

2.3.1.2 Real-time PCR

Real-time PCR is a PCR technique that can be used for the detection and expression analysis of gene(s) in real-time. It has found numerous applications in areas such as quantitative genotyping, genetic variation of inter and intra organisms, forensics and others (Deepak et al. 2007). Real-time PCR can be used to accurately quantify specific nucleic acids in mixtures of high complexity using fluorescent detection of labeled PCR products. The PCR products can be detected using both specific and nonspecific fluorescent probes (Fraga et al. 2008). The real-time PCR makes use of fluorescent reporter molecules for monitoring the generation of amplification products during each PCR cycle. This involves a combination of DNA amplification and detection steps into one homogeneous assay, while eliminating the need for gel electrophoresis for detecting amplification products. Suitable data analysis and/or use of apposite chemistries obviate the need for Southern blotting or DNA sequencing to identify amplicons. The method is simple, specific and very sensitive, with high throughput potential (Bustin 2005).

Real-time PCR has been used to establish traceability of fruits in food matrices (Palmieri et al. 2009, An et al. 2019), to detect plant species in food matrices (Campos et al. 2018, Curtis and Coyle 2014) as well as for the detection of a variety of foods of plant origin such as wheat-based foods (Sonnate et al. 2009, Pasqualone et al. 2007, Terzi et al. 2003), cassava (Wang et al. 2019) basmati rice (Lopez 2008), coffee (Ferreira et al. 2016, Couto et al. 2019), olive oil (Wu et al. 2008), vegetable oils (Su et al. 2019), peanuts (López-Calleja et al. 2013), saffron (Jiang et al. 2014), etc.

The method appears to be a promising alternative in food diagnostics. Amongst its main advantages over conventional culturing, Rodríguez-Lázaro et al. (2013) referred to its time effectiveness, high sensitivity and selectivity and its potential for quantification. They also mentioned, however, the need for expensive equipment, reagents, qualified personnel and the absence of standardized protocols that can hinder its wide implementation in the industry.

2.3.1.3 DNA barcoding

DNA barcoding is one of the most accurate molecular approaches used for assessing the authenticity of food materials. There are however a number of analytical constraints that frequently render its use unsuitable for the food industry (Valentini et al. 2017). It has however been widely recognized as a universal and versatile technique (Galimberti et al. 2014).

Parvathy et al. (2015) employed the method to detect adulterants of plant origin in turmeric (*Curcuma longa* L.). Three barcoding loci, namely *ITS*, *rbcL*, and *matK* were used for PCR amplification. The amplification with the loci *ITS* and *rbcL* was 100% successful. Techen et al. (2014), de Boer et al. (2015) and Selvaraj et al. (2012) used it also for the authentication of herbs, while Bosmali et al. (2012) used microsatellite and DNA-barcode regions typing in combination with high resolution melting (HRM) for the identification of different varieties of lentils.

Coming back to the area of spices, an existing problem in the cinnamon market is the presence and identification of spurious species in traded barks of cinnamon. DNA markers can now be employed to allow the detection of this type of adulteration. A DNA barcoding method was used for detecting these adulterants in real samples of true cinnamon obtained from the market using the loci *rbcL*, *matK* and *psbA-trnH*. The rate of success of the PCR, the sequencing efficiency, inter and intraspecific divergence, and the presence of SNPs were all used for assessing whether each barcode loci can be used for the authentication of *C. verum* from its adulterants. The amplification and sequencing were 100% successful in the cases of *rbcL* and *psbA-trnH* while *matK* was unsuccessful. *rbcL* allowed for higher interspecific divergence in comparison to *psbA-trnH*. *C. cassia*-specific SNPs were found in *rbcL* locus in seven out of the ten market samples. Amongst the 3 loci used, *rbcL* was proved to be the most effective for this particular purpose. In another study, Parvathy et al. (2014) used DNA barcoding aiming at detecting chili adulteration in black pepper powder. The loci *psbA-trnH*, *rbcL* and *rpoC1* were employed in this study. Sequence analysis and BLAST results indicated chili adulteration in two out of nine market samples, which was probably a result of the use of chili to fortify the product. Chili adulteration in black pepper could most effectively be detected using the locus *psbA-trnH*. The method was very effective in detecting adulteration at levels as low as 0.5%.

Ganopoulos et al. (2012a) used Barcode-DNA high-resolution melting (with the locus *ITS2*) to identify and quantify adulterants of main pasture plant species. The method allowed the detection of *Medicagolupulina* adulterants in *Trifoliumpratense* seeds at levels as low as 1:100, thus indicating that it could be used to provide fast, reliable and cost-effective results to authenticate forage and pasture species and quantitatively assess the purity of their seeds. Furthermore, in the study of Chen et al. (2015), amplification and analysis of *ITS* sequences of 43 maca samples, obtained from several different sources, were carried out. *ITS* sequences of 19 potential adulterants of maca were also obtained and subjected to analysis. It was proved that the *ITS* sequence of maca was consistent across all samples and could be uniquely identified when compared with its adulterants. It was therefore indicated that the method can be used to reliably identify maca and its adulterants.

Finally, DNA barcoding was also chosen to easily, reliably and accurately distinguish *Radix Astragali* from its adulterants. Thirty-seven samples covering four genera within the family Fabaceae were analyzed in total. A comparison of four DNA barcoding markers followed, using *ITS*, *matK*, *rbcL*, and *coxI* sequences from nuclear, chloroplast, and mitochondrial genomes. It was indicated that there were higher divergences among genera and interspecies in the case of *ITS*, while *matK* showed sufficient divergence at the intraspecific level. Furthermore, two indels found in the *matK* sequence could be used in PCR studies to distinguish *R. astragali* from its adulterants. It was therefore indicated that a combination of the barcoding regions of *ITS* and *matK* are of high importance for the detection of adulteration of *R. astragali* but was also suggested that the method may be equally effective when used for other herbs (Guo et al. 2010).

2.3.1.4 HRM

HRM is a high accurate and effective method for SNP genotyping, mutation scanning and sequence scanning of DNA. Its development has recently become possible due to the fact that better double-stranded DNA (dsDNA)–binding dyes and next-generation real-time PCR instrumentation have become available. HRM analysis relies on PCR melting (dissociation) curve techniques. Specifically, the method characterizes nucleic acid samples based on their disassociation behavior and allows the detection of small sequence differences in PCR amplified sequences via direct melting. The discrimination of samples can also be achieved based on sequence length, G-C content and strand complementarity (http://www.premierbiosoft.com/tech_notes/high_resolution_melting_analysis.html). The method is easy to use for gene scanning and genotyping of individual and multiple SNPs. The technique can rapidly, conveniently and inexpensively provide very accurate results, while can be easily transitioned to high-throughput scale (Słomka et al. 2017). HRM is based on quantitative analysis of the melt curve of a DNA fragment following PCR amplification, and combined with qPCR it allows genetic variation in nucleic acid sequences to be identified by the controlled melting of a double-stranded PCR amplicon. In contrast to standard melt curves that are used in qPCR experiments, HRM melt curves include the collection of melt data in certain temperature increments. The elimination though of background fluorescence from any HRM analysis is of paramount importance for the identification of minor nucleotide changes. New developments in real time PCR instrumentation, software and reagents (e.g., fluorescent dyes) have rendered HRM a valuable analytical technique with the capability to detect very small proportion of variant DNA at sensitivities that approach 5% (Canadian BioTechnologist 2.0 2010).

The technique has been used successfully in a number of studies conducted on foods of plant origin. Ballin et al. (2019) examined the use of the method for the identification of species of herbs. As explained in this study there has been a number of fraud incidents involving herbs, and therefore the development of a method that allows the simultaneous extraction of information from different barcode regions is important from both a universality and discrimination point of view. As a result, DNA amplification from five different barcode regions was conducted in a multiplexed PCR format and HRM was subsequently used. This multiplexed Bar-HRM approach was initially applied to plants at a plant kingdom level and was then gradually narrowed down to cover genetic variability within the Lamiaceae and the Solanaceae families and finally closely related cultivars. The distinct melting profiles obtained for species that originated from 29 different families indicated that the technique has universal use, while effective discrimination was possible for species, sub-species, and a few cultivars via the use of multivariate statistics to HRM. It was therefore proved that the technique is capable of discriminating a vast amount of species within the whole plant kingdom, without the need for prior knowledge of the species' DNA sequence and can take place in a closed system within 2.5 h. In the work of Batista et al. (2009), a method relying on real-time PCR combined with HRM was employed for the discrimination of *Cynara scolymus* from other Cynara species and was used to analyze herbal mixtures and food supplements labelled as containing artichoke. A number of Cynara voucher species (*C. scolymus*, *C. cardunculus*, *C. humilis* and *C. syriaca*) and samples of herbal infusions and supplements were obtained. DNA from plant samples and supplements was extracted using the commercial NucleoSpin Plant II kit. New primers were developed on a microsatellite region of *C. cardunculus* to allow qualitative PCR and real-time PCR assays to be developed in order to differentiate *Cynara* spp. Prior to the specific PCR assays, DNA extracts were positively tested aiming at a universal eukaryotic sequence (18S rRNA gene). The qualitative PCR results were specific to Cynara genus. When the method was further developed, it was indicated that the tested *Cynara* spp. were grouped in three distinct clusters with a level of confidence higher than 99.4%, thus allowing *C. scolymus* to be discriminated from the others. With regards to the plant products and supplements, it was shown that all samples, apart from one food supplement sample, contained a eukaryotic gene. The analysis confirmed the presence of *C. scolymus* with a high level of confidence (> 98.8%) in the tested samples.

Ganopoulos et al. (2012b) reported that as a result of its high price, "Fava Santorinis" is often adulterated with other legume products derived from other Lathyrus or Vicia and Pisum species. Using plant DNA barcoding regions (trnL and rpoC) coupled with Bar-HRM, these researchers developed a method that allowed the detection and authentication of "Fava Santorinis". Bar-HRM was once again proved to be very sensitive and was capable of genotyping Lathyrus and its closed relative species as well as detecting admixtures. The sensitivity of the method was high enough to detect 1:100 of non-"Fava Santorinis" in "Fava Santorinis" commercial products.

Sea buckthorn (Hippophae) is another crop that has been analyzed using this method due to its growing demand that has given rise to a number of adulteration incidents and attempts. AnHRM assay, targeting a DNA barcoding region of the internal transcribed spacer 2 (ITS2) (Bar-HRM) was used, aiming at identifying seven native Chinese Hippophae species and authenticating commercial sea buckthorn products. Melting data from the HRM assay indicated that all Hippophae species could be effectively differentiated.

The method has also been used to verify the authenticity of products such as olive oil (Ganopoulos et al. 2012a, Gomes et al. 2018, Pereira et al. 2018, Vietina et al. 2013), wine (Pereira et al. 2017, Pereira et al. 2018).

2.4 Enzyme-linked immunosorbent assay (ELISA)

ELISA is an immunological assay employed widely for measuring antibodies, antigens, proteins and glycoproteins in samples of biological origin (https://www.immunology.org/public-information/bitesized-immunology/experimental-techniques/enzyme-linked-immunosorbent-assay). The basic ELISA can be differentiated from other antibody-based assays due to the fact that specific and non-specific separation interactions take place via serial bindings to a solid surface, frequently a multiwell plate made of polystyrene, and also because quantitative results can be obtained. The ELISA produces a colored end product that is related to the level of analyte present in the original sample. The assay is characterized by speed and simplicity and since a high number of samples can be processed at the same time, it is very popular amongst researchers. The development of ELISAs in the 1970s aimed at replacing radioimmunoassays. They are still widely used in their original format and with several modifications allowing the use of several analytes per well, very sensitive readouts, as well as direct cell-based output (https://www.bio-rad-antibodies.com/an-introduction-to-elisa.html).

As ELISAs involve multistep reactions,measuring analyte levels can be very complicated when complex samples are examined, due to the presence of a high amount of analytes interfering with the measurement (https://www.mybiosource.com/learn/ELISA). Three types of ELISA exist:

a) Direct ELISA—this is the least complicated variation of ELISA in which passive adsorption is used for binding the antigen to the solid phase. Washing removes any unbound molecules and direct incubation with a conjugated antibody then follows. After incubation and additional washing, a substrate is inserted,producing a signal. The substrate reaction is terminated after a certain period of time and the signal is quantified. Direct ELISA is usually employed to titter conjugated secondary antibodies and is very effective for the estimation of antigen cross-reactivity (https://rockland-inc.com/elisa.aspx).

b) Indirect ELISA begins by coating a multi-well plate. An unlabeled primary antibody, specific to an antigen, is applied. Next, an enzyme-labeled secondary antibody is bound to the first antibody. The secondary antibody is often an anti-species polyclonal antibody (https://www.bosterbio.com/protocol-and-troubleshooting/elisa-principle).

c) Sandwich ELISA—It is highly sensitive and can be employed for measuring the antigen concentration of samples. Quantification of the antigen of interest between two layers of antibodies (the capture and the detection antibody) is carried out. These antibodies have to bind to non-overlapping epitopes on the antigen. Monoclonal or affinity-purified polyclonal antibodies can be employed to play the role of capture and detection antibodies, a decision that is usually taken based on how sensitive the assay needs to be (https://www.genwaybio.com/services/sandwich-elisa).

Some examples of studies in which DNA-based methods were used for the assessment of the authenticity of foods of plant origin are presented in Table 2.

2.5 Electronic noses

When certain foods are considered, their aroma is specific enough to allow the discrimination of an authentic product from its fraudulent counterpart. Electronic nose (e-nose) is a quick and effective technique that can be used for the determination of the aroma of a product. No complex sample preparation process is required (Gliszczyńska-Świgło and Chmielewski 2017). Electronic-nose instrumentation, adopted from a number of different types of aroma-sensor technologies, can be used in a variety of applications in agriculture and forestry. Latest advances in this technology concerning applications in plant sciences include improvements in gas-sensor designs, innovative data analysis methods and pattern-recognition algorithms, as well as advances in material science and systems integration methods. The use of electronic noses has been reported in various industries in the wide field of agriculture, such as agronomy, biochemical processing, botany, cell culture, pesticide detection, plant physiology and others. These aroma-detection technologies have led to improvements in characteristics of products of plant origin and have rendered certain manufacturing processes more efficient and effective (Wilson 2013).

'Electronic nose' systems cover a variety of partially specific electronic chemical gas sensors but also suitable statistical methods and allow complex odors to be recognized. As commercial instruments have been developed, the research around the use of electronic noses has increased significantly. The gas sensor technologies used commercially cover metal

Table 2. Use of DNA-based techniques for the assessment of the authenticity of foods of plant origin.

Food Type/Target	Adulterant/purpose	Method of Analysis	Effectiveness	References
Durum wheat-based pasta and bread	Quantification of soft wheat adulteration	Real-Time PCR	Reliable quantification, suitable for verification of labeling	Sonnante et al. 2009
Basmati rice	Non-basmati rice		Very accurate, detection limit 1%	Lopez 2008
Semolina and durum wheat bread	Soft wheat		Detection of wheat in semolina and breads with a threshold of 2.5%	Pasqualone et al. 2007
Commercial coffees	Cereals		Sensitive, specific, quantification down to 0.6 pg, 14 pg and 16 pg of barley, corn and rice DNA, respectively	Ferreira et al. 2016
Olive oil	Detection of olive oil		Detection limit of 0.07 ng olive DNA	Wu et al. 2008
Processed foods	Detection of peanut (allergen)		Detection of 0.1 ppm	López-Calleja et al. 2013
Berry fruits (aronia, blackberry, cranberry, and strawberry)	Detection of berries in commercial foods		Highly efficient, suitable	An et al. 2019
Coffea arabica and *C. canephora*	Discrimination between the two		Highly specific for identification and quantification	Couto et al. 2019
Edible vegetable oils (soybean, peanut, rapeseed, and cottonseed oils and their blends)	Identification of species and determination of ratio in mixed vegetable oils blended from two or three oil-species		Simple, sensitive and efficient	Su et al. 2019
Sweet potato (*Ipomoea batatas*) noodles	Cassava (*Manihot esculenta*)		Very specific (detection limit 1%)	Wang et al. 2019

Poaceae	Simultaneous detection of adlay, barley, maize, rice, wheat	Multiplex PCR	Effective, broad applications possible	Moon et al. 2016
Saffron	Safflower		Effective, simple, rapid	Babaei et al. 2014
Soya	GMO		Sensitive, accurate, quick, quantitative determination, results comparable to real-time PCR technique	Burns et al. 2003
			Very sensitive, time and cost-effective	Dainese et al. 2004
Soya and Maize				Forte et al. 2005
Anemarrhena asphodeloides Bunge (Jimo)	Sagan, Chog-chag-yag, and Shig-ban-pung		Rapid, effective	Jigden et al. 2010
Lentinula edodes	Other strains		Suitable, time and cost-effective	Wu et al. 2010
Turmeric powder	Detection of plant-based adulterants	DNA Barcoding	Efficient, 100% success in PCR and sequencing of loci rbcL and ITS, matK showed no amplification	Parvathy et al. 2015
Medicinal plants (herbs)			Reliable regardless of other environmental factors	Techen et al. 2014
			Effective and powerful method	De Boer et al. 2015
Boerhavia diffusa			DNA Barcode ITS was found effective in detecting adulterants	Selvaraj et al. 2012
Black pepper (*Piper nigrum* L.)	Detection of adulteration with other pepper		Very accurate, *psbA-trnH* provided optimum results.	Parvathy et al. 2014

Table 2 contd. ...

...Table 2 contd.

Food Type/Target	Adulterant/purpose	Method of Analysis	Effectiveness	References
Lentils (*Lens culinaris*)	Distinction and identification of different varieties	Microsatellite and DNA-barcode regions typing combined with High Resolution Melting	Very effective identification and quantification of adulteration	Bosmali et al. 2012
Cinnamomum species		DNA barcoding	*rbcL* locus proved to be efficient in tracing out adulterants in traded cinnamon.	Swetha et al. 2014
Medicago lupulina adulterants in *Trifolium pratense* seeds	Identification and quantification of adulterant		*Bar-HRM* (barcoding region *ITS2*) detected the adulterant when present at levels as low as 1:100.	Ganopoulos et al. 2012b
Radix Astragali	Distinguishing *Radix Astragali* from its adulterants		Very effective when the combined barcoding regions of *ITS* and *matK* we used.	Guo et al. 2010
Angiosperms, gymnosperm, mosses, and liverwort (Marchantiophyta).	Discriminate between species	HRM	Effective, inexpensive.	Ballin et al. 2019
Artichoke (*Cynara scolymus* L.)	*C. scolymus*, *C. cardunculus*, *C. humilis* and *C. syriaca*		Highly accurate (level of confidence > 98.8%), rapid.	Batista et al. 2016
Lathyrus clymenum	Detection and quantification of other *Lathyrus* or *Vicia* and *Pisum* species		Rapid, cost effective with high resolution.	Ganopoulos et al. 2012a

Olive oil	Canola oil	Barcode DNA HRM	Limit of detection of 1% (w/w)	Ganopoulos et al. 2013
	Discrimination between varieties of olive oil	Microsatellite HRM	Reliable, time- and cost-effective.	Gomes et al. 2018
Sea buckthorn (*Hippophae*)	Authentication of commercial products containing *Hippophae* species	Barcode DNA HRM	Effective identification of species	Liu et al. 2018
Wine	Genotype discrimination	HRM	Rapid, efficient, suitable for wine	Pereira et al. 2017
Extra virgin olive oil	Hazelnut, maize, sunflower, peanut, sesame, soybean, rice and pumpkin		Cost effective, accurate	Vietina et al. 2013

oxide semiconductors, metal oxide semiconductor field effect transistors, organic conducting polymers, and piezoelectric crystal sensors as well as fibreoptic, electrochemical sensors and others. The technology can be used in processes such as quality control, process monitoring, freshness assessment, shelf-life evaluation and assessment of authenticity (Schaller et al. 1998) as well as the determination of geographical origin of products, contamination and spoilage. In the majority of cases, successful classification of samples has been above 85%, however, the proper assessment of several performance characteristics (e.g., drift, humidity influence, selectivity and signal to noise ratio) of electronic noses can be very challenging (Berna 2010).

2.6 Chemometric methods

"Chemometrics" as a term was reported several decades ago to describe the techniques related to the mathematical manipulation and interpretation of chemistry-related data. It has only been a few decades though, that chemometrics has been widely adopted and attracted particular interest. Scientists in modern laboratories frequently face the need to handle, analyze and interpret a huge amount of information in the form of numerical and graphical data. It is therefore reasonable to expect that the use of these data can limit the operational efficiency and effectiveness of a laboratory. More sophisticated analytical instrumentation has also been used lately for direct on-line or in-line process monitoring purposes, which also increases the need for the use of computerized decision making. Chemometrics, as a subject, complements laboratory automation. Similarly, to the fact that automation is related to the tools used to handle the mechanics and chemistry within a laboratory, chemometrics is concerned with the application of mathematical and statistical operations to assist in data handling (Adams 1995). In the last decade, numerous multivariate statistical techniques have been employed widely in food science and technology. It is common that chemometrics is employed in cases where a large and complex dataset needs to be handled, usually when this is related to sample numbers, types, and responses. The results obtained are used to authenticate the geographical origin, farming systems or detect adulteration of certain food commodities. Overall, chemometrics can significantly facilitate the extraction of results in cases of extensive, multiple, and complex problems, in a multifactorial and holistic frame. It can therefore be a tool employed by governmental bodies and industries whenever there is a need for monitoring the quality of foods, raw materials, and processes via the manipulation of high-dimensional data sets (Granato et al. 2018).

3. Conclusions

A number of methods and techniques have been developed to detect fraud and adulteration in foods of plant origin. Innovative techniques are being continuously designed and their effectiveness, practical use, cost of use and accuracy are tested against different ranges of plant species and complex plant matrices or commercial products. The selection of a method is highly dependent on the type and complexity of material tested and/or the adulterant or characteristic of interest, the relevant costs as well as the sophistication and instrumentation needed for the preparation of samples. The majority of spectroscopic techniques (MIR, NIR, FT-NIR, Raman, UV-Vis and Fluorescence spectroscopy, IRMS, NMR, ICP-MS, LC-MS, etc.) commonly used for analyses in the food industry have been used to test materials of plant origin, detect fraud or verify their authenticity. Chromatographic techniques are also often the methods of choice due to their reliability and accuracy in detecting substitution of ingredients with cheaper alternatives even at trace levels. The development of DNA-based methods relying on PCR and ELISA as well as electronic noses have more recently been introduced in the industry and are used mostly in combination with chemometrics providing highly accurate results and offering solutions when very complex datasets need to be processed. Multiplex-PCR, DNA barcoding and HRM have also been used in a number of studies and have recently been introduced to the industry improving the accuracy, reliability and reproducibility of the results. Although the sophistication of the methods used has commonly been increasing to allow the detection of more and more sophisticated adulteration methods, there is now a trend towards the development of particularly sensitive techniques, yet characterized by high universality and time-effectiveness that can be performed following relatively simple methodologies and inexpensive instrumentation and reagents. Although full automation of testing is not yet practical, advancements in these techniques could allow that to happen in the near future.

Abbreviations

Charged coupled device	CCD
Enzyme-linked immunosorbent assay	ELISA
Electrospray ionisation	ESI
European Union	EU
Food and Drug Administration	FDA
Flame-ionization detection	FID
Fourier Transform Near-infrared	FT-NIR
Fourier transform infrared	FTIR
Gas chromatography	GC

Genetically modified organisms	GMO
High Performance Liquid Chromatography	HPLC
High Resolution Melting	HRM
Inductively Coupled Plasma-Mass Spectrometry	ICP-MS
Infrared	IR
Isotope ratio mass spectrometer	IRMS
Liquid chromatography-mass spectrometry	LC-MS
Mid-infrared	MIR
Mass spectrometer/spectrometry	MS
Multivariate analysis	MVA
Near-infrared	NIR
Nuclear Magnetic Resonance	NMR
Randomly amplified polymorphic DNA	RAPD
Stable isotope ratio mass spectrometry	SIRMS
Single nucleotide polymorphism	SNP
Sequence-related amplified polymorphism	SRAP
Polymerase chain reaction.	PCR
Sequence characterized amplified region	SCAR
United States	US
Ultraviolet-visible	UV-Vis
Volatile organic compound	VOC

References

Adams, M.J. 1995. Chemometrics in Analytical Spectroscopy. N.W. Barnett (Ed.). Wolverhampton, UK.

Al-Jowder, O., E.K. Kemsley and R.H. Wilson. 1997. Mid-infrared spectroscopy and authenticity problems in selected meats: a feasibility study. Food Chemistry 59(2): 195–121.

An, J., J.-C. Moon, J.H. Kim, G.S. Kim and C.S. Jang. 2019. Development of DNA-based species-specific real-time PCR markers for four berry fruits and their application in commercial berry fruit foods. Applied Biological Chemistry 62: 1–7.

Andrikopoulos, N.K., I.G. Giannakis and V. Tzamtzis. 2001. Analysis of olive oil and seed oil triglycerides by capillary gas chromatography as a tool for the detection of the adulteration of olive oil. Journal of Chromatographic Science 39: 137–145.

Antolovich, M., X. Li and K. Robards. 2001. Detection of adulteration in australian orange juices by Stable Carbon Isotope Ratio Analysis (SCIRA). Journal of Agricultural and Food Chemistry 49: 2623–2626.

Aparicio, R. and R. Aparicio-Ruíz. 2000. Authentication of vegetable oils by chromatographic techniques. Journal of Chromatography A 881: 93–104.

Babaei, S., M. Talebi and M. Bahar. 2014. Developing an SCAR and ITS reliable multiplex PCR-based assay for safflower adulterant detection in saffron samples. Food Control 35: 323–328.

Baeten, V. and R. Aparicio. 2000. Edible oils and fats authentication by Fourier transform Raman spectrometry. Biotechnology, Agronomy, Society and Environment 4: 196–203.

Baeten, V. and M. Meurens. 1996. Detection of virgin olive oil adulteration by Fourier Transform Raman Spectroscopy. Journal of Agricultural and Food Chemistry 44(8): 2225–2230.

Baeten, V., J.A.F. Pierna, P. Dardenne, M. Meurens, D.L. García-González and R. Aparicio-Ruiz. 2005. Detection of the presence of hazelnut oil in olive oil by FT-Raman and FT-MIR Spectroscopy. Journal of Agricultural and Food Chemistry 53: 6201–6206.

Ballin, N.Z., J.O. Onaindia, H. Jawad, R. Fernandez-Carazo and A. Maquet. 2019. High-resolution melting of multiple barcode amplicons for plant species authentication. Food Control 105: 141–150.

Barth, A. and P. Haris. 2009. Infrared spectroscopy-past and present. pp. 1–52. *In*: Barth, A. and P. Haris [eds.]. Advances in Biomedical Spectroscopy, volume 2. Biological and Biomedical Infrared Spectroscopy. Lancaster, UK.

Bartlett, J.M.S. and D.A. Stirling. 2003. A short history of the polymerase chain reaction. pp. 3–6. *In*: Bartlett, J.M.S. and D.A. Stirling [eds.]. PCR Protocols. Humana Press Inc. Totowa, NJ, USA.

Batista, A., J. Costa, T.J.R. Fernandes, J.S. Amaral, M.B.P.P. Oliveira and I. Mafra. 2016. High resolution melting analysis as a new tool to authenticate plant food supplements: the case of artichoke (*Cynara Scolymus*). PROGRAM & BOOK OF ABSTRACTS Assuring the integrity of the food chain: FIGHTING FOOD FRAUD April 6–7, 2016 Prague, Czech Republic.

Bazilio, A. and J. Weinrich. 2012. The Easy Guide to: Inductively Coupled Plasma-Mass Spectrometry (ICP-MS) pp. 1–11.

Berg, J.M., J.L. Tymoczko and L. Stryer. 2002. Exploring genes and genomes. pp. 134–163. *In*: Berg, J.M., J.L. Tymoczko and L. Stryer [eds.]. Biochemistry 6th Ed. W. H. Freeman and Company, USA.

Berna, A. 2010. Metal oxide sensors for electronic noses and their application to food analysis. Sensors 10: 3882–3910.

Bosmali, I., I. Ganopoulos, P. Madesis and A. Tsaftaris. 2012. Microsatellite and DNA-barcode regions typing combined with High Resolution Melting (HRM) analysis for food forensic uses: A case study on lentils (*Lens culinaris*). Food Research International 46: 141–147.

Branch, S., S. Burke, P. Evans, B. Fairman and C.S.J.W. Briche. 2003. A preliminary study in determining the geographical origin of wheat using isotope ratio inductively coupled plasma mass spectrometry with ^{13}C, ^{15}N mass spectrometry. Journal of Analytical Atomic Spectrometry 18: 17–22.

Burns, M., D. Shanahan, H. Valdivia and N. Harris. 2003. Quantitative event-specific multiplex PCR detection of Roundup Ready soya using LabChip technology. European Food Research and Technology 216: 428–433.

Bustin, S.A. 2005. Real-time PCR. pp. 1117–1125. *In*: Fuchs, J. and M. Podda [eds.]. Encyclopedia of Diagnostic Genomics and Proteomics. Marcel Dekker, Inc. New York, USA.

Cabañero, A.I., J.L. Recio and M. Rupérez. 2010. Simultaneous stable carbon isotopic analysis of wine glycerol and ethanol by liquid chromatography coupled to isotope ratio mass spectrometry. Journal of Agricultural and Food Chemistry 58: 722–728.

Campos, C., L. Morello, L. Braglia, D. Breviario and H.G. Cardoso. 2018. A TaqMan real-time PCR method based on alternative oxidase genes for detection of plant species in animal feed samples. PLOS ONE 13: 1–13.

Canadian BioTechnologist 2.0. 2010. A Practical Guide to High Resolution Melt Analysis Genotyping. https://cbt20.wordpress.com/2010/11/03/a-practical-guide-to-high-resolution-melt-analysis-genotyping. Accessed 14.03.2020.

Chen, H., C. Tan and Z. Lin. 2018. Authenticity detection of black rice by near-infrared spectroscopy and support vector data description. International Journal of Analytical Chemistry 1–8.

Chen, J-J., Q-S. Zhao, Y-L. Liu, S-H. Zha and B. Zhao. 2015. Identification of maca (Lepidium meyenii Walp.) and its adulterants by a DNA-barcoding approach based on the ITS sequence. Chinese Journal of Natural Medicines 13(9): 0653–0659.

Chen, Q., J. Zhao and H. Lin. 2009. Study on discrimination of Roast green tea (*Camellia sinensis* L.) according to geographical origin by FT-NIR spectroscopy and supervised pattern recognition. Spectrochimica Acta Part A: Molecular and Biomolecular Spectroscopy 72: 845–850.

Chromedia. 2018. FLUORESCENCE SPECTROMETRY. http://www.chromedia.org/chromedia?waxtrapp=mkqjtbEsHiemBpdmBlIEcCArB&subNav=cczbdbEsHiemBpdmBlIEcCArBP. Accessed 25.12.2018.

Contreras, U., O. Barbosa-García, J.L. Pichardo-Molina, G. Ramos-Ortíza, J.L. Maldonado, M.A. Meneses-Nava, N.E. Ornelas-Soto and P.L. López-de-Alba. 2010. Screening method for identification of adulterate and fake tequilas by using UV–VIS spectroscopy and chemometrics. Food Research International 43: 2356–2362.

Cotte, J.F., H. Casabianca, S. Chardon, J. Lheritier and M.F. Grenier-Loustalot. 2003. Application of carbohydrate analysis to verify honey authenticity. Journal of Chromatography A 1021: 145–155.

Couto, C.C., T.F. Santos, A.M.G.N. Mamede, T.C. Oliveira, A.M. Souza, O. Freitas-Silva and E.M.M. Oliveira. 2019. *Coffeaarabica* and *C. canephora* discrimination in roasted and ground coffee from reference material candidates by real-time PCR. Food Research International 115: 227–233.

Cozzolino, D., H.E. Smyth and M. Gishen. 2003. Feasibility study on the use of visible and near-infrared spectroscopy together with chemometrics to discriminate between commercial white wines of different varietal origins. Journal of Agricultural and Food Chemistry 51: 7703–7708.

Cozzolino, D., M. Holdstock, R.G. Dambergs, W.U. Cynkar and P.A. Smith. 2009. Mid infrared spectroscopy and multivariate analysis: A tool to discriminate between organic and non-organic wines grown in Australia. Food Chemistry 116: 761–765.

Cozzolino, D., E. Restaino and A. Fassio. 2010. Discrimination of yerba mate (*Ilex paraguayensis* St. Hil.) samples according to their geographical origin by 2014. Means of near infrared spectroscopy and multivariate analysis. Sensing and Instrumentation for Food Quality and Safety 4: 67–72.

Cozzolino, D. 2015. Infrared spectroscopy as a versatile analytical tool for the quantitative determination of antioxidants in agricultural products, foods and plants. Antioxidants (Basel) 4: 482–497.

Cserháti, T., E. Forgács, Z. Deyl and I. Miksik. 2005. Chromatography in authenticity and traceability tests of vegetable oils and dairy products: a review. Biomedical Chromatography 19: 183–190.

Cunha, S.C. and M.B.P.P. Oliveira. 2006. Discrimination of vegetable oils by triacylglycerols evaluation of profile using HPLC/ELSD. Food Chemistry 95: 518–524.

Curtis, K. and H.M. Coyle. 2014. Development of a quantitative real-time polymerase chain reaction (RTPCR) assay for plant species. Journal of Forensic Research and Criminal Studies 1: 1–5.

Dainese, E., C. Angelucci, P. De Santis, M. Maccarrone and I. Cozzani. 2004. A Multiplex PCR-Based assay for the detection of genetically modified soybean. Analytical Letters 37(6): 1139–1150.

De Boer, H.J., M.C. Ichim and S.G. Newmaster. 2015. DNA barcoding and pharmacovigilance of herbal medicines. Drug Safety 38: 611–620.

Deepak, S.A., K.R. Kottapalli, R. Rakwal, G. Oros, K.S. Rangappa, H. Iwahashi, Y. Masuo, and G.K. Agrawal. 2007. Real-time PCR: Revolutionizing detection and expression analysis of genes. Current Genomics 8: 234–251.

Defernez, M., E. Wren, A.D. Watson, Y. Gunning, I.J. Colquhoun, G. Le Gall, D. Williamson, and E.K. Kemsleya. 2017. Low-field ^{1}H NMR spectroscopy for distinguishing between arabica and robusta ground roast coffees. Food Chemistry 216: 106–113.

Dennis, M.J. and P.R. Ashurst. 1996. An introduction to food authentication. pp. 1–14. *In*: Dennis, M.J. and P.R. Ashurst [eds.]. Food Authentication. Chapman & Hall. London, UK.

Dionisi, F., J. Prodolliet and E. Tagliaferri. 1995. Assessment of olive oil adulteration by reversed-phase high-performance liquid chromatography/amperometric detection of tocopherols and tocotrienols. Journal of the American Oil Chemists' Society 72: 1505–1511.

Downey, G., P. McIntyre and A.N. Davies. 2002. Detecting and quantifying sunflower oil adulteration in extra virgin olive oils from the eastern mediterranean by visible and near-infrared spectroscopy. Journal of Agricultural and Food Chemistry 50: 5520–5525.

Duraipandian, S., J.C. Petersen and M. Lassen. 2019. Authenticity and concentration analysis of extra virgin olive oil using spontaneous raman spectroscopy and multivariate data analysis. Applied Sciences 9(12): 1–11.

Edith, I.N. and E.M. Ochubiojo. 2012. Food Quality Control: History, Present and Future. Dr. Benjamin Valdez [ed.], ISBN: 978-953-307-916-5, InTech, Available from: http://www.intechopen.com/books/scientific-health-and-social-aspects-of-the-food-industry/food-quality-control-history-present-and-future.

Ehling, S. and S. Cole. 2011. Analysis of organic acids in fruit juices by liquid chromatography–mass spectrometry: an enhanced tool for authenticity testing. Journal of Agricultural and Food Chemistry 59: 2229–2234.

Emwas, A.H. 2015. The strengths and weaknesses of NMR spectroscopy and mass spectrometry with particular focus on metabolomics research. Methods in Molecular Biology 1277: 161–193.

Esteki, M., Z. Shahsavari and J. Simal-Gandara. 2018. Use of spectroscopic methods in combination with linear discriminant analysis for authentication of food products. Food Control 91: 100–112.

European Commission. Food Law General Requirements. https://ec.europa.eu/food/safety/general_food_law_en. Accessed 25.12.2018.

Farrukh, M.A. 2012. Advanced Aspects of Spectroscopy. DOI: 10.5772/2757.

Ferreira, T., A. Farah, T.C. Oliveira, I.S. Lima, F. Vitório and M.M. Oliveirac. 2016. Using Real-Time PCR as a tool for monitoring the authenticity of commercial coffees. Food Chemistry 199: 433–438.

Feudo, G.L., A. Naccarato, G. Sindona and A. Tagarelli. 2010. Investigating the origin of tomatoes and triple concentrated tomato pastes through multielement determination by inductively coupled plasma mass spectrometry and statistical analysis. Journal of Agricultural and Food Chemistry 58: 3801–3807.

Fifield, F.W. and D. Kealey. 2000. Principles and Practice of Analytical Chemistry. Blackwell Science Ltd, Oxford, UK.

Filho, E.G.A. 2016. Nuclear magnetic resonance applied to food analysis. Journal of Experimental Food Chemistry. International Conference on Food Chemistry and Hydrocolloids. August 11–12, 2016. Toronto, Canada.

Forte, V.T., A. Di Pinto, C. Martino, G.M. Tantillo, G. Grasso and F.P. Schema. 2005. A general multiplex-PCR assay for the general detection of genetically modified soya and maize. Applied Biological Chemistry 62: 1–7.

Fraga, D., T. Meulia and S. Fenster. 2008. Real-time PCR. Current Protocols 00: 10.3.1–10.3.34.

Galimberti, A., M. Labra, A. Sandionigi, A. Bruno, V. Mezzasalma and F. De Mattia. 2014. DNA barcoding for minor crops and food traceability. Advances in Agriculture, 1–8.

Gamache, P., E. Ryan and I.N. Acworth. 1993. Analysis of phenolic and flavonoid compounds in juice beverages using high-performance liquid chromatography with coulometric array detection. Journal of Chromatography A 1: 143–150.

Ganopoulos, I., P. Madesis and A. Tsaftaris. 2012a. Universal ITS2 barcoding DNA region coupled with high-resolution melting (HRM) analysis for seed authentication and adulteration testing in leguminous forage and pasture species. Plant Molecular Biology Reporter 30: 1322–1328.

Ganopoulos, I., P. Madesis, N. Darzentas, A. Argiriou and A. Tsaftaris. 2012b. Barcode high resolution melting (Bar-HRM) analysis for detection and quantification of PDO "Fava Santorinis" (*Lathyrusclymenum*) adulterants. Food Chemistry 133: 505–512.

Ganopoulos, I., C. Bazakos, P. Madesis, P. Kalaitzis and A. Tsaftaris. 2013. Barcode DNA high-resolution melting (Bar-HRM) analysis as a novel close-tubed and accurate tool for olive oil forensic use. Journal of the Science of Food and Agriculture 93: 2281–2286.

Gliszczyńska-Świgło, A. and J. Chmielewski. 2017. Electronic nose as a tool for monitoring the authenticity of food. A Review. Food Analytical Methods 10: 1800–1816.

Gomes, S., R. Breia, T. Carvalho, V. Carnide and P. Martins-Lopes. 2018. Microsatellite High-Resolution Melting (SSR-HRM) to track olive genotypes: from field to olive oil. Journal of Food Science, 83: 2415–2423.

Gordon, M.H. and R. Macrae. 1987a. Infrared spectroscopy. pp. 133–145. *In*: Gordon and Macrae [eds.]. Instrumental Analysis in the Biological Sciences. Blackie Academic & Professional, Glasgow, UK.

Gordon, M.H. and R. Macrae. 1987b. Nuclear magnetic resonance spectroscopy. pp. 146–166. *In*: Instrumental Analysis in the Biological Sciences. Blackie Academic & Professional, Glasgow, UK.

Gordon, M.H. and R. Macrae. 1987c. UV-visible spectrophotometry. pp. 92–116. *In*: Gordon and Macrae [eds.]. Instrumental Analysis in the Biological Sciences. Blackie Academic & Professional, Glasgow, UK.

Gottlieb, S. 2018. We're Committed to Guarding Against the Intentional Adulteration of Food and Implementing the New Rule Efficiently. FDA U.S> Food & Drug Administration. https://blogs.fda.gov/fdavoice/index.php/2018/03/were-committed-to-guarding-against-the-intentional-adulteration-of-food-and-implementing-the-new-rule-efficiently/. Accessed 25.12.2018.

Grassi, S. and C. Alamprese. 2018. Advances in NIR spectroscopy applied to process analytical technology in food industries. Current Opinion in Food Science 22: 17–21.

Guimet, F., R. Boqué and J. Ferré. 2004. Cluster analysis applied to the exploratory analysis of commercial spanish olive oils by means of excitation–emission fluorescence spectroscopy. Journal of Agricultural and Food Chemistry 52: 6673–6679.

Guo, H., W. Wang, N. Yang, B. Guo, S. Zhang, R. Yang, Y. Yuan, J. Yu, S. Hu, Q. Sun, and J. Yu. 2010. DNA barcoding provides distinction between *Radix Astragali* and its adulterants. Science China Life Sciences 53: 992–999.

Gurdeniz, G., F. Tokatli and B. Ozen. 2007. Differentiation of mixtures of monovarietal olive oils by mid-infrared spectroscopy and chemometrics. European Journal of Lipid Science and Technology 109: 1194–1202.

Gurdeniz, G. and B. Ozen. 2009. Detection of adulteration of extra-virgin olive oil by chemometric analysis of mid-infrared spectral data. Food Chemistry 116: 519–525.

Hammerstone, J.F., S.A. Lazarus, A.E. Mitchell, R. Rucker and H.H. Schmitz. 1999. Identification of procyanidins in cocoa (Theobroma cacao) and chocolate using high-performance liquid chromatography/mass spectrometry. Journal of Agricultural and Food Chemistry 47: 490–496.

Haris, P.I. 2009. Series editor preface. pp. ii. *In*: Harris (Ed.). Advances in Biomedical Spectroscopy. Biological and Biomedical Infrared Spectroscopy. IOS Press BV, Amsterdam, The Netherlands.

Hewlett-Packard Company. 1998. Basics of LC/MS. USA.

Horacek, M., K. Hansel-Hohl, K. Burg, G. Soja, W. Okello-Anyanga and S. Fluch. 2015. Control of origin of sesame oil from various countries by stable isotope analysis and DNA based markers—A pilot study. PLoS One. doi: 10.1371/journal.pone.0123020. eCollection 2015.

http://hiq.linde-gas.com/en/analytical_methods/liquid_chromatography/high_performance_liquid_chromatography.html. Accessed 08.09.2018.

http://orgchem.colorado.edu/Spectroscopy/irtutor/IRtheory.pdf Accessed 31.08.2018.

http://orgchem.colorado.edu/Spectroscopy/irtutor/IRtheory.pdf. Accessed 18.07.2018.

http://www.aga.is/is/images/AGA%20HIQ%20Flame%20Ionization%20Detector%20Datasheet%20UK_tcm648-102263.pdf. Accessed 09.09.2018.

http://www.cbrnetechindex.com/Explosives-Detection/Technology-ED/Molecular-Spectroscopy-ED-T/Mid-infrared-Spectroscopy-ED-MS Accessed 31.08.2018.
http://www.chem.ucalgary.ca/courses/351/Carey5th/Ch13/ch13-ir-1.html. Accessed 31.08.2018.
http://www.impublications.com/content/introduction-near-infrared-nir-spectroscopy Accessed 31.08.2018.
http://www.kosi.com/na_en/products/raman-spectroscopy/raman-technical-resources/raman-tutorial.php. Accessed 31.08.2018.
http://www.premierbiosoft.com/tech_notes/high_resolution_melting_analysis.html. Accessed 14.03.2020.
http://www.premierbiosoft.com/tech_notes/multiplex-pcr.html. Accessed 15.09.2018.
http://www.renishaw.com/en/a-basic-overview-of-raman-spectroscopy--25805. Accessed 31.08.2018.
http://www.spectroscopyonline.com/sample-preparation-problem-solving-inductively-coupled-plasma-mass-spectrometry-liquid-introduction. Accessed 08.09.2018.
http://www.waters.com/waters/en_GB/How-Does-High-Performance-Liquid-Chromatography-Work%3F/nav.htm?cid=10049055&locale=en_GB. Accessed 31.08.2018.
https-//chem.libretexts.org/Textbook_Maps/Physical_and_Theoretical_Chemistry_Textbook_Maps/Supplemental_Modules_(Physical_and_Theoretical_Chemistry)/Spectroscopy/Vibrational_Spectroscopy/Infrared_Spectroscopy/Infrared%3A_Interpretation. Accessed 06.08.2018.
https-//www.gov.uk/government/groups/food-authenticity-steering-group#contact-details. Accessed 05.08.2018.
https://chem.libretexts.org/Textbook_Maps/Analytical_Chemistry/Supplemental_Modules_(Analytical_Chemistry)/Instrumental_Analysis/Chromatography/Gas_Chromatography. Accessed 09.09.2018
https://rockland-inc.com/elisa.aspx. Accessed 23.09.2018.
https://www.airproducts.com/industries/Analytical-Laboratories/analytical-lab-applications/product-list/gc-with-flame-ionization-detector-gc-fid-analytical-laboratories.aspx?itemId=D6D6641C668A47139A6F1960D9441B93. Accessed 09.09.2018.
https://www.azom.com/materials-equipment.aspx?cat=125. Accessed 31.08.2018.
https://www.bio-rad-antibodies.com/an-introduction-to-elisa.html. Accessed 23.09.2018.
https://www.bosterbio.com/protocol-and-troubleshooting/elisa-principle. Accessed 23.09.2018.
https://www.bruker.com/products/infrared-near-infrared-and-raman-spectroscopy/ft-nir.html. Accessed 31.08.2018.
https://www.chromacademy.com/lms/sco10/Theory_and_Instrumentation_Of_GC_Introduction.pdf Accessed 09.09.2018.
https://www.genome.gov/10000207/polymerase-chain-reaction-pcr-fact-sheet/. Accessed 09.09.2018.
https://www.genwaybio.com/services/sandwich-elisa. Accessed 23.09.2018.
https://www.globalspec.com/learnmore/labware_scientific_instruments/spectrometers_analytical_photometers/infrared_spectrometers. Accessed 31.08.2018.
https://www.immunology.org/public-information/bitesized-immunology/experimental-techniques/enzyme-linked-immunosorbent-assay. Accessed 23.09.2018.
https://www.jeol.co.jp/en/products/category_nmr.html. Accessed 08.09.2018.
https://www.jove.com/science-education/10187/gas-chromatography-gc-with-flame-ionization-detection. Accessed 09.09.2018.
https://www.mybiosource.com/learn/ELISA. Accessed 23.09.2018.
https://www.ncbi.nlm.nih.gov/probe/docs/techpcr/. Accessed 09.09.2018.
https://www.ru.nl/science/gi/facilities-activities/elemental-analysis/icp-ms/. Accessed 08.09.2018.
https://www.studyread.com/gas-chromatography-principle/ Accessed 09.09.2018.

https://www.thermofisher.com/uk/en/home/industrial/spectroscopy-elemental-isotope-analysis/spectroscopy-elemental-isotope-analysis-learning-center/molecular-spectroscopy-information/nir-technology.html. Accessed 31.08.2018.

Hu, Y., S. Wang, S. Wang and X. Lu. 2017. Application of nuclear magnetic resonance spectroscopy in food adulteration determination: the example of Sudan dye I in paprika powder. Scientific Reports 7: 2637: 1–9.

Huang, W.E., M. Li, R.M. Jarvis, R. Goodacre and S.A. Banwart. 2010. Shining light on the microbial world: The application of Raman microspectroscopy. pp. 153–186. *In*: Laskin, A.I., S. Sariaslani and G.M. Gadd [eds.]. Advances in Applied Microbiology. Elsevier Inc. London, UK.

Husted, S., B.F. Mikkelsen, J. Jensen and N.E. Nielsen. 2004. Elemental fingerprint analysis of barley (*Hordeumvulgare*) using inductively coupled plasma mass spectrometry, isotope-ratio mass spectrometry, and multivariate statistics. Analytical and Bioanalytical Chemistry 378: 171–182.

Jamin, E., J. Gonzalez, G. Remaud, N. Naulet, G.G. Martin, D. Weber, A. Rossmann and H.-L. Schmidt. 1997. Improved detection of sugar addition to apple juices and concentrates using internal standard ^{13}C IRMS. AnalyticaChimicaActa 347: 359–368.

Jiang, C., L. Cao, Y. Yuan, M. Chen, Y. Jin and L. Huang. 2014. Barcoding melting curve analysis for rapid, sensitive, and discriminating authentication of saffron (*Crocus sativus* L.) from its adulterants. BioMed Research International 809037: 1–10.

Jigden, B., H. Wang, N. Samdan and D.-C. Yang. 2010. Molecular identification of oriental medicinal plant Anemarrhena asphodeloides Bunge ('Jimo') by multiplex PCR. Molecular Biology Reports 37: 955–960.

Kaur, P. and P.B. O'Connor. 2007. Quantitative determination of isotope ratios from experimental isotopic distributions. Analytical Chemistry 79: 1198–1204.

Kelly, S.D. and C. Rhodes. 2002. Emerging techniques in vegetable oil analysis using stable isotope ratio mass spectrometry. Grasas y Aceites 53: 34–44.

Kelly, S., M. Baxter, S. Chapman, C. Rhodes, J. Dennis and P. Brereton. 2002. The application of isotopic and elemental analysis to determine the geographical origin of premium long grain rice. European Food Research and Technology 214: 72–78.

Kelly, S., K. Heaton and J. Hoogewerff. 2005. Tracing the geographical origin of food: The application of multi-element and multi-isotope analysis. Trends in Food Science and Technology 16: 555–567.

Kunz, M.R., J. Ottaway, J.H. Kalivas, C.A. Georgiou and G.A. Mousdis. 2011. Updating a synchronous fluorescence spectroscopic virgin olive oil adulteration calibration to a new geographical region. Journal of Agricultural and Food Chemistry 59: 1051–1057.

Li, W., C. Gu, H. Zhang, D.V.C. Awang, J.F. Fitzloff, H.H.S. Fong and R.B. van Breemen. 2000. Use of high-performance liquid chromatography–tandem mass spectrometry to distinguish *Panax ginseng* C.A. Meyer (Asian Ginseng) and *Panaxquinquefolius* L. (North American Ginseng). Analytical Chemistry 72: 5417–5422.

Lísa, M., M. Holčapek and M. Boháč. 2009. Statistical evaluation of triacylglycerol composition in plant oils based on high-performance liquid chromatography–atmospheric pressure chemical ionization mass spectrometry data. Journal of Agricultural and Food Chemistry, 57: 6888–6898.

Liu, L., D. Cozzolino, W.U. Cynkar, M. Gishen and C.B. Colby. 2006. Geographic classification of spanish and australian tempranillo red wines by visible and near-infrared spectroscopy combined with multivariate analysis. Journal of Agricultural and Food Chemistry 54: 6754–6759.

Liu, Y., L. Xiang, Y. Zhang, X. Lai, C. Xiong, J. Li, Y. Su, W. Sun and S. Chen. 2018. DNA barcoding based identification of *Hippophae* species and authentication of commercial products by high resolution melting analysis. Food Chemistry 242: 62–67.

López-Calleja, I.M., S. de la Cruz, N. Pegels, I. González, T. García and R. Martín. 2013. Development of a real time PCR assay for detection of allergenic trace amounts of peanut (*Arachishypogaea*) in processed foods. Food Control 30: 480–490.

López-Díez, E.C., G. Bianchi and R. Goodacre. 2003. Rapid quantitative assessment of the adulteration of virgin olive oils with hazelnut oils using Raman spectroscopy and chemometrics. Journal of Agricultural and Food Chemistry 51: 6145–6150.
Lopez, S.J. 2008. TaqMan based real time PCR method for quantitative detection of basmati rice adulteration with non-basmati rice. European Food Research and Technology 227: 619–622.
Määttä, K.R., A. Kamal-Eldin and A.R. Törrönen. 2003. High-performance liquid chromatography (HPLC) analysis of phenolic compounds in berries with diode array and electrospray ionization Mass Spectrometric (MS) Detection: *Ribes* Species. Journal of Agricultural and Food Chemistry 51: 6736–6744.
Mafra, I., I.M.P.L.V.O. Ferreira and M.B.P.P. Oliveira. 2008. Food authentication by PCR-based methods. European Food Research and Technology 227: 649–665.
Manley, M. and V. Baeten. 2018. Spectroscopic technique: near infrared (NIR) spectroscopy. pp. 51–102. *In*: Sun, D.-W. [ed.]. Modern Techniques for Food Authentication. Elsevier Inc. London, UK.
Martelo-Vidal, M.J., F. Domínguez-Agis and M. Vázquez. 2012. Ultraviolet/visible/near-infrared spectral analysis and chemometric tools for the discrimination of wines between subzones inside a controlled designation of origin: a case study of RíasBaixas. Australian Journal of Grape and Wine Research 19: 62–67.
Mertz, C., V. Cheynier, Z. Günata and P. Brat. 2007. Analysis of phenolic compounds in two blackberry species (*Rubusglaucus* and *Rubusadenotrichus*) by high-performance liquid chromatography with diode array detection and electrospray ion trap mass spectrometry. Journal of Agricultural and Food Chemistry 55: 8616–8624.
Mesko, M.F., C.A. Hartwig, C.A. Bizzi, J.S.F. Pereira, P.A. Mello and E.M.M. Flores. 2011. Sample preparation strategies for bioinorganic analysis by inductively coupled plasma mass spectrometry. International Journal of Mass Spectrometry 307: 123–136.
Monakhova, Y.B., D.N. Rutledge, A. Roßmann, H.U. Waiblinger, M. Mahler, M. Ilse, T. Kuballa and D.W. Lachenmeier. 2014. Determination of rice type by ^{1}H NMR spectroscopy in combination with different chemometric tools. Journal of Chemometrics 28: 83–92.
Monteiro, P.I., J.S. Santos, V.R.A. Brizola, C.T.P. Deolindo, A. Koot, R. Boerrigter-Eenling, S. van Ruth, K. Georgouli, A. Koidis and D. Granato. 2018. Comparison between proton transfer reaction mass spectrometry and near infrared spectroscopy for the authentication of Brazilian coffee: A preliminary chemometric study. Food Control 91: 276–283.
Moon, J.-C., J.-H. Kim and C.S. Jang. 2016. Development of multiplex PCR for species-specific identification of the *Poaceae* family based on chloroplast gene, *rpo*C2. Applied Biological Chemistry 59: 201–207.
Nanophoton. 2016. https://www.nanophoton.net/raman/index.html. Accessed 23.09.2018.
Osborne, B.G. 2006. Near-infrared Spectroscopy in Food Analysis. Encyclopedia of Analytical Chemistry. 2006 by John Wiley & Sons, Ltd. DOI: 10.1002/9780470027318.a1018.
Palmieri, L., E. Bozza and L. Giongo. 2009. Soft fruit traceability in food matrices using real-time PCR. Nutrients 1: 316–328.
Paradkar, M.M., S. Sivakesava and J. Irudayaraj. 2003. Discrimination and classification of adulterants in maple syrup with the use of infrared spectroscopic techniques. Journal of the Science of Food and Agriculture 83: 714–721.
Parasuraman, S., R. Anish, R. Balamurugan, S. Muralidharan, S.J. Kumar and V. Vijayan. 2014. An overview of liquid chromatography-mass spectroscopy instrumentation. Pharmaceutical Methods 5: 47–55.
Park, Y.W., P.-S. Chang and J.H. Lee. 2010. Application of triacylglycerol and fatty acid analyses to discriminate blended sesame oil with soybean oil. Food Chemistry 123: 377–383.
Parvathy, V.A., V.P. Swetha, T.E. Sheeja, N.K. Leela, B. Chempakam and B. Sasikumar. 2014. DNA Barcoding to detect chilli adulteration in traded blck pepper powder. Food Biotechnology 1: 25–40.

Parvathy, V.A., V.P. Swetha, T.E. Sheeja and B. Sasikumar. 2015. Detection of plant-based adulterants in turmeric powder using DNA barcoding. Pharmaceutical Biology 53(12): 1774–1779.

Pasqualone, A., C. Montemurro, A. Grinn-Gofron, G. Sonnante and A. Blanco. 2007. Detection of soft wheat in semolina and durum wheat bread by analysis of DNA microsatellites. Journal of Agricultural and Food Chemistry 55: 3312–3318.

Pasquini, C. 2018. Near infrared spectroscopy: A mature analytical technique with new perspectives—A review. AnalyticaChimicaActa 8–36.

Pellati, F., G. Orlandini, K.A. van Leeuwen, G. Anesin, D. Bertelli, M. Paolini, S. Benvenuti, and F. Camin. 2013. Gas chromatography combined with mass spectrometry, flame ionization detection and elemental analyzer/isotope ratio mass spectrometry for characterizing and detecting the authenticity of commercial essential oils of *Rosa damascena* Mill. Rapid Communications in Mass Spectrometry 27: 591–602.

Peng, D., Y. Bi, X. Ren, G. Yang, S. Sun and X. Wang. 2015. Detection and quantification of adulteration of sesame oils with vegetable oils using gas chromatography and multivariate data analysis. Food Chemistry 188: 415–421.

Pereira, L., S. Gomes, C. Castro, J.E. Eiras-Dias, J. Brazão, A. Graça, J.R. Fernandes and P. Martins-Lopes. 2017. High Resolution Melting (HRM) applied to wine authenticity. Food Chemistry 216: 80–86.

Pereira, L., S. Gomes, S. Barrias, J.R. Fernandes and P. Martins-Lopes. 2018. Applying high-resolution melting (HRM) technology to olive oil and wine authenticity. Food Research International 103: 170–181.

Petrakis, E.A., L.R. Cagliani, M.G. Polissiou and R. Consonni. 2015. Evaluation of saffron (*Crocus sativus* L.) adulteration with plant adulterants by ^{1}H NMR metabolite fingerprinting. Food Chemistry 173: 890–896.

Pitt, J.J. 2009. Principles and applications of liquid chromatography-mass spectrometry in clinical biochemistry. The Clinical Biochemist Reviews 30: 19–34.

Pröfrock, D. and A. Prange. 2012. Inductively Coupled Plasma–Mass Spectrometry (ICP-MS) for quantitative analysis in environmental and life sciences: a review of challenges, solutions, and trends. Applied Spectroscopy 66: 843–868.

Radboud University. 2018. https://www.ru.nl/science/gi/facilities-activities/elemental-analysis/irms/. Accessed 06.09.2018.

Rodríguez-Lázaro, D., N. Cook and M. Hernández. 2013. Current challenges in real-time PCR diagnostics in food science. pp. 21–26. *In*: Rodríguez-Lázaro, D. [ed.]. Real-Time PCR in Food Science-Current Technology and Applications. Caister Academic Press, Norfolk, UK.

Rodushkin, I., F. Ödman and P.K. Appelblad. 1999. Multielement determination and lead isotope ratio measurement in alcoholic beverages by high-resolution inductively coupled plasma mass spectrometry. Journal of Food Composition and Analysis 12: 243–257.

Rohman, A. and Y.B.C. Man. 2011. The use of Fourier transform mid infrared (FT-MIR) spectroscopy for detection and quantification of adulteration in virgin coconut oil. Food Chemistry 129: 583–588.

Royal Society of Chemistry. 2009. Ultraviolet - Visible Spectroscopy (UV), pp. 1–7.

Schaller, E., J.O. Bosset and F. Escher. 1998. 'Electronic Noses' and their application to food. LWT - Food Science and Technology 31: 305–316.

Scotter, C.N.G. 1997. Non-destructive spectroscopic techniques for the measurement of food quality. Trends in Food Science & Technology 8: 285–292.

Selvaraj, D., D. Shanmughanandhan, R.K. Sarma, J.C. Joseph, R.V. Srinivasan and S. Ramalingam. 2012. DNA barcode *ITS* effectively distinguishes the medicinal plant *Boerhaviadiffusa* from its adulterants. Genomics, Proteomics & Bioinformatics 10: 364–367.

Seo, H.-Y., J. Ha, D.-B. Shin, S.-L. Shim, K.-M. No, K.-S. Kim, K.-B. Lee and S.-B. Han. 2010. Detection of corn oil in adulterated sesame oil by chromatography and carbon isotope analysis. Journal of the American Oil Chemists' Society 87: 621–626.

Shafiee, S. and S. Minaei. 2018. Combined data mining/NIR spectroscopy for purity assessment of lime juice. Infrared Physics & Technology 91: 193–199.

Shaw, R.A. and H.H. Mantsch. 2006. Infrared spectroscopy in clinical and diagnostic analysis. Encyclopedia of Analytical Chemistry, by John Wiley & Sons, Ltd. Chichester. ISBN 0471 97670 9.

Shui, G. and L.P. Leong. 2002. Separation and determination of organic acids and phenolic compounds in fruit juices and drinks by high-performance liquid chromatography. Journal of Chromatography A 977: 89–96.

Sinelli, N., E. Casiraghi, D. Tura and G. Downey. 2008. Characterisation and Classification of Italian Virgin Olive Oils by Near- and Mid-Infrared Spectroscopy 16: 335–342.

Sinija, V.R.N. and H.N. Mishra. 2009. Characterisation and classification of italian virgin olive oils by near- and mid-infrared spectroscopy. Fresh Produce 3: 76–80.

Słomka, M., M. Sobalska-Kwapis, M. Wachulec, G. Bartosz and D. Strapagiel. 2017. High resolution melting (HRM) for High-throughput genotyping—limitations and caveats in practical case studies. International Journal of Molecular Sciences 18(11): 1–21.

Sonnante, G., C. Montemurro, A. Morgese, W. Sabetta, A. Blanco and A. Pasqualone. 2009. DNA microsatellite region for a reliable quantification of soft wheat adulteration in durum wheat-based foodstuffs by real-time pCR. Journal of Agricultural and Food Chemistry 57: 10199–10204.

Stuart, B. 2004. Infrared Spectroscopy—Fundamentals and Applications. John Wiley & Sons, Ltd, West Sussex. England.

Su, T., P. Wei, L. Wu, Y. Guo, W. Zhao, Y. Zhang, Z. Chi and L. Qiu. 2019. Development of Nucleic Acid Isolation by Non-silica-based Nanoparticles and Real-time PCR Kit for edible Vegetable oil Traceability 300: 1–6.

Swetha, V.P., V.A. Parvathy, T.E. Sheeja and B. Sasikumar. 2014. DNA barcoding for discriminating the economically important *Cinnamomumverum* from its adulterants. Food Biotechnology 28: 183–194.

Tafvizi, F. and M. Hashemzadegan. 2016. Specific identification of chicken and soybean fraud in premium burgers using multiplex-PCR method. Journal of Food Science and Technology 53: 816–823.

Techen, N., I. Parveen, Z. Pan and I. Khan. 2014. DNA barcoding of medicinal plant material for identification. Current Opinion in Biotechnology 25: 103–110.

Terzi, V., M. Malnati, M. Barbanera, A.M. Stanca and P. Faccioli. 2003. Development of analytical systems based on real-time PCR for *Triticum* species-specific detection and quantitation of bread wheat contamination in semolina and pasta. Journal of Cereal Science 38: 87–94.

The Forensics Library. 2018. Available at https://aboutforensics.co.uk/isotope-ratio-mass-spectrometry/. Accessed 16.08.2021.

Thermo Fisher Scientific Inc. 2006. Advantages of Fourier-Transform Near-Infrared Spectroscopy. http://www.nicoletcz.cz/upload/kc/files/nir_spektrometry/Fourier TransformNIR.pdf. Accessed 20.08.2018.

Tsao, R., R. Yang, J.C. Young and H. Zhu. 2003. Polyphenolic profiles in eight apple cultivars using high-performance liquid chromatography (HPLC). Journal of Agricultural and Food Chemistry 51: 6347–6353.

USGS. 2013. https://crustal.usgs.gov/laboratories/icpms/intro.html. Accessed 08.09.2018.

Vaclavik, L., O. Lacina, J. Hajslova and J. Zweigenbaum. 2011. The use of high performance liquid chromatography–quadrupole time-of-flight mass spectrometry coupled to advanced data mining and chemometric tools for discrimination and classification of red wines according to their variety. AnalyticaChimicaActa 685: 45–51.

Valentini, P., A. Galimberti, V. Mezzasalma, F. De Mattia, M. Casiraghi, M. Labra and P.P. Pompa. 2017. DNA barcoding meets nanotechnology: development of a universal colorimetric test for food authentication. AngewandteChemie International Edition 56: 8094–8098.

Vietina, M., C. Agrimonti and N. Marmiroli. 2013. Detection of plant oil DNA using high resolution melting (HRM) post PCR analysis: A tool for disclosure of olive oil adulteration. Food Chemistry 141: 3820–3826.

Wang, D., Y. Wang, K. Zhu, L. Shi, M. Zhang, J. Yu and Y. Liu. 2019. Detection of cassava component in sweet potato noodles by real-time loop-mediated isothermal amplification (Real-time LAMP) method. Molecules 24: 1–8.

Wang, H., E.J. Race and A.J. Shrikhande. 2003. Characterization of anthocyanins in grape juices by ion trap liquid chromatography–mass spectrometry. Journal of Agricultural and Food Chemistry 51: 1839–1844.

Wang, L., F.S.C. Lee, X. Wang and Y. He. 2006. Feasibility study of quantifying and discriminating soybean oil adulteration in camellia oils by attenuated total reflectance MIR and fiber optic diffuse reflectance NIR. Food Chemistry 95: 529–536.

Warwick Scientific Services. 2017. Fluorescence Spectroscopy. https://warwick.ac.uk/services/ris/impactinnovation/impact/analyticalguide/fluorescence. Accessed 06.09.2018.

Weinert, B., M. Ulrich and A. Mosandl. 1999. GC-IRMS analysis of black Ceylon, Assam and Darjeeling teas. Zeitschriftfür Lebensmitteluntersuchung und-Forschung A 208: 277–281.

Wilke, W.W., R.N. Jones and L.D. Sutton. 1995. Automation of polymerase chain reaction tests. Reduction of human errors leading to contamination. Diagnostic Microbiology and Infectious Disease 21: 181–185.

Wilson, A.D. 2013. Diverse applications of electronic-nose technologies in agriculture and forestry. Sensors (Basel) 13: 2295–2348.

Wu, X., H. Li, W. Zhao, L. Fu, H. Peng, L. He, J. Cheng, H. Wei and Q. Wu. 2010. SCAR makers and multiplex PCR-based rapid molecular typing of *Lentinula edodes* strains. Current Microbiology 61: 381–389.

Wu, Y., Y. Chen, Y. Ge, J. Wang, B. Xu, W. Huang and F. Yuan. 2008. Detection of olive oil using the Evagreen real-time PCR Method 227: 1117–1124.

Yabsley, W., S. Homer-Vanniasinkam and J. Fisher. 2012. Nuclear magnetic resonance spectroscopy in the detection and characterisation of cardiovascular disease: Key Studies 1–11.

Zalacain, A., S.A. Ordoudi, E.M. Díaz-Plaza, M. Carmona, I. Blázquez, M.Z. Tsimidou and G.L. Alonso. 2005. Near-infrared spectroscopy in saffron quality control: determination of chemical composition and geographical origin. Journal of Agricultural and Food Chemistry 53: 9337–9341.

Zhang, L., J.-r. Pan and C. Zhu. 2012. Determination of the geographical origin of Chinese teas based on stable carbon and nitrogen isotope ratios. Journal of Zhejiang University SCIENCE B 13: 824–83.

4

Authenticity of Tomato

Konstantinos V Kotsanopoulos[1,*] and *M Nazim Uddin*[2]

1. Introduction

The tomato fruit (*Solanum lycopersicum*) is a highly popular product that is consumed across the world. Tomato is originally perennial and is grown year-round in several parts of the world. The fruit of tomato—botanically considered to be a berry—has been consumed for centuries. It possesses hollow spaces that are filled with moisture and seeds. In the last few decades, this originally South American plant has spread throughout the world, becoming one of the most valuable crops on the planet. Tomatoes are the main dietary source of lycopene, an antioxidant linked to a variety of health benefits, including a lower risk of heart disease and cancer (Matri 2016). Specifically, research has indicated that there are strong correlations between the consumption of tomatoes and the reduction of the risk of a number of cancers and cardiovascular diseases. Due to the fact that tomato is the second most widely consumed vegetable worldwide, following only potato, it is a vital source of numerous nutrients, including minerals and antioxidants, such as vitamins C and E, lycopene, b-carotene, lutein and flavonoid compounds (e.g., quercetin). Selective cultivation, environmental parameters, agronomic practices, the stage of ripeness when the fruits are harvested, and certain methods of handling the fruits throughout the supply chain can all be used to improve the level of their phytochemicals (Dorais et al. 2008).

[1] School of Agricultural Sciences, Department of Ichthyology and Aquatic Environment, University of Thessaly, Fytoko St, 38446 Nea Ionia Magnesias, Volos, Hellas, Greece.
[2] Olericulture Division, Horticulture Research Center, Bangladesh Agriculture Research Institute, Joydebpur, Gazipur-1701, Bangladesh.
* Corresponding author: kostaskot@yahoo.gr

Tomato has been extensively used both as food and for research purposes. Tomato plants are characterised by fleshy fruits, a sympodial shoot and compound leaves, attributes that are not seen in other model plants such as rice and the Arabidopsis genus. The majority of these characteristics are of high importance and cannot be examined when other model plants are used. There are a total of 13 known wild tomato species covering a great variety of phenotypes that can be crossed with the cultivated tomato. These species can be used for breeding purposes, to source desirable traits, as well as for the conduction of evolutionary studies. Recent advancements on the research around the tomato genome sequencing project have provided a wealth of information that can be used to study the plant. Furthermore, as the tomato is part of the large family of Solanaceae, its study and results obtained from it are also relevant for the study of other related plants, such as eggplants, tobacco, etc. It is therefore considered to be a valuable plant from a research perspective and is often used as the model organism for the Solanaceae family and particularly for fleshy-fruited plants (Kimura and Sinha 2008). The species of tomato (*S. lycopersicum*) that is commonly cultivated is a member of the large and diverse family, Solanaceae. It is one of a group of 13 closely related species that all emerged in arid habitats in the western coast of South America and then migrated to the north before being domesticated in Mesoamerica (in between Mexico and northern Costa Rica) (Peralta and Spooner 2000). Jenkins (1948) had noted that the ancestral form of the tomato that is now cultivated originates from the Peru-Ecuador area. Following its spread towards the north, most probably as a weed, it only started being domesticated when it eventually reached Mexico.

As mentioned above, tomatoes are known as a very good source of lycopene, β-carotene, flavonoid compounds, vitamin C hydroxycinnamic acid derivatives and others. Data indicating that lycopene has anti-oxidative and anti-cancer properties have led to a significant increase in the interest around tomatoes, while advancements in genetic engineering and biotechnology have shown the potential of tomato plants (Gerszberg et al. 2015).

Food adulteration attempts are not a new phenomenon. Although it first appeared when humans started trading food, adulteration practices have now become very sophisticated. Foodstuffs most prone to adulteration are usually high-value products. Adulteration attempts aim mainly at gaining additional profit. To shield themselves against fraudsters, food manufacturers implement a number of quality controls to their ingredients, requiring the use of techniques such as high-performance liquid chromatography (HPLC), thin-layer chromatography (TLC) or tests based on enzymes and physical characteristics (Cozzolino 2012).

2. Authenticity of Tomato and Products Thereof

Authenticity and traceability are two highly important elements for the food industry and are essential for maintaining consumer trust and fair market terms, while authenticity can also have a direct effect on food safety and quality. As in the case of many other foodstuffs, tomato can be the subject of adulteration attempts that can take the form of falsified geographical origin claims or methods or production (Arvanitoyannis and Vaitsi 2007). The trace elements' content of tomatoes can vary considerably but distinct differences are seen when organic and conventional methods of production are used. When multivariate analysis was therefore used to differentiate between tomatoes produced via organic and conventional methods, based on their levels of trace elements and stable nitrogen isotope ($\delta^{15}N$‰), it was shown that there were clear differences in a number of elements (Mn, Ca, Cu, Zn). These differences were attributed to the presence of higher levels of arbuscular mycorrhizal fungi (AMF) in the soil used for organic cultivation. These elements could therefore be good markers in such authenticity assessments, although such an approach could not be used when hydroponic methods of cultivation are followed (Kelly and Bateman 2010). The level and type of nutrients provided to tomato plants can have a significant effect on both the yield and the fruit's quality characteristics during storage. There are certain elements, such as N, P, K, Ca, Mg, and S that are required by the plants in larger amounts as they are used for growth and reproduction, while elements such as Fe, Cu, Zn, Mn, B, Mo, and Cl, are required in smaller amounts. Fertilizers are therefore commonly used in tomato cultivation, involving supplements of N, P, and K and less frequently Ca and Mg. Different elements can also be provided if the plants show signs of deficiencies. It is important to note though that when excessive amounts of nutrients are applied (exceeding the levels needed), reduction in yield is seen, while the cost of production unavoidably increases. It is therefore important that the soil and plant samples are periodically analysed to ensure appropriate levels of fertilisers are used (Sainju et al. 2003). The development of advanced analytical techniques for the authentication of tomatoes and products produced using tomatoes is thus very important. Tomatoes are used to produce various foodstuffs such as sauces and pastes, while are also used extensively as part of recipes for raw, semi-cooked or ready to eat foods and ingredients, in which cases any visual and in certain cases even organoleptic characteristics cannot be determined. Various methods have therefore been used to assess the authenticity and quality of tomatoes and detect adulterants or falsified/incorrect claims. A summary of the main techniques reported in the literature is presented in Table 1.

Table 1. Overview of key techniques used for the authentication of tomato.

Technique	Scope	Effectiveness	References
IR-MALDESI (MS imaging)	Cherry tomatoes	Effective imaging (100–1200 m/z,), covering a range of metabolites, operates at ambient conditions, no need for sample derivatization	Bagley et al. 2020
Multispectral imaging and chemometrics	Measurement of lycopene and phenolics in tomato	Effective grading, non-destructive	Liu et al. 2015
	Detection of sucrose adulteration in tomato paste	Detection of 1% adulteration, no misclassification using chemometrics	Liu et al. 2017
Backscattering and multispectral imaging	Development of prediction models for tomato quality factors (elasticity, moisture content, soluble solid content, titratable acidity)	Good prediction using wavelengths found using fusion of genetic algorithm and t-Score techniques; Models using artificial neural network led to good correlation for estimation of quality factors	Mollazade et al. 2015
Multispectral imaging and chemometrics	Identification to tomato varieties	Non-destructive, up to 100% discrimination	Shrestha et al. 2015
ICP-MS (Al, As, Ba, Be, Ca, Cd, Ce, Cu, Dy, Fe, K, La, Lu, Mg, Mn, Na, Nd, Pb, Rb, Sm, Sr, Th, U, V, Zn) and chemometrics	Differentiation of tomatoes and triple concentrated tomato pastes based on geographical origin	> 75% accurate	Feudo et al. 2010
Isotope ratio MS	Organic vs conventional tomatoes	Effective differentiation based on differences in mean $\delta^{15}N$ value but no unequivocal evidence	Bateman et al. 2007
GC-C-IRMS (analysis of C and N isotope ratios)		Effective discrimination based on $\delta^{15}N$ ratio	Bontempo et al. 2020
HPLC-HRAMS		Effective classification using $\delta^{15}N$ data and chemometrics	Bueno et al. 2018
NMR and chemometrics		^{1}H NMR profiling and LDA led to 100% correct classification	Hohmann et al. 2014
NMR, MIR and Stable Isotope analysis		Combinations of the 3 methods led to higher accuracy than use of methods individually	Hohmann et al. 2015

Whole genome re-sequencing analysis	Differentiation between two genotypes and genetic mapping	Successful differentiation of genotypes	Cambiaso et al. 2019b
SSRs	Evaluation of phenotypic and genetic components for fruit weight and shelf life	24 markers associated to weight, LC gene (chromosome 2) and other six markers in chromosomes 1, 2, 6, and 11. Genomic regions in chromosomes 9, 10 and 12 determined shelf life	Cambiaso et al. 2019a
	Assessment of authenticity of PDO tomato varieties	Highly accurate and reliable	Scarano et al. 2015
ISSR	Analysis of genetic diversity in different tomato genotypes	Polymorphism percentage: 50 to 90%; average genetic diversity index 0.61; data obtained sufficient for discrimination of genetic distances among genotypes	Sharifova et al. 2017
Metal oxide semiconductor e-nose and extreme learning machine	Prediction of quality of cherry tomato treated with high pressure argon post-harvest	Clear categorisation into 4 quality grades, reliable and effective	Feng et al. 2018
E-nose and e-tongue (combinations) and chemometrics	Detection of adulteration in cherry tomato juices	Better performances when the methods are used combined	Hong and Wang 2014
		Better authentication based on direct (no desiccants) e-nose measurement. (2) Best authentication ability using fusion approach based on ANOVA selected variables	Hong et al. 2014
E-nose (gas sensors array) and chemometrics	Adulteration of tomato paste with pumpkin, potato, and starch	Up to 99% detection of adulteration	Mohammad-Razdari et al. 2019

2.1 Spectroscopic methods

Spectroscopic methods involving ultraviolet (UV), visible (VIS), near infrared (NIR), and mid-infrared (MIR) spectroscopy have found applications in the verification of authenticity of foods of plant origin. Adulterants frequently added in foodstuffs include sugars, acids, volatile oils, juices and others, but adulteration can also take the form of over-dilution, addition of concentrates in fresh product as well as the use of lower quality materials than what is expected/labeled (Cozzolino 2012). Moreover, the qualitative and quantitative determination of chemical contaminants can only be effectively conducted using sensitive methods for verifying the composition, qualitative characteristics and authenticity of food. Methods using spectroscopy have been developed and employed in the past few decades for the assessment of adulteration and the verification of authenticity of foodstuffs. UV–Vis, fluorescence and vibrational spectroscopy, hyperspectral imaging and nuclear magnetic resonance (NMR) in combination with chemometrics have been employed to obtain chemical and structural information in order to characterize and classify foodstuffs and detect and quantify contaminants. These techniques have found various applications in food research and the industry, mainly due to their ability to analyze a wide range of substances, their speed, accuracy and high throughput. The spectra that are obtained represent certain matrix profiles and can therefore be used as fingerprints for a specific (bio)chemical composition. The combination of these methods though with linear discriminant analysis (LDA) has revealed their enormous potential for analyzing complex mixtures and extracting multispectral data for discriminating and classifying food (Esteki et al. 2018).

Sequi et al. (2007) described magnetic resonance imaging (MRI) spectroscopy as a particularly promising, non-invasive and non-destructive technique that can be used to accurately determine the geographical origin of fresh cherry tomatoes. In their study, these researchers succeeded in distinguishing PGI (Protected Geographical Indication) cherry tomatoes from non-PGI alternatives and cv. Naomi from cv. Shiren samples. The method they used allowed the determination of physical and morphological characteristics, transverse relaxation times and thicknesses, which were combined into a total of four equations (with two of which used for the determination of varieties and two for determining the plants' geographical origin). Successful determination of varieties was achieved in approximately 90% of the cases, while the geographical origin of the samples was accurately determined in approximately 80% of samples (Sequi et al. 2007).

2.1.1 NIR

The use of non-destructive analytical methods by the food industry to detect adulteration is becoming more and more important as routine controls are applied in factories (Posudin et al. 2015). NIR spectroscopy is frequently used to analyse the composition, function and sensory characteristics of foods of plant origin. It usually requires no sample preparation and produces quick results. Amongst the most important benefits this technology offers is that it can simultaneously measure more than one ingredient (Osborne 2006). The technique is not destructive and the fact that no sample preparation is required also means that it requires no handling of hazardous chemicals. It can be employed for both quantitative and qualitative analysis, while it is highly accurate (https://www.bruker.com/products/infrared-near-infrared-and-raman-spectroscopy/ft-nir.html). In the study of Vitalis et al. (2020), NIR spectroscopy and electronic tongue (e-tongue) were employed for the detection of low levels of paprika seed and cornstarch (0.5, 1, 2, 5, 10%) as well as sucrose and salt (0.5, 1, 2, 5%), in tomato paste. For comparison purposes, both conventional (soluble solid content, consistency) and sophisticated, highly sensitive techniques (NIR spectroscopy, e-tongue) were used to analyze the samples. Univariate statistics (ANOVA) was used to analyze the data obtained via conventional methods, while the results obtained with NIR and e-tongue were analyzed with Principal component analysis (PCA), LDA and partial least squares regression (PLSR). The conventional methods were not capable of detecting low concentrations (< 5%) of adulterants, while NIR spectroscopy and e-tongue were very accurate and able to identify adulteration levels as low as 0.5%. NIR spectroscopy though was found less difficult to implement and more accurate when estimations of the adulterant levels were given with R^2 above 0.96 and root mean square error (RMSE) below 1%. NIR [visible and short-wave NIR (Vis/SW NIR) diffuse reflectance] was also used in the study of Xie et al. (2009) for the classification of tomatoes of different genotypes. A comparison was also made between the discrimination performance of different chemometrics techniques, such as least-squares support vector machines (LS-SVM), discriminant analysis (DA), soft independent modeling of class analogy (SIMCA) and discriminant partial least-squares (DPLS). It was proved that LS-SVM performed the same as DA with 100% success, but was quicker when bigger varieties of tomatoes were to be differentiated based on genotypes. The technique in combination with chemometrics was therefore found very effective in differentiating tomatoes' genotypes.

2.2 Mass spectrometry

2.2.1 Isotope ratio mass spectrometry (IRMS)

Bontempo et al. (2020) carried out a stable isotope ratio analysis of hydrogen (H, δ^2H), carbon (C, $\delta^{13}C$), nitrogen (N, $\delta^{15}N$), oxygen (O, $\delta^{18}O$) and sulfur (S, $\delta^{34}S$) using samples of organic and conventional tomatoes originating from two Italian regions over a period of two years. A gas chromatography (GC)-combustion-IRMS method was designed and employed to analyze the C and N isotope ratios in amino acids. Amongst the bulk isotope ratios, $\delta^{15}N$ was of the highest importance for the discrimination between organic and conventional tomatoes. A significant improvement in the classification power was allowed using compound-specific isotope analysis irrespective of years of production and geographical regions. It was concluded that isotope analysis of amino acids can be effectively used to facilitate control procedures related to the assessment of organic tomato products.

IRMS was also used in the study of Bateman et al. (2007), in which a number of organic and conventional tomatoes', lettuces', and carrots' samples were subjected to analysis to determine their $\delta^{15}N$ composition. The data from the analyses were then used to build datasets that would allow any systematic differences in nitrogen isotope composition as a result of different production methods to be established. Based on the tomato and lettuce datasets, the distinct types of fertilizers usually used in organic and conventional systems lead to different nitrogen isotope compositions in these crops. In the case of organic tomatoes, the mean $\delta^{15}N$ value was 8.1‰, while tomatoes grown with conventional methods had a mean $\delta^{15}N$ value of −0.1‰. An overlap was found between the $\delta^{15}N$ values of organic and conventional datasets indicating that the use of a statistical method is required for classifying random samples as per their cultivation method. All in all, it was demonstrated that nitrogen isotope analysis can be employed to allow the detection of adulteration of organic crops with conventionally grown plants. Nevertheless, it became clear that a random $\delta^{15}N$ analysis of a sample cannot be used to prove whether synthetic fertilizers were used in the field but could to some extent be used as supporting evidence.

2.2.2 Inductively coupled plasma mass spectrometry (ICP-MS)

ICP-MS was used in the study of Feudo et al. (2010) to determine the levels of Al, As, Ba, Be, Ca, Cd, Ce, Cu, Dy, Fe, K, La, Lu, Mg, Mn, Na, Nd, Pb, Rb, Sm, Sr, Th, U, V and Zn in tomatoes originating from a total of four Italian regions as well as in triple concentrated tomato pastes of Italian, Chinese, Greek and US origin. Chemometrics were then used to process the extracted multi elemental profiles, aiming at discriminating between regions of origin. The samples were digested in a closed-vessel microwave and, following dilution, were subjected to analysis using Dynamic Reaction Cell (DRC)-ICP-MS with CH_4 as reaction gas. A total of three supervised pattern recognition procedures, LDA, SIMCA and K-nearest neighbors (KNN) were used to assess whether the samples were of Italian origin or not. The method was found to be sufficiently accurate (correct classification in the range of 75–120%) in the case of 25 out of the 35 elements (Feudo et al. 2010).

Effective differentiation of samples according to their geographical origin was also achieved in the study of Fragni et al. (2018) using a similar approach. Specifically, the construction of a new mineral elements database was proposed for authenticating processed tomatoes of Italian origin, and differentiating them from Chinese, American and Spanish tomatoes. ICP orthogonal acceleration Time-of-Flight MS and ICP Optical Emission Spectrometry were employed to quantify 26 elements, namely Li, Be, Na, Mg, Al, K, Ca, V, Cr, Mn, Co, Cu, Zn, Ga, As, Rb, Sr, Ag, Cd, In, Cs, Ba, Tl, Pb, Bi and U. A total of 183 samples of Italian, Chinese, US and Spanish origin produced in 2013, 2015 and 2017 were used in this study. LDA based on 28 variables (single elements and ratios) clearly differentiated the Italian samples from the rest. The ratios Li/Cu, Co/Rb and Sr/Cd were found to be highly successful in the identification of Italian samples (100% prediction ability, 98.8% in cross-validation), indicating that rations of elements can be more robust markers than single elements when it comes to discriminating samples.

2.2.3 LC-MS

Juices extracted from common red tomato and a tangerine tomato variety were compared with regard to their content in phytochemicals other than carotenoids. An LC high-resolution MS metabolomics-based method was employed to allow a broad evaluation of the differences in carotenoids and certain phytochemicals between the two types of tomatoes to be made. This untargeted approach can be effectively used to rapidly detect and extensively characterise different types of phytochemical substances. The tomato juices examined presented significant differences in numerous phytochemical compounds such as carotenoids, chlorophylls, lipids, and substances derived from cinnamic acid. The most important differences were related to the samples' contents in lycopene, phytoene, phytofluene, neurosporene, and ζ-carotene. Differences though were also found in polar phytochemicals (i.e., chlorogenic acid, hydroxyferulic acid, phloretin-di-C-glycoside, and isopropylmalic acid) (Morgan et al. 2017).

Moreover, in the study of Bueno et al. (2018), methodologies relying on HPLC-high-resolution accurate MS were followed to identify markers that could be used to differentiate between organic and conventional tomatoes grown in greenhouses. The samples were analysed, and the data extracted were processed using databases (mzCloud, ChemSpider and PubChem). Furthermore, the use of stable nitrogen isotope analysis ($\delta^{15}N$) was examined as a potential market that could allow samples to be categorized in accordance with their production systems, based on crop/fertilizer correlations. Pesticide residue analyses were also used to assess whether or not the samples were organically produced. It was proved that the use of high-resolution accurate MS and $\delta^{15}N$ data and chemometrics can effectively classify organic and conventional samples. PCA clearly clustered samples in accordance with the cultivation method followed and the samples were also found to differ significantly in their levels of L-tyrosyl-L-isoleucyl-L-threonyl-L-threonine, trilobatin, phloridzin, tomatine, phloretin and echinenone. HPLC was also used in the study of Cheng et al. (2011) but this time to allow a number of substances present in tomatoes during ripening to be determined. These researchers noted that during ripening, tomatoes' physico-chemical characteristics change. These changes impact the color, texture, aroma, and sugar concentrations. Changes in color can be used to define the ripening stages, while they can also be related to lycopene metabolism. The detection of internal changes in the levels of sugar and lycopene in tomatoes during ripening was achieved using chemical shift imaging (CSI) and HPLC. The imaging experiments were carried out in a Bruker 7T MRI system employed with a gradient-echo sequence for scanning. The spatial distribution of the sugar and lycopene levels in the fruits was therefore examined and a comparison was made before and after ripening. The CSI indicated that CSI intensity and sugar solutions concentrations in the range of 0.1–1.0 M were linearly correlated. It was also indicated though that there was no consistent trend in how the sugar signal intensity varied during ripening. The lycopene levels on the other hand increased considerably in the outer pericarp and columella at the red ripe stage. These outcomes were verified using HPLC analyses and it was noted that the CSI technique is not only nondestructive but is also highly effective in analyzing physiological changes in tomatoes.

2.2.4 Matrix-assisted laser desorption/ionization–mass spectrometry imaging (MALDI–MSI)

The presence and levels of metabolites are of paramount importance for a deep understanding of the physiological changes occurring as a result of ripening and wounding stress in tomatoes. Metabolomics were therefore used to examine tissue samples of tomato (S. *lycopersicum* L.) with the use of a MALDI–MSI technique in combination with a matrix sublimation/recrystallization method. The method effectively allowed the distribution patterns of about 30 metabolites to be examined at the same time. Changes in metabolites in the tissue of the fruit as physiological changes take place were examined using MALDI–MSI, taking into account the ripening phenotypes of mature green and red tomatos. Even though several metabolites did not seem to present differences in their location and levels between the two types of tomatoes, the levels of glutamate and adenosine monophosphate increased in both mesocarp and locule during ripening. On the other hand, malate decreased in both regions. MALDI–MSI was also used for the evaluation of other metabolic changes occurring as a result of wounding stress. A glycoalkaloid, tomatine, and limited amounts of esculeoside A, were detected in the wound region, while their inverse levels were found in undamaged regions. Moreover, the levels of these two substances were different at different developmental stages. This MALDI–MSI technique therefore provided useful information related to physiological changes in tomatoes involving changes in metabolic compounds (Nakamura et al. 2017).

Bagley et al. (2020) also reported that they used a MALDI–MSI technique [(infrared matrix-assisted laser desorption electrospray ionization (IR-MALDESI)] to study the metabolic profile of cherry tomatoes. This technique can be followed at ambient condition, requiring no samples derivatization. The samples were flash frozen and cryosectioned and imaging followed to allow a distinguishment to be made between principal tissues (e.g., skin, mesocarp, endocarp, locular tissue, etc.). Metabolites were scanned at 100–1200 m/z, thus covering a big range of metabolites, such as amino acids, lipids, terpenes, phenolics, glycosides, and alkaloids. The metabolic profiling led to the detection of endogenous carotenoid hydrocarbons, such as lycopene or β-carotene, ionized as radical cations.

2.3 NMR

NMR spectroscopy is a highly effective method, capable of very quickly analyzing mixtures down to molecular level, while there is no need for any separation and/or purification of samples in advance of analysis, rendering the technique very suitable for analyzing food. Although increasingly popular among scientists, NMR is not widely used in the food industry principally because of its cost, relatively low sensitivity, and the need for technicians with NMR expertise (Hatzakis 2019).

As it is known that ^{1}H NMR can be effectively used for differentiating between organic and conventional tomatoes, an optimization of ^{1}H NMR classification models that was reported, attempted to take into account data obtained using IRMS ($\delta^{13}C$, $\delta^{15}N$, and $\delta^{18}O$) and MIR spectroscopy. A number of methods (^{1}H NMR + MIR, ^{1}H NMR + IRMS, MIR + IRMS, and ^{1}H NMR + MIR + IRMS) were assessed either individually or in combinations using PCA, partial least squares discriminant analysis (PLS-DA), LDA, and common components and specific weight analysis. It was proved that fused data of ^{1}H NMR + MIR + IRMS are effective (95.0–100.0%

correct classification) in classifying samples in comparison to individual methods (^{1}H NMR, 91.3–100%; MIR, 75.6–91.7%), thus indicating that when analytical profiles are jointly examined, organic tomatoes can be more accurately authenticated (Hohmann et al. 2015).

^{1}H NMR was also used in the study of Hohmann et al. (2014) to authenticate organically cultivated tomatoes (*S. lycopersicum*). A total of 361 samples belonging 22 different cultivars and were subjected to analysis during a period of seven months. PCA was employed for the separation of tomatoes as per their production method with very satisfactory results ($p < 0.001$ using the t test). The use of LDA allowed the cultivation regiments to be discriminated, while external validation proved that correct classification had been achieved in all cases. Validation studies that followed, however, indicated that there were certain differences between different producers that were not predicted and interfered with the prediction of the cultivation method. Nevertheless, it was clear that ^{1}H NMR spectra significantly deferred between organic and conventional tomatoes.

^{1}H NMR spectroscopy was also used for discriminating triple concentrated tomato paste originating from China and Italy. Discrimination could be achieved regardless of cultivars and ripening level when a combination of NMR and PCA was used. Supervised orthogonal projection to latent structure discriminant analysis (OPLS-DA) was employed to construct effective classification models, while S-plot was used for identifying statistically significant variables that allow class separation. Citrate was the most critical substance in the discrimination between Chinese and Italian samples. Furthermore, in an effort to identify additional substances that can be used as markers to differentiate samples, citrate content was excluded, and a new classification model was constructed, which proved that aspartate, glutamine, and sugars can also be significant variables that can be used to differentiate tomato products (Consonni et al. 2009).

3. Electronic Noses

The flavor of numerous foodstuffs is a major indicator of quality and authenticity. Aroma is in many cases sufficient to allow the discrimination of authentic from adulterated food. Electronic nose (e-nose) is a quick and robust technique that does not require any particular sample preparation and can be used for the determination of a food's aroma (Gliszczyńska-Świgło and Chmielewski 2017). E-noses-based methods are constructed with the aim of mimicking a nose. E-noses incorporate sensors that allow the non-selective interaction with odor molecules, leading to the production of signals that are received by a computer. Multivariate statistics are then employed to detect patterns in the data. These patterns can indicate whether two samples are similar or different on the basis of headspace volatiles. Various e-nose sensors have been designed, including but not limited to organic polymers, metal oxides and quartz crystal microbalance. GC or a combination of GC and MS can also be employed as e-noses that use chemical mass or GC patterns (Baldwin et al. 2011).

In the study of Hong and Wang (2014), a total of seven approaches were taken in an effort to authenticate cherry tomato juice containing different concentrations of overripe tomato juices (0–30%) as an adulterant. PCA, factor *F* and stepwise selection were employed to analyze the data. A quantitative determination of adulteration levels was principally made using Canonical Discriminant Analysis (CDA) and Library Support Vector Machines (Lib-SVM). A quantitative calibration based on pH and soluble solids content was conducted using principle components regression. Good classification rates were given with all approaches while better predictions could be obtained using fusion approaches than when e-nose or e-tongue were solely used. It is important to note though that classification and prediction performances relying on different fusion approaches can vary. It was therefore suggested that using both instruments simultaneously can improve the performance of the classification compared to using e-nose or e-tongues individually, but it is important that suitable data fusion approaches are followed. In the study of Hong et al. (2014) a total of six approaches (two e-nose and an e-tongue approaches as well as three fusion approaches followed using both instruments) were followed to recognise and quantitatively analyse four different tomato juice products, one authentic and three juices containing different levels of adulterant. Recognition of the products was achieved using PCA and cluster analysis. A quantitative calibration based on pH and soluble solids content was conducted using (1) principle components regression with stepwise selection, (2) multiple linear regression (MLR) based on raw feature vector, (3) forward selection and (4) stepwise selection features. A comparison of cluster analysis based on different data standardization and distance calculation approaches was made and precision-recall measure was used for the quantification of the clustering outcomes. The data extracted indicated that using the optimum standardization and distance calculation methods for all datasets prior to cluster analysis is of paramount importance. The humidity factor was also taken into account and it was proved that there was no improvement when desiccants are used during e-nose measurements. The fusion dataset consisting of variables chosen by ANOVA was the most effective in authenticating samples, while the quality indices showed a high correlation to this dataset.

In another study, a gas sensor array was developed to assess the authenticity of tomato paste. To this regard, an olfactory machine system equipped with five gas sensors was assessed for its ability to detect adulteration with pumpkin, potato, and starch (0, 5, 10, 15, and 20%). PCA,

LDA, SVM, and PLS were employed to classify and analyse the sensors' responses. PCA was capable of explaining 99% of the variance in the data. The sensors TGS2610, MQ3, TGS2620, and TGS2600 are commonly employed to classify samples. LDA was 79.07, 83.87, and 92.22% precise in the classification of pumpkin, potato and starch, respectively. In the SVM method, C-SVM function as polynomial function and Nu-SVM as radial base function were employed to classify pumpkin samples with 98.84 and 88.14% accuracy, respectively. In the case of potato adulteration classification, the polynomial function was 90.32 and 83.87% precise in C-SVM and 96.77 and 86.02% in Nu-SVM. It was also proved that TGS2610 and MQ3 delivered the best results when PLS was used. When the different methods of classification were compared, PCA had the best performance when used to categorise potato and starch adulteration (Mohammad-Razdari et al. 2019).

The flavor of tomato is the result of numerous physicochemical changes and is a very important indicator of overall quality and freshness. An effort was therefore made to assess how reliable electronic noses (e-noses) are in evaluating the quality and freshness of cherry tomatoes following high pressure treatments with argon. Extreme learning machine (ELM) and PLS were employed to predict a number of quality parameters. It was shown that the freshness of cherry tomatoes as perceived using flavor as the marker during storage allowed the classification of samples into fresh, acceptable, stale and very stale. High pressure treatment with argon at 0.8 MPa played a crucial role in indicating the freshness and quality of the fruits during cold storage. ELM models performed better than PLS when used to evaluate the firmness, soluble solids concentration and pH based on their higher fitting correlation coefficients ($R^2 > 0.9500$). It was finally proved that e-nose technology in combination with ELM allowed the evaluation of quality and freshness of tomatoes during cold storage with high reliability (Feng et al. 2018).

4. Multispectral Imaging

Multispectral imaging has recently emerged and has gained recognition due to its non-destructive nature and effectiveness in discriminating and identifying varieties of the same fruit species. The use of the method was reported in the assessment of tomato cultivars from Nepal. Two sample sets were compiled to carry out this study, the first one with two parents and their crosses and the second one involving eleven cultivars, aiming at examining parents and offspring relationship and identifying varieties. Normalized CDA (nCDA) and PCA were employed for analyzing and comparing the results for parents and offspring. The results of the study indicated that parents and offspring could be clearly discriminated. nCDA was also employed to allow a pairwise discrimination of the eleven cultivars examined, accurately discriminating up to 100% of pairs. PLS-DA was also employed for the classification of all cultivars, achieving a classification accuracy of 82%, while the use of stepwise PLS-DA allowed additional improvements to be made and delivered classification accuracies of up to 96%. It was therefore proved that multispectral imaging could be used with high success for identifying and discriminating between tomato varieties (Shrestha et al. 2015).

Liu et al. (2017) examined the use of multispectral imaging (405–970 nm) in combination with chemometrics in rapidly detecting sucrose adulteration in tomato paste. PLS, LS-SVM, and back propagation neural network (BPNN) were all employed to construct quantitative models. It was shown that LS-SVM had a significantly better prediction performance in comparison to PLS and BPNN, with coefficient of determination in prediction of 0.936 to 0.966, RMSE of prediction of 0.521% to 0.445%, and residual predictive deviation of 5.014 to 5.865. Moreover, the use of multispectral imaging allowed the detection of sucrose in tomato paste at concentrations as low as 1%, while, using chemometrics, no samples were misclassified. It was therefore proved that multispectral imaging is highly effective in rapidly determining sucrose adulteration in tomato paste.

Liu et al. (2015) assessed the use of multispectral imaging in combination with chemometrics for the simultaneous and nondestructive measurements of the levels of lycopene and phenolics in tomatoes. PLS, LS-SVM BPNN were employed for developing the models. It was shown that BPNN had a significantly better performance, in comparison to PLS and LS-SVM with coefficient of determination in prediction (R^2_p) = 0.938 and 0.965, and residual predictive deviation of 4.590 and 9.335 for lycopene and total phenolics content prediction, respectively. It was therefore proved that multispectral imaging can be effectively used instead of standard methods for determining the levels of bioactive compounds in tomatoes.

Mollazade et al. (2015) conducted a study that aimed at predicting quality factors of tomato during storage with the use of backscattering and multispectral imaging. The data required to develop the prediction models were collected from 200 tomatoes (cv. Pannovy) of two different levels of maturity. The fruits were stored for a period of up to four weeks. During this period, elasticity, moisture concentration, soluble solids, titratable acidity, hyperspectral data, and backscattering images were measured/obtained from 40 tomatoes at regular intervals of one week. Spectral data from 40 points on each sample were obtained. Following that, a number of techniques such as filter-(Relief F, Fisher-Score, and t-Score) and wrapper-based (genetic algorithm) methods were employed to identify the sensitive wavelengths for each quality factor. The wavelengths determined by the fusion of the genetic algorithm and t-Score showed a good prediction performance in the case of all qualitative parameters. Verification of the importance of selected wavelengths was possible using backscattering and multispectral imaging techniques. The artificial neural network allowed the production of calibration models which showed a quite good correlation for the estimation of elasticity, soluble solids and titratable acidity at 660 nm and moisture at 830 nm of tomato from backscattering images. The correlation coefficients between the multispectral and backscattering imaging prediction results and reference measurements were 0.952 and 0.891 in the case of elasticity, 0.727 and 0.539 in the case of moisture, 0.736 and 0.561 in the case of soluble solids and 0.811 and 0.706 in the case of titratable acidity.

5. Molecular Procedures

DNA markers rely on the use of nucleotide sequences for identifying different species of plants. These markers have a very high discriminating power. Thus, characterising species using this type of markers is a very effective approach that can be taken to identify not only plant species, but also varieties belonging to the same species (Ganie et al. 2015). Tomato is a fruit of high commercial importance that is commonly used as an experimental model for the Solanaceae family of plants as well as for other fruiting plants. To render breeding more efficient, tomato has been the subject of many studies employing DNA markers. These markers have been used to construct genetic linkage maps which have allowed the development of quantitative trait locus (QTL) and gene mappings for certain highly important traits, and the development of the field of comparative genomics around Solanaceae. The determination of the whole genome sequence of tomato facilitated the development of high numbers of DNA markers, as well as the promotion of new genotyping techniques that do not rely on DNA markers. Furthermore, the construction of databases for genomes, DNA markers, linkage maps and data obtained using proteomics, metabolomics, phenome information, etc., can all contribute to advancements in the field of molecular breeding of tomatoes (Shirasawa and Hirakawa 2013).

Incorrect food labeling (or mislabeling) is a serious concern worldwide. Fish and meat adulteration with foreign/cheaper species has been documented numerous times. It is true though that there is not so much information when it comes to mislabeling plant products. Molecular markers could be used to verify the authenticity of foods of plant origin and detect adulterants. As "San Marzano" is one of the most important varieties of tomatoes worldwide, there is an EU Protected Designation of Origin (PDO) scheme that protects authenticity and origin claims. Considering this, single sequence repeats (SSRs) were employed to test 20 tomato varieties with the aim of subsequently using them to determine the genetic identity of 25 commercial tins labeled as "San Marzano". SSR fingerprint revealed that these products contained tomatoes that did not have the genetic profile of 'San Marzano sel. 2' or 'Kiros' varieties, which are the two varieties that can be used to produce PDO labeled foodstuffs. The effectiveness of DNA based markets in verifying the authenticity of premium quality tomatoes was therefore demonstrated. It was also shown that SSRs can support the industry in meeting the ever-increasing consumer demands for authentic products (Scarano et al. 2015). Also, in the study of Sharifova et al. (2017), inter simple sequence repeat (ISSR) markers were employed to analyze the variations among 41 tomato accessions. Fifty scorable bands were produced in total with 32 of them (63.3% of all the amplified loci) being polymorphic. The percentages of polymorphism were between 50 and 90%. The average genetic diversity index was 0.61. The primers UBC860 and UBC825 showed the highest diversity indexes of 0.89 and 0.85, respectively, while the lowest diversity index was 0.34 identified by primer UBC808. The genetic similarity within the genotypes taken into account ranged from 0.52 to 0.98. Cluster analysis relying on Jaccard's similarity coefficient allowed the division of genotypes into 6 clearly distinct clusters. The lowest genetic distances were detected between 'Gronastiy' and 'AG1224' (0.52), 'Orange' and 'AG1224' (0.54), and 'Evgeniya' and 'AG1224' (0.55) accessions.

Furthermore, Cambiaso et al. (2019a) noted that the weight of tomatoes and their shelf-life are very important characteristics from a commercial point of view. Development of tomatoes of recombinant inbred lines was performed via antagonistic and divergent selection for two traits from an interspecific cross between the *S. lycopersicum* L. cv. "Caimanta" and the *S. pimpinellifolium* L. accession "LA0722". The purpose of this study was the evaluation of phenotypic and genetic components related to fruit weight and shelf-life. The collection of phenotypic data from recombinant inbred lines was conducted during a period of three years. Sixteen SSRs were designed based on the sequences of "Caimanta" and "LA0722", and a total of 4 functional markers for fruit size-determining genes were employed. Both fruit weight and shelf-life presented significant genetic variability. When tomatoes were characterised at a molecular level, it was shown that their genetic structure followed fruit weight. Marker data were examined to investigate changes on allelic frequencies at *loci* within the phenotypic extreme group of recombinant inbred lines for fruit weight and shelf-life. A total of 24 markers were related to fruit weight and the *LC* gene in chromosome 2 and other six markers in chromosomes 1, 2, 6, and 11 presented the most important associations. Three new genomic regions in chromosomes 9, 10 and 12 were found to be associated with tomatoes' shelf-life.

Very recently, next generation sequencing was also used in trials involving tomatoes. The reduction of the cost of next generation sequencing technologies has allowed the technique to be considered for use in more plant breeding programmes. An effort was therefore made to sequence the entire genome of the *S. lycopersicum* L. cultivar Caimanta and *Solanum pimpinellifolium* L. accession LA0722 with assembly relative to the Heinz 1706 reference version SL2.50. The level and distribution of polymorphism in "Caimanta" and "LA0722" were analysed and alleles in candidate genes related to disease resistance, fruit shape, weight and quality were examined. Molecular markers were also developed for the construction of a genetic linkage map based on a F2 population. In total, the detection of 1,397,518 polymorphisms was reported when "Caimanta" and "LA0722" were compared. A resistant allele for Rx4/Xv3 was found by sequence and a confirmation was given through inoculation. Insertion/deletion DNA markers were designed that could be multiplexed and scored via the use of genotyping platforms. These markers were employed for constructing a genetic linkage map, finally demonstrating that the whole genome sequencing of parental lines is capable of revealing phenotypes and characterizing reference populations via following standard genotyping strategies (Cambiaso et al. 2019b).

6. Conclusions

Tomato is one of the most important foods of plant origin from both a commercial and nutritional point of view. Its extensive use in research and wide cultivation across the world has triggered the conduction of numerous studies on the plant and the physicochemical and organoleptic characteristics of the fruits. Adulteration in tomato and products thereof (pastes, sauces, etc.) is common and its detection is very rarely possible without the use of advanced analytical techniques, especially when tomato products are used as ingredients for food preparation. Tomato adulteration can take the form of incorrect geographical origin claims, substitution of PDO or higher-quality varieties with cheaper alternatives, use of conventional grown tomatoes in organic cultivation claims, etc. Various techniques ranging from spectrometry-based methods, DNA markers e-noses/e-tongues-based approaches, multispectral imaging, etc., usually combined with chemometrics have been assessed for their effectiveness in detecting adulteration and verifying the authenticity of tomato fruits and tomato-derived products. Spectrometric methods allow the quantitative and qualitative determination of contaminants and in combination with chemometrics can be used to accurately detect adulteration. Techniques based on NIR have the additional benefit that are nondestructive and are therefore suitable for analyses in the industry. Techniques analyzing isotopes and their ratios such as ICP-MS can be used to determine the geographical origin of samples and are therefore suitable for supporting geographical origin claims. NMR on the other hand is highly accurate and suitable for analyzing mixtures and therefore valuable considering the enormous variety of commercial tomato uses. E-noses have found applications in the food industry due to the significance of aromas in determining the quality of foodstuffs. Studies have shown that they can effectively classify samples and detect different levels of adulterants in tomato products. Multispectral imaging has also been employed to analyze tomato and products thereof, mainly assisting in differentiating between different varieties, providing very accurate and reliable results. Finally, molecular markers, such as microsatellites, allow the unequivocal identification of tomato varieties, but can also be used to study qualitative characteristics. It is possible that future research will examine further the use of whole genome sequencing thus allowing our in depth understanding of the genetic characteristics of tomatoes.

Abbreviations

Arbuscular mycorrhizal fungi	AMF
Analysis of variance	ANOVA
Back propagation neural network	BPNN
Canonical Discriminant Analysis	CDA
Chemical shift imaging	CSI
Discriminant analysis	DA
Discriminant partial least-squares	DPLS
Dynamic Reaction Cell	DRC
Electronic nose	e-nose
Electronic tongue	e-tongue
Extreme learning machine	ELM
Gas chromatography	GC
High-performance liquid chromatography	HPLC
Inductively coupled plasma mass spectrometry	ICP-MS
Isotope Ratio Mass Spectrometry	IRMS
Inter simple sequence repeat	ISSR
K-nearest neighbors	KNN
Linear discriminant analysis	LDA
Library Support Vector Machines	Lib-SVM
Least-squares support vector machines	LS-SVM
Matrix-assisted laser desorption/ionization–mass Spectrometry imaging	MALDI–MSI
Mid-infrared spectroscopy	MIR
Multiple linear regression	MLR
Magnetic resonance imaging	MRI
Near infrared spectroscopy	NIR
Nuclear magnetic resonance	NMR
Orthogonal projection to latent structure discriminant analysis	OPLS-DA

Principal component analysis	PCA
Protected Designation of Origin	PDO
Protected Geographical Indication	PGI
Partial least squares discriminant analysis	PLS-DA
Partial least squares regression	PLSR
Quantitative trait locus	QTL
Root mean square error	RMSE
Soft independent modeling of class analogy	SIMCA
Single sequence repeat	SSR
Thin-layer chromatography	TLC
Ultraviolet	UV
Visible	VIS

References

Arvanitoyannis, I.S. and O.B. Vaitsi. 2007. A review on tomato authenticity: quality control methods in conjunction with multivariate analysis (Chemometrics). Crit. Rev. Food Sci. Nutr. 47: 675–699.

Bagley, M.C., C.L., Pace, M. Ekelöfa and D.C. Muddiman. 2020. Infrared matrix-assisted laser desorption electrospray ionization (IR-MALDESI) mass spectrometry imaging analysis of endogenous metabolites in cherry tomatoes. Analyst 16: 1–8.

Baldwin, E.A., J. Bai, A. Plotto and S. Dea. 2011. Electronic noses and tongues: applications for the food and pharmaceutical industries. Sensors 11: 4744–4766.

Bateman, A.S., S.D. Kelly and M. Woolfe. 2007. Nitrogen isotope composition of organically and conventionally grown crops. J. Agric. Food Chem. 55: 2664–2670.

Bontempo, L., K.A. van Leeuwen, M. Paolini, K.H. Laursen, C. Micheloni, P.D. Prenzler, D. Ryan and F. Camin. 2020. Bulk and compound-specific stable isotope ratio analysis for authenticity testing of organically grown tomatoes. Food Chemistry 318(126426): 1–11.

Bueno, M.J.M., F.J. Díaz-Galiano, L. Rajski, V. Cutillas and A.R. Fernández-Alba. 2018. A non-targeted metabolomics approach to identify food markers to support discrimination between organic and conventional tomato crops. J. Chromatogr. A 1546: 66–76.

Cambiaso, V., M.D. Gimenez, J.H.P. da Costa, D.V. Vazquez, L.A. Picardi, G.R. Pratta and G.R. Rodríguez. 2019a. Selected genome regions for fruit weight and shelf life in tomato RILs discernible by markers based on genomic sequence information. Breeding Science 69: 447–454.

Cambiaso, V., G.R. Pratta, J.H.P. da Costa, R. Zorzoli, D.M. Francis and G.R. Rodríguez. 2019b. Whole genome re-sequencing analysis of two tomato genotypes for polymorphism insight in cloned genes and a genetic map construction. Sci. Hortic. 247: 58–66.

Cheng, Y.C., T.T. Wang, J.H. Chen and T.T. Lin. 2011. Spatial–temporal analyses of lycopene and sugar contents in tomatoes during ripening using chemical shift imaging. Postharvest Biol. Technol. 62: 17–25.

Consonni, R., L.R. Cagliani, M. Stocchero and S. Porretta. 2009. Triple concentrated tomato paste: discrimination between italian and chinese products. J. Agric. Food Chem. 57: 4506–4513.

Cozzolino, D. 2012. Recent trends on the use of infrared spectroscopy to trace and authenticate natural and agricultural food products. Appl. Spectrosc. Rev. 47: 518–530.

Dorais, M., D.L. Ehret and A.P. Papadopoulos. 2008. Tomato (*Solanum lycopersicum*) health components: from the seed to the consumer. Phytochem. Rev. 7: 231–250.

Esteki, M., Z. Shahsavari and J. Simal-Gandara. 2018. Use of spectroscopic methods in combination with linear discriminant analysis for authentication of food products. Food Control 91: 100–112.

Feng, L., M. Zhang, B. Bhandari and Z. Guo. 2018. A novel method using MOS electronic nose and ELM for predicting postharvest quality of cherry tomato fruit treated with high pressure argon. Comput. Electron Agric. 154: 411–419.

Feudo, G.L., A. Naccarato, G. Sindona and A. Tagarelli. 2010. Investigating the origin of tomatoes and triple concentrated tomato pastes through Multielement determination by inductively coupled plasma mass spectrometry and statistical analysis. J. Agric. Food Chem. 58: 3801–3807.

Fragni, R., A. Trifirò, A. Nucci, A. Seno, A. Allodi and M. Di Rocco. 2018. Italian tomato-based products authentication by multi-element approach: A mineral elements database to distinguish the domestic provenance. Food Control. 93: 211–218.

Ganie, S.H., P. Upadhyay, S. Das and M.P. Sharma. 2015. Authentication of medicinal plants by DNA markers. Plant Gene 4: 83–99.

Gerszberg, A., K. Hnatuszko-Konka, T. Kowalczyk and A.K. Kononowicz. 2015. Tomato (*Solanum lycopersicum* L.) in the service of biotechnology. Plant Cell Tissue Organ Cult. 120: 881–902.

Gliszczyńska-Świgło, A. and J. Chmielewski. 2017. Electronic nose as a tool for monitoring the authenticity of food. A Review. Food Anal. Methods 10: 1800–1816.

Hajšlová, J., V. Schulzová, P. Slanina, K. Janné, K.E. Hellenäs and C.H. Andersson. 2005. Quality of organically and conventionally grown potatoes: four-year study of micronutrients, metals, secondary metabolites, enzymic browning and organoleptic properties. Food Addit. Contam. 22: 514–534.

Hatzakis, E. 2019. Nuclear Magnetic Resonance (NMR) spectroscopy in food science: a comprehensive review. Comprehensive Reviews in Food Science and Food Safety 18: 189–220.

Hohmann, M., N. Christoph, H. Wachter and U. Holzgrabe. 2014. ^{1}H NMR profiling as an approach to differentiate conventionally and organically grown tomatoes. J. Agric. Food Chem. 62: 8530–8540.

Hohmann, M., Y. Monakhova, S. Erich, N. Christoph, H. Wachter and U. Holzgrabe. 2015. Differentiation of organically and conventionally grown tomatoes by chemometric analysis of combined data from proton nuclear magnetic resonance and mid-infrared spectroscopy and stable isotope analysis. J. Agric. Food Chem. 63: 9666–9675.

Hong, X. and J. Wang. 2014. Detection of adulteration in cherry tomato juices based on electronic nose and tongue: Comparison of different data fusion approaches. J. Food Eng. 126: 89–97.

Hong, X., J. Wang and S. Qiu. 2014. Authenticating cherry tomato juices—Discussion of different data standardization and fusion approaches based on electronic nose and tongue. Food Res. Int. 60: 173–179.

https://www.bruker.com/products/infrared-near-infrared-and-raman-spectroscopy/ft-nir.html. Accessed 17.05.2021.

Jenkins, J.A. 1948. The origin of the cultivated tomato. Economic Botany 2: 379–392.

Kelly, S.D. and A.S. Bateman. 2010. Comparison of mineral concentrations in commercially grown organic and conventional crops-Tomatoes (*Lycopersicon esculentum*) and lettuces (*Lactuca sativa*). Food Chemistry 119: 738–745.

Kimura, S. and N. Sinha. 2008. Tomato (*Solanum lycopersicum*): a model fruit-bearing crop. CSH Protoc 3: 1–9.

Liu, C., W. Liu, W. Chen, J. Yang and L. Zheng. 2015. Feasibility in multispectral imaging for predicting the content of bioactive compounds in intact tomato fruit. Food Chemistry 173: 482–488.

Liu, C., G. Hao, M. Su, Y. Chen and L. Zheng. 2017. Potential of multispectral imaging combined with chemometric methods for rapid detection of sucrose adulteration in tomato paste. J. Food Eng. 215: 78–83.

Mohammad-Razdari, A., M. Ghasemi-Varnamkhasti, S.H. Yoosefian, Z. Izadi and M. Siadat. 2019. Potential application of electronic nose coupled with chemometric tools for authentication assessment in tomato paste. J. Food Process Eng. 42: 1–12.

Mollazade, K., M. Omid, F.A., Tab, Y.R. Kalaj and S.S. Mohtasebi. 2015. Data mining-based wavelength selection for monitoring quality of tomato fruit by backscattering and multispectral imaging. Int. J. Food Prop. 18(4): 880–896.

Nakamura, J., T. Morikawa-Ichinose, Y. Fujimura, E. Hayakawa, K. Takahashi, T. Ishii, D. Miura and H. Wariishi. 2017. Spatially resolved metabolic distribution for unraveling the physiological change and responses in tomato fruit using matrix-assisted laser desorption/ionization-mass spectrometry imaging (MALDI-MSI). Anal. Bioanal. Chem. 409: 1697–1706.

Osborne, B.G. 2006. Near-infrared spectroscopy in food analysis. Encyclopedia of Analytical Chemistry 1: 1–14.

Peralta, I.E. and D.M. Spooner. 2000. Classification of wild tomatoes: a review. Kurtziana 28: 45–54.

Posudin, Y.I., K.S. Peiris and S.J. Kays. 2015. Non-destructive detection of food adulteration to guarantee human health and safety. Ukrainian Food Journal 4: 207–260.

Sainju, U.M., R. Dris and B. Singh. 2003. Mineral nutrition of tomato. Journal of Food, Agriculture and Environment 1: 176–183.

Scarano, D., R. Rao, P. Masi and G. Corrado. 2015. SSR fingerprint reveals mislabeling in commercial processed tomato products. Food Control 51: 397–401.

Sequi, P., M.T. Dell'Abate and M. Valentini. 2007. Identification of cherry tomatoes growth origin by means of magnetic resonance imaging. J. Sci. Food Agric. 87: 127–132.

Sharifova, S.S., S.P. Mehdiyeva and M.A. Abbrasov. 2017. Analysis of genetic diversity among different tomato genotypes using ISSR DNA Marker. Genetika 49: 31–42.

Shirasawa, K. and H. Hirakawa. 2013. DNA marker applications to molecular genetics and genomics in tomato. Breeding Science 63: 21–30.

Shrestha, S., L.C. Deleuran, M.H. Olesen and R. Gislum. 2015. Use of multispectral imaging in varietal identification of tomato. Sensors 15(2): 4496–4512.

Vitalis, F., J.-L.Z. Zaukuu, Z. Bodor, B. Aouadi, G. Hitka, T. Kaszab, V. Zsom-Muha, Z. Gillay and Z. Kovacs. 2020. Detection and quantification of tomato paste adulteration using conventional and rapid analytical methods. Sensors 20: 1–22.

Xie, L., Y. Ying and T. Ying. 2009. Classification of tomatoes with different genotypes by visible and short-wave near-infrared spectroscopy with least-squares support vector machines and other chemometrics. J. Food Eng. 94(1): 34–39.

5

Authenticity of Banana

Konstantinos V Kotsanopoulos

1. Introduction

There are indications that organised agricultural activities in countries like New Guinea had begun by 6950–6440 cal BP or even earlier. Banana plants were most probably domesticated and cultivated initially in New Guinea and their cultivation was subsequently extended to Asia (Denham et al. 2004). Bananas are not trees but monocotyledonous herbaceous plants, growing rapidly. Their fruits can be eaten raw when ripe while in the case of plantains, which are highly starchy even when fully ripe, the fruits need to be cooked before consumption. The banana plant is part of the class Musa (family Musaceae) and its cultivation aims mainly at producing fruits for food but also fibers for other uses and also products for ornamental purposes (Rubatzky and Yamaguchi 1997). Nadal-Medina (2009) noted that how the germplasms are classified and the genotypes identified in bananas can be challenging as when traditional methods (morphology) are used the process can be quite hard and lead to mistakes. Morphological characteristics in combination with molecular identification markers has allowed the accurate taxonomy and determination of banana genotypes. The morphological and genetic characterisation of 22 accessions of bananas and plantains belonging to several subgroups such as Cavendish, Red, Plantain, Ibota and Silk, as well as hybrids was carried out by these researchers using random amplified polymorphic DNA (RAPD). The accessions were obtained from the germplasm bank of the Instituto Nacional de Investigaciones Forestales Agrícolas y Pecuarias of Colima in

School of Agricultural Sciences, Department of Ichthyology and Aquatic Environment, University of Thessaly, Fytoko St, 38446 Nea Ionia Magnesias, Volos, Hellas, Greece.

México. A total of 90 markers from 22 accessions were determined, with 83 being found to be polymorphic. It was shown that classification based on morphology followed the results obtained using molecular markers. A dendrogram was developed indicating that there were two groups (A and B): Group A covered 10 accessions belonging to the "Cavendish" subgroup and the hybrid 'FHIA 03', while Group B covered accessions of the subgroups "Red", "Plantain" and "Cavendish", and the hybrids 'FHIA 01', 'FHIA20', 'FHIA 21' and 'SH-3640'.

2. Distribution and Ecology

Bananas (*Musa* spp., Musaceae) are rhizomatous perennial monocotyledonous herbaceous plants originating from Southeast Asia and Melanesia. Their domestication was achieved via selection for sterility and parthenocarpy. The origin species of the edible banana are the wild species *M. acuminata* and *M. balbisiana*. The many variations of the plant are a result of its genomic origin, the level of ploidy and somatic mutations. Bananas and plantains are grown across both the tropic and some subtropic regions. Their global production was estimated at 145 million tonnes in 2013, with a large amount (85%) sold in the regions of Africa, Asia, and Latin America (where most production takes place), and exports are done to Latin America, Africa, and Philippines (15%) (Israeli and Lahav 2017). Banana plants are highly sensitive to environmental changes. Monteith's analysis of plant responses to intercepted radiation can be used to analyse the productivity of bananas and plantains. Bananas have high sensitivity to soil water deficits, and the effects can be rapidly seen on emerging leaves and fruits. While soil dries, stomata close keeping leaves hydrated. Productivity though is considerably reduced. A more efficient use of water in plantations can be achieved by examining the water used versus the amount of water the plants need. Scientists and plantation personnel can quantify the water that becomes available in a particular location and estimate if and how much extra water is needed to be provided. The duration of sunlight can influence the rate of bunch initiation with periods of light availability of less than 12 hours reducing it noticeably (Turner et al. 2007).

3. Origin and Evolution

Banana is one of the world's most important crops from a production and trading perspective, following rice, wheat and maize when it comes to gross value of production. It is a significant food source for millions of people and an income source via both local and international trade. Bananas can be categorised into dessert and cooking types. Dessert bananas are eaten raw when ripe, while cooking (starchy) bananas can be eaten once boiled, fried brewed, roasted or turned into powder. Plantains are the most widely known type of banana comprising one-third of all bananas produced. Bananas and plantains are not clearly distinguished, botanically or genetically, but plantains have often a higher dry matter content (Nayar 2010).

The wild bananas contain seeds and are diploid, while cultivated bananas and plantains are categorised by parthenocarpy, sterility, and polyploidy, mainly triploidy. Parthenocarpy is a result of complementary dominant genes which are present in wild *M. acuminata*. Parthenocarpic triploids come from unreduced gametes (2n+n) and do not have seeds. It is believed that they have been directly chosen and multiplied by offshoots. Hybrids of diploid *M. acuminata* (AA genomes) and *M. bulbisiana* (BB genomes) led to the production of diploid AB hybrids and AAB and ABB triploids, which are plantains and cooking types of banana. There are also several known tetraploid hybrids (AAAB, AABB, and ABBB) (Janick 2005).

4. Cultivation Type Factors and Environment

Bananas and plantains are therefore a very important produce from a commercial point of view and, depending on the region where they are cultivated and factors such as the variety, type of planting material, season and method of cultivation, they can produce high yields. The banana varieties that are generally preferred due to their particularly high yield potential and marketability are those belonging to the Cavendish group such as Grand Naine, Williams and Robusta. Planting sword suckers is preferred over water suckers, butts and bits of rhizomes. Nevertheless, the *in vitro* Dwarf Cavendish and Robusta bananas are of higher quality compared to conventional suckers due to their fast growth, early flowering (around 19 days) and quick crop production (within 29 and 22 days, respectively). When plant spacing and planting density are considered, various factors should be taken into account, such as the banana variety, soil fertility, prevailing climatic conditions, etc. It has been observed that maintenance at densities of 625–1000 plants ha^{-1} can lead to low productivity of < 30 t ha^{-1}, while a high density of 5000 plants ha^{-1} can lead to the production of 120 t ha^{-1}. It is important to mention though that it can take particularly longer for plants to mature in higher densities. Bunch grade and the quality of fruits are however highly dependent on pre-harvest bunch management practices. The methods of irrigation, use of fertilizers and the levels of elements provided, the use of organic farming methods as well as the use of bunch sleeves can all affect yield and fruit quality (Mustaffa and Kumar 2012). Banana is cultivated commercially in around 130 countries and the global production of the fruit amounts to about 120 million tonnes/year. An excess of 800,000 tonnes are imported into the UK every year and each banana is estimated to have a carbon footprint of 100–200 rams. Interestingly, over a million bananas are wasted every day in Great Britain, which leads to emissions of > 30,000 tonnes on an annual basis. Disease is the most important threat to producers as almost all bananas come from the same variety (Cavendish). Topical Race 4, a banana disease attributed to a fungus is currently responsible for the destruction of large numbers of banana plants worldwide. Also, damages due to storms, floods and drought can also be very detrimental to banana production. The careful irrigation and use of nutrients can boost the health of the plants and make them more resilient to both climate changes and diseases and can support effective pest control (Reay 2019).

5. Conventional vs GMO

Bananas and plantains are large herbaceous plants that belong to the genus Musa. They are characterised by sterility and parthenocarpy and produce seedless fruits. The hybrids and species that are commonly cultivated are mainly triploid ($2n = 3x = 33$; a limited number are diploid or tetraploid), and many of them have been propagated from their wild mutants. In 2007 the global annual production of bananas was estimated at 100 million tonnes. As diseases attributed to viruses, abiotic stresses and new requirements for sustainability, quality, transport, and yield are amongst the main challenges, genetic resistance to such stress factors becomes very important. Therefore, genomic approaches for *Musa* are now becoming a very important field of research, assisting also in targeted breeding, transformation and taking full advantage of the plants' biodiversity (Heslop-Harrison and Schwarzacher 2007). From a genetic point of view, the varieties cultivated worldwide are very few and therefore the risk of certain characteristics disappearing is high as new diseases emerge. Breeding plants resistant to Fusarium wilt and Sigatoka leaf spot diseases (SLSDs) is the priority. New strains of Fusarium wilt are detected in several parts of the world, including Australia, South Africa and Asia and Cavendish varieties are significantly affected. In Central America, the control of BLSD with chemicals means becomes less and less effective, and increased doses of pesticides are required to give the same results. Furthermore, the pathogen seems to also have affected a number of natural varieties mainly in the Pacific region. Ensuring the effective production of banana fruits, while protecting the environment will most probably be where future research will focus on. Induced mutation and genetic transformation can bring a lot of benefits, including the breeding of shorter plants with better bunch and fruit shapes. Historically, improvements in banana have been made almost exclusively via hybridization (Bakry et al. 2009). Conventional breeding to improve banana varieties is therefore not new, but since the majority of cultivated bananas are triploids and produce seedless fruits, the process becomes more challenging and time-consuming (Hwang and Ko 2004). An effective and reproducible system for genetically transforming bananas using embryogenic cell suspensions (ECS) was suggested by Tripathi et al. (2015). In this study, two different types of explants (immature male flowers and multiple buds) were examined for their ability to develop ECS in different banana varieties. ECS of "Cavendish Williams" and "Gros Michel" varieties were given using multiple buds, while in the case of "Sukali Ndiizi", immature male flowers were used. The regeneration efficiency of ECS was found to range from 20,000 to 50,000 plantlets per ml of settled cell volume (SCV). ECS of three varieties were transformed using *Agrobacterium* and the gusA reporter gene. The regeneration of twenty to seventy independent transgenic events per ml SCV of ECS on selective medium was observed. The fact that the gene was present in transgenic plants was verified using polymerase chain reaction (PCR), dot blot, and Southern blot analysis as well as histochemical β-glucuronidase (GUS) assays. The method was successful in generating hundreds of transgenic lines resistant to diseases.

Paul et al. (2011) examined whether the transgenic expression of apoptosis-inhibition-related genes in banana can be employed to render the plant resistant to Fusarium wilt. ECS of 'Lady Finger' cultivars were subjected to transformation with animal genes that negatively regulate apoptosis, namely Bcl-xL, Ced-9 and Bcl-2 3′ UTR, and regeneration of independently transformed plant lines was carried out for testing purposes. After 12 weeks of exposure to Foc race 1 in small-plant glasshouse bioassays, a total of 7 transgenic lines (2 × Bcl-xL, 3 × Ced-9 and 2 × Bcl-2 3′ UTR) presented distinctly less internal and external disease symptoms than the control samples. Out of these, one Bcl-2 3′ UTR line was found to be as resistant to disease as the wild-type Cavendish bananas that were considered in this study as resistant negative controls. Moreover, the resistance of this line lasted for a period of 23 weeks after the inoculation at which time the experiment ended. The use of TUNEL assays indicated that Foc race 1 induced apoptosis-like features in the roots of wild-type 'Lady Finger' plants known to correspond to the necrotrophic phase in the life cycle of the pathogen. This evidence was also enforced by the fact that these features were also reduced in the roots of the resistant Bcl-2 3′ UTR-transgenic line.

In another study, Shekhawat et al. (2012) investigated whether intron-hairpin-RNA (ihpRNA) transcripts that correspond to viral master replication initiation protein (Rep) can be used for generating banana bunchy top virus (BBTV)-resistant transgenic bananas. Two ihpRNA constructs named ihpRNA-Rep and ihpRNA-ProRep that were formed via Rep full coding sequence or Rep partial coding sequence together with its 5′ upstream regulatory region, respectively, and castor bean catalase intron were effectively subjected to transformation into banana embryogenic cells. ihpRNA-Rep- and ihpRNA-ProRep-derived transgenic bananas that were chosen following preliminary screening for successful reporter gene expression, showed high resistance to BBTV infection (i.e., no symptoms after a period of 6 months of viruliferous aphid inoculation). The resistance to BBTV infection was further proved when no cDNAs coding for viral coat protein, movement protein and Rep protein could be detected using reverse transcription polymerase chain reaction (RT-PCR) on inoculated transgenic leaf extracts. Southern analysis of the two groups of transgenics proved the stable integration of ihpRNA transgene into the plant's genome. Small interfering RNAs (siRNAs) derived from the ihpRNA transgene sequence that were detected in transformed BBTV-resistant plants proved that RNA was linked with the resistance to BBTV. The successful screening of optimal transformants in bananas proved that all the transgenic plants examined had resistance to BBTV infection.

Finally, it is known that the A20/AN1 zinc finger domain that contains Stress Associated Proteins (SAP) is related to stress response pathways in plants. The novel banana SAP gene, MusaSAP1, was characterized by overexpression in transgenic bananas. The study of expression profiles in bananas indicated that MusaSAP1 was up-regulated by factors such as lack of water, salt, cold, elevated temperatures, oxidative stress and treatment with abscisic acid. A cellular localization assay was run by making a MusaSAP1::GFP fusion protein and showed that MusaSAP1 is not completely translocated to nucleus. Copy number analysis carried out using real time PCR and Southern blotting proved that MusaSAP1 is found in the banana genome in a single copy per 11 chromosome set. Transgenic banana plants constitutively overexpressing MusaSAP1 were more stress resistant in comparison to controls in both *in vitro* and *ex vivo* assays. Reduced membrane damage attributed to reduced malondialdehyde levels in transgenic leaves that had been exposed to drought, salt or oxidative stress indicated that MusaSAP1 plays an important part in stress resistance pathways of banana. Strong up-regulation of a polyphenol oxidase (PPO) coding transcript in MusaSAP1 overexpressing plants in conjunction with induction of MusaSAP1 by wounding and methyl jasmonate treatment showed that MusaSAP1 is probably related to biotic stress responses where PPOs acts as a major factor in defense pathways (Sreedharan et al. 2012).

6. Genetically Modified Bananas

As regards genetically modified bananas, Sági et al. (1995) suggested the use of a simple protocol that can be used to produce transgenic banana plants. Foreign genes were delivered into embryogenic suspension cells with the use of accelerated particles coated with DNA. An optimization of the bombardment parameters was carried out for a modified particle gun that resulted in elevated levels of transient expression of the beta-glucuronidase gene in banana and plantain cells. The selection of bombarded banana cells was done with hygromycin and their regeneration into plants followed. Molecular and histochemical characterization of transformants proved the stable integration of the transferred genes into the banana genome.

In another study it was noted that banana Xanthomonas wilt (BXW), a diseased caused as a result of *Xanthomonas campestris* pv. Musacearum infection is considered to be the most damaging disease of bananas in certain regions of Africa. The bacterium is a threat to the livelihood of millions of Africans. The disease can be catastrophiç for many banana varieties, such as the East African Highland bananas. As bananas exhibit no natural resistance to the disease, the constitutive expression of the hypersensitivity response-assisting protein (Hrap) gene from sweet pepper (*Capsicum annuum*) was assessed for its potential to render bananas resistant to BXW. The generation of transgenic lines that express the Hrap gene under the regulation of the constitutive CaMV35S promoter was carried out using ECS of the cultivars: 'Sukali Ndiizi' and 'Mpologoma'. Molecular analysis was used to characterize the lines and the bacterium was used to assess the efficacy of the Hrap gene against BXW. Most transgenic lines (75%) expressing Hrap exhibited no symptoms of infection following artificial inoculation of plants samples, while control nontransgenic plants exhibited extensive symptomatology that resulted in total wilting. It was therefore demonstrated that the constitutive expression of the sweet pepper Hrap gene in banana leads to improved banana plant resistance to BXW (Tripathi et al. 2010). Vishnevetsky et al. (2011) also proposed a transformation system for banana expressing the endochitinase gene ThEn-42 from *Trichoderma harzianum* in conjunction with the grape stilbene synthase (StSy) gene in transgenic bananas, controlled by the 35S promoter and the inducible PR-10 promoter, respectively. The tomato-derived superoxide dismutase gene Cu,Zn-SOD, controlled by the ubiquitin promoter, was used for improving scavenging of free radicals formed during fungal infections. A trial that was conducted during a period of 4 years showed that a number of transgenic banana lines became more tolerant to Sigatoka. Since the genes responsible for rendering the plants Sigatoka tolerant may be involved in a wide spectrum of anti-fungal attributes, inoculation of the regenerated bananas with *Botrytis cinerea* was also trialled. The most Sigatoka tolerant lines were also tolerant against *B. cinerea* under laboratory conditions.

Atkinson et al. (2004) evaluated whether protein engineering could be employed to control banana infection with nematodes. Cavendish banana was transformed using *Agrobacterium tumefaciens* to express a protein engineered rice cystatin (OcIdeltaD86) aiming at controlling infection by these parasites. Expression for each line was aiming at controlling infection by these parasites. Expression for each line was controlled by a constitutive promoter obtained from the maize ubiquitin gene (UBI-1), a constitutive, chimeric promoter obtained from the octopine and mannopine synthase genes of *A. tumefaciens* or a promoter obtained from a tubulin gene of Arabidopsis (TUB-1). The selection of the lines was carried out following 8 weeks of trials. Sixteen lines were chosen in total as potentially resistant and Western blots indicated that eight out of these expressed cystatin with a mean of 0.08 +/− 0.04% tsp. Only two out of 19 negatively selected lines from bioassays were found not to express cystatin. It was calculated that the mean resistance level was 69 +/− 17%. Enzyme-linked immunosorbent assay (ELISA) determined that the positive lines under control of UBI led to considerably higher expression levels of OcIdeltaD86 compared to when the other two promoters were used. A confirmation of lines of interest was given by determining the ones that produce a transcript for OcIdeltaD86 by RT-PCR. Eight plants of one UBI promoter line that expressed only 0.1 +/− 0.004% tsp as cystatin were re-tested with *R. similis* demonstrating that they were 70 +/− 10% resistant. Repeat western blotting that followed indicated that cystatin was produced by this line even after the trial. Finally, in the study of Roderick et al. (2012), transgenic plantain cv. Gonja manjaya (Musa AAB) plants, that were found to express a maize cystatin effective in the inhibition of nematode digestive cysteine proteinases and a synthetic peptide effective in the disruption of nematode chemoreception, were examined for their potential in resisting to nematode infection. Lines were used in this experiment, expressing each gene separately or at the same time in a stacked defence. Testing followed with a single nematode species or mixed populations and the results showed that 10 lines were highly resistant to infection with resistance levels against *Radopholus similis* as high as 84% ± 8% for the cystatin, 66% ± 14% for the peptide and 70% ± 6% for the dual defence. In the case of mixed population, resistance was also demonstrated against *Helicotylenchus multicinctus*. A chemodisruptive peptide labelled with fluorescent underwent retrograde transport along certain sensory dendrites of *R. similis* in a way that disrupted chemoreception. Degradation of the peptide was observed after 30 minutes in simulated intestinal fluid or boiling water and after 1 hour in nonsterile soil.

7. Physicochemical Properties

The biochemical characteristics (moisture, protein, ash, ascorbic acid, sucrose, glucose, fructose) in Pequena Enana and Gran Enana bananas obtained from cultivars in Tenerife and Ecuador (South America) were studied to evaluate how the biochemical composition and region of origin are related. The biochemical compositions in fruits from the two different geographical regions differed significantly. The cultivar though did not have an effect on the nutritional composition of the fruits, with fibers being the only exception. The biochemical composition of fruits from Tenerife differed when different cultivation methods (greenhouse and field, conventional and organic) were used but differences were also observed between bananas cultivated in the north and south of Tenerife. Clear and almost absolute differentiation (91.7%) was achieved between fruits from Tenerife and Ecuador using protein, ash, and ascorbic acid as markers and employing stepwise analysis to analyze the data. The use of various analysis also allowed bananas to be differentiated based on region of production and farming method (Forster et al. 2002). In the study of Cabrera and Saúco (2005), plant height (Grande Naine > Zelig > Gruesa) and pseudostem thickness (Gruesa > Grande Naine = Zelig) were found to differ considerably between banana cultivars. Each cultivar showed a stable (height:thickness) ratio that did not change, regardless of trial conditions. Gruesa cycle length was longer than those of the other cultivars. The production of hands per bunch and fingers per hand was higher in the case of Gruesa. In all three cultivars, finger length was directly linked to pseudostem height. The organoleptic characteristics of the fruits were similar among cultivars. A slightly reduced cycle length and increased productivity were seen though in the northern plot under the polyethylene cover in comparison to the plots under other covers and outdoors.

Studies have also focused on defining the post-harvest characteristics of new hybrids. Yomeni et al. (2004) examined how the stage of ripening of French Sombre, Dwarf Kalapua and Bluggoe affects the organoleptic and physico-chemical attributes of processed fruits. Chips made with fruits of these cultivars at different ripeness stages were found to contain less than 20 g kg^{-1} water for fruits at stages 1 and 3 and less than 60 g kg^{-1} for fruits at stages 4 and 5. There was no statistically significant difference ($p > 0.05$) between the ash levels of chips at different stages of ripeness, while protein levels increased in all cases with increasing ripeness. In contrast, the fat levels decreased with increasing ripeness, but variations were seen between cultivars. The energy content of the chips was calculated at > 4840 kcal kg^{-1} for all cultivars and stages of ripeness. The highest quality chips were obtained when fruits at stages 1 and 3 were used. Flour produced from these fruits at different stages of ripeness was found to contain less than 60 g kg^{-1} water. The fat, ash and protein levels were also low and were compensated by high levels of carbohydrates. It was also found that the yields of chips and flour were higher in the case of the plantain (French Sombre) in comparison to the rest (bananas), regardless of the state of ripeness. The cakes and chips that were trialled by consumers for their organoleptic characteristics were considered all acceptable, although there were differences. Fruits at stages 1 and 3 of ripeness were more suitable for drying and were therefore the best options for producing flour for cakes.

Another study was carried out for assessing the physicochemical and qualitative attributes of different value-added banana products. Banana flour was employed to produce dough, chapati and banana kheer. For comparison purposes, chappati was also produced using rice and basin flour. Unripe banana fruit, banana flour and products produced using them were tested for pH, titratable acidity, moisture (%), total soluble solids (TSS) (Brix) and vitamin C. It was shown that the physico-chemical and organoleptic characteristics of banana flour and their products were considerably affected by value added banana products. The maximum pH (7.68) and TSS (26.30 brix) values were recorded in banana kheer, while the highest ash (0.8260) and vitamin C (18.35) level of chapatti did not differ significantly for those of unripe banana fruit (17.54) and banana kheer (15.79). The maximum moisture content recorded in unripe bananas was (72.08%), while the minimum pH (6.79) and titratable acidity (0.02) were found in unripe banana fruit. The minimum moisture and ash recorded in banana were 7.47% and 0.2363%, respectively, while the minimum vitamin C (12.83) and TSS (5.30) were recorded in chapati and banana and basin flour, respectively. It was therefore concluded that processing banana fruits to produce value added products significantly affects the physico-chemical characteristics of banana flour (Farooq et al. 2018).

8. Conclusion

Bananas are therefore very important from both a nutritional and a trade/marketing perspective, being part of the diet of billions of people worldwide and a significant source of income for millions. Classification of bananas based simply on morphological characteristics can be very problematic and therefore advanced analytical and molecular techniques have been developed to distinguish between varieties, geographical origins, and methods of cultivation. Molecular markers have been proved very useful in this respect. A lot of research has also been carried out on the effect of environmental factors on yield and fruit quality, with a particular interest being paid on the amount of water that the plants require and how this can be controlled and adjusted in a way that enables optimum production, while preserves this natural resource. It is also important to mention that since it is known that the global banana production is threatened by the spread of certain diseases caused mainly by species of bacteria and fungi, a lot of research has focused on producing genetically modified banana plants that are more resistant to infections by such microorganisms. Although significant steps have been made towards achieving this goal, attention needs to be paid on the correct labelling of such products so that consumers know they have been genetically modified and are allowed to make informed purchase decisions. The future of research in the field of banana and plantain production will therefore unavoidably focus on achieving four main goals: (a) maximum yield, (b) improved quality of fruits, (c) resistance to diseases and (d) environmentally friendly methods of cultivation.

Abbreviations

Banana bunchy top virus	BBTV
Embryogenic cell suspensions	ECS
Enzyme-linked immunosorbent assay	ELISA
Hypersensitivity response-assisting protein	Hrap
Polymerase chain reaction	PCR
Polyphenol oxidase	PPO
Random Amplified Polymorphic DNA	RAPD
Reverse transcription polymerase chain reaction	RT-PCR
Settled cell volume	SCV
Sigatoka leaf spot diseases	SLSDs
Stilbene synthase	StSy
Stress Associated Proteins	SAP
Total soluble solids	TSS
Xanthomonas wilt	BXW
β-glucuronidase	GUS

References

Atkinson, H.J., S. Grimwood, K. Johnston and J. Green. 2004. Prototype demonstration of transgenic resistance to the nematode *Radopholussimilis* conferred on banana by a cystatin. Transgenic Res. 13(2): 135–142.

Bakry, F., F. Carreel, C. Jenny and J-P. Horry. 2009. Genetic improvement of banana. pp. 3–50. *In*: Mohan, J.S. and P.M. Priyadarshan [Eds.]. Breeding Plantation Tree Crops: Tropical Species. Springer, New York, NY.

Cabrera, J.C. and V.G. Saúco. 2005. Evaluation of the banana cultivars Zelig, Grande Naine and Gruesa under different environmental conditions in the Canary Islands. Fruits 60(6): 357–369.

Denham, T., S. Haberle and C. Lentfer. 2004. New evidence and revised interpretations of early agriculture. Highland New Guinea. Antiquity 78(302): 839–857.

Farooq, M., S.A. Sheikh, T.F. Miano and N-u-N. Memon. 2018. Physicochemical characteristics of value added banana products. J. Agric. Res. 56(4): 265–269.

Forster, M.P., E.R. Rodríguez and C.D. Romero. 2002. Differential characteristics in the chemical composition of bananas from Tenerife (Canary Islands) and Ecuador. J. Agric. Food Chem. 50(26): 7586–7592.

Heslop-Harrison, J.S. and T. Schwarzacher. 2007. Domestication, genomics and the future for banana. Ann. Bot. 100(5): 1073–1084.

Hwang, S-C. and W-H. Ko. 2004. Cavendish banana cultivars resistant to Fusarium wilt acquired through somaclonal variation in Taiwan. Plant Dis. 88(6): 580–588.

Israeli, Y. and E. Lahav. 2017. Banana. *In*: Encyclopedia of Applied Plant Sciences (Second Edition). Volume 3. Elsevier, pp. 363–381.

Janick, J. 2005. The origins of fruits fruit growing and fruit breeding. Plant Breed Rev. 25: 255–326.

Mustaffa, M.M. and V. Kumar. 2012. Banana production and productivity enhancement through spatial, water and nutrient management. J. Hortic. Sci. 7(1): 1–28.

Nadal-Medina, R., G. Manzo-Sánchez, J. Orozco-Romero, M. Orozco-Santos and S. Guzmán-González. 2009. Genetic diversity of bananas and plantains (*Musa* spp.) determined by RAPD markers. Rev. Fitotec. Mex. 32(1): 1–7.

Nayar, N.M. 2010. The bananas: Botany, origin, dispersal. Horticult. Rev. 36(3): 117–164.

Paul, J.Y., D.K. Becker, M.B. Dickman, R.M. Harding, H.K. Khanna and J.L. Dale. 2011. Apoptosis-related genes confer resistance to Fusarium wilt in transgenic 'Lady Finger' bananas. Plant Biotechnol. J. 9(9): 1141–1148.

Reay, D. 2019. Climate-smart bananas. pp. 81–91. *In*: Reay, D. [ed.]. Climate-Smart Food. Palgrave Pivot, Cham.

Roderick, H., L. Tripathi, A. Babirye, D. Wang, J. Tripathi, P.E. Urwin and H.J. Atkinson. 2012. Generation of transgenic plantain (*Musa* spp.) with resistance to plant pathogenic nematodes. Mol. Plant Pathol. 13(8): 842–851.

Rubatzky, V.E. and M. Yamaguchi. 1997. Plantain, starchy banana, breadfruit, and jackfruit. pp. 253–276. *In*: Rubatzky, V.E. and M. Yamaguchi [eds.]. World Vegetables. Springer, Boston, MA.

Sági, L., B. Panis, S. Remy, H. Schoofs, K. De Smet, R. Swennen and B.P. Cammue. 1995. Genetic transformation of banana and plantain (*Musa* spp.) via particle bombardment. Biotechnology 13(5): 481–485.

Shekhawat, U.K.S., T.R. Ganapathi and A.B. Hadapad. 2012. Transgenic banana plants expressing small interfering RNAs targeted against viral replication initiation gene display high-level resistance to banana bunchy top virus infection. J. Gen. Virol. 93(8): 1804–1813.

Sreedharan, S., U.K.S. Shekhawat and T.R. Ganapathi. 2012. MusaSAP1, a A20/AN1 zinc finger gene from banana functions as a positive regulator in different stress responses. Plant Mol. Biol. 80: 503–517.

Tripathi, J.N., R.O. Oduor and L. Tripathi. 2015. A high-throughput regeneration and transformation platform for production of genetically modified banana. Front. Plant Sci. 6: 1025.

Tripathi, L., H. Mwaka, J.N. Tripathi and W.K. Tushemereirwe. 2010. Expression of sweet pepper Hrap gene in banana enhances resistance to *Xanthomonas campestris* pv. musacearum. Mol. Plant Pathol. 11(6): 721–731.

Turner, D.W., J.A. Fortescue and D.S. Thomas. 2007. Environmental physiology of the bananas (*Musa* spp.). Brazilian Journal of Plant Physiology 19(4): 463–484.

Vishnevetsky, J., T.L. White, A.J. Palmateer, M. Flaishman, Y. Cohen, Y. Elad, M. Velcheva, U. Hanania, N. Sahar, O. Dgani and A. Perl. 2011. Improved tolerance toward fungal diseases in transgenic Cavendish banana (*Musa* spp. AAA group) cv. Grand Nain. Transgenic Res. 20(1): 61–72.

Yomeni, M.O., J. Njoukam and J.T. Tchango. 2004. Influence of the stage of ripeness of plantains and some cooking bananas on the sensory and physicochemical characteristics of processed products. J. Sci. Food Agric. 84(9): 1069–1077.

6

Detection of Genetically Modified Organisms in Foods of Plant Origin

Spiros Paramithiotis, Maria K Syrokou* and *Eleftherios H Drosinos*

1. Introduction

The definition of genetically modified organism (GMO) is clearly stated in article 2(2) of the directive 2001/18/EC of the European Parliament and of the Council: *'an organism, with the exception of human beings, in which the genetic material has been altered in a way that does not occur naturally by mating and/or natural recombination'*. In order to remove any doubt that may occur, the techniques that lead to genetic modification are listed in Annex 1 A, part 1, while the ones that are not considered to result in genetic modification are listed in Annex 1 A, part 2 of the same directive.

There are currently two generations of genetically modified crops that have been developed and made commercially available. The first generation aims to assist the farmer during production; crops have been modified to offer resistance to insects, or other adverse environmental conditions and facilitate the application of certain herbicides. The second generation aims to provide benefits to the consumer, mainly through an enhanced nutritional content.

Laboratory of Food Quality Control and Hygiene, Department of Food Science and Human Nutrition, Agricultural University of Athens, Greece.
* Corresponding author: sdp@aua.gr

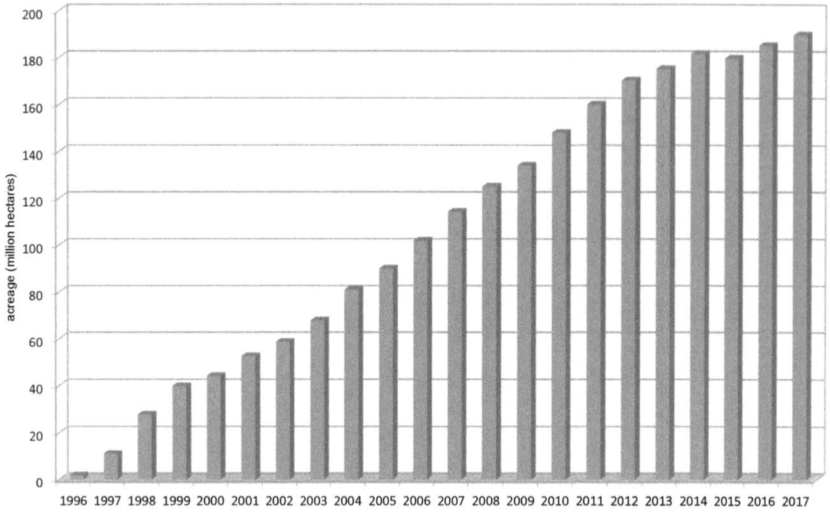

Figure 1. Worldwide acreage (in million hectares) covered by genetically modified crops, from 1996 to 2017.

The worldwide acreage covered by genetically modified crops, from 1996 to 2017, is presented in Fig. 1. This period may be further subdivided into three terms. The first one, from 1996 to 1999, is characterized by a rapid increase of the total cultivation area every year, from the 1.7 million hectares of 1996 to the 39.9 million hectares of 1999. The second term is from 2000 to 2010, during which an annual increase of more than 10% on average was noted. The total area in 2000 was 44.2 million hectares and reached 148 million hectares in 2010. Finally, a third period, from 2011 to 2017, is observed. During the latter, an annual increase of less than 5% on average was noted. The USA is the top cultivation country, in which 40% of the worldwide cultivation area is located. Brazil is in second place with 26% followed by Argentina (12%), Canada (7%) and India (6%). Soybeans are the top cultivated crop, accounting for 49.6% of the cultivated genetically modified crops, worldwide. Maize is in second place with 31.8%, followed by cotton (13.1%) and canola (4.9%) (ISAAA 2017).

In Fig. 2 the top three genetically modified crops per country are exhibited along with the respective acreage covered in 2017. Soybean, maize and cotton are the mostly cultivated crops worldwide. The first is the most common crop in the USA, Argentina, Brazil, Paraguay and Uruguay. Moreover, it is the only genetically modified crop cultivated in Bolivia. Maize is the most common crop in Chile, Colombia and South Africa and the only genetically modified one cultivated in Portugal, Spain, Honduras, Vietnam and Philippines. Cotton is the most common crop in China, Mexico and Costa Rica and the only genetically modified one cultivated in India, Pakistan, Myanmar and Sudan. Finally, canola is the

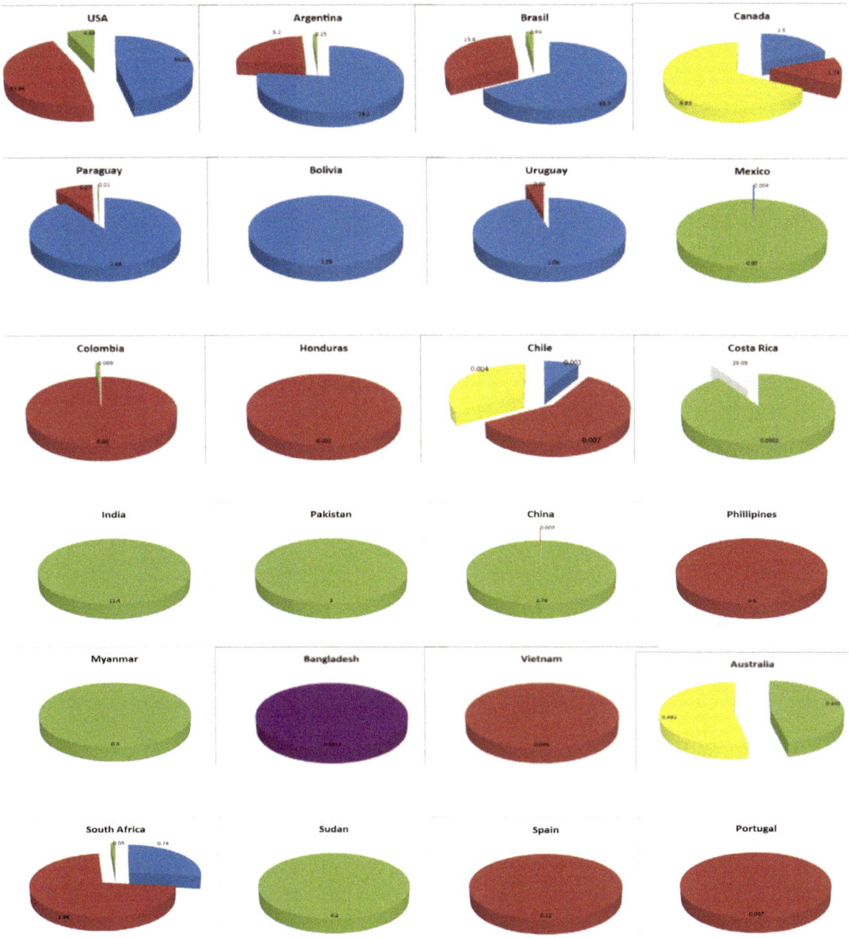

Figure 2. Top three genetically modified crops per country. The labels indicate the respective acreage (in million hectares) covered in 2017 [soybean (blue), maize (red), cotton (green), canola (yellow), papaya (orange), pink pineapple (white), eggplant (purple)].

most common crop in Canada and Australia while eggplant is the only genetically modified one cultivated in Bangladesh.

Labeling of food containing genetically modified material has been the subject of intense controversy. In the EU it is regulated by Regulation (EC) 1830/2003, which states that any food or feed containing genetically modified material more than 0.9% has to be labeled. A labeling legislation was also expected to come into force in 2018 in the USA; further details are not currently known. The enforcement of the respective legislation implies the existence of analytical techniques that may effectively detect the genetically modified material, both qualitatively and quantitatively.

The aim of the present chapter is to present in a comprehensive way the approaches currently used and discuss their advantages and limitations.

2. Introducing a Trait—Essential Background

The purpose of this paragraph is to explain the rationale behind genetic modifications in order to facilitate understanding of the detection strategies and approaches. Production of a GMO essentially involves the detection of the gene or genes that will provide with the desired property, the construction of a gene cassette that will ensure the efficient transcription of the gene(s) of interest and finally the introduction of the gene cassette into the organism of interest. The development of GMOs along with all related technical aspects have been extensively reviewed (Fedoroff 2010, Hefferon 2015, Raman 2017, Moghissi et al. 2016, Brophy et al. 2018, Nuccio 2018). From a GMO monitoring point of view, the most important aspect is the composition of the gene cassette. In Fig. 3 an example of a gene cassette encoding for two properties is presented. Upstream from the gene coding region, a promoter, a leader and/or an intron may be present, whereas a transcript termination sequence is located downstream. The promoter more frequently used is the *35S* of the cauliflower mosaic virus (CaMV *35S*). In addition, the use of opine, maize ubiquitin, rice actin 1 as well as maize alcohol dehydrogenase 1 promoters has been reported. Presence of leader or intron sequences is not very often; the leader sequence most often used is the one of the wheat *cab* gene, encoding for the major chlorophyll a/b binding protein, while the respective introns are from the rice actin gene or the maize heat shock protein Hsp70. A variety of transcript termination sequences is also used, including the ones from nopaline synthase of *Agrobacterium tumefaciens*, wheat heat shock protein Hsp17, *35S* CaMV, potato proteinase inhibitor 2.

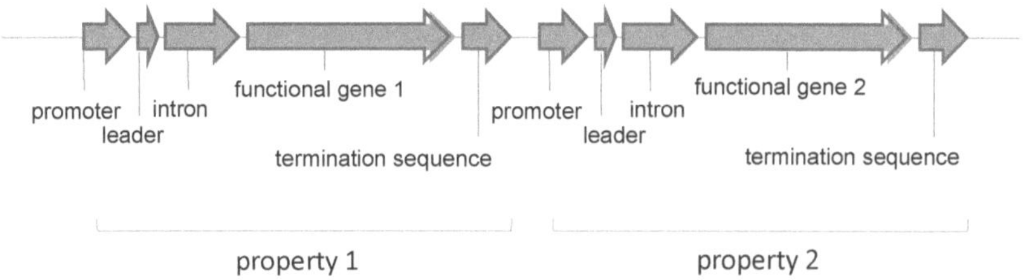

Figure 3. Example of a gene cassette encoding for two properties.

3. GMO Detection

The majority of the GMO detection protocols rely on the detection of the respective DNA sequences and only a few on the detection of the protein that is encoded, because of the stability of the former and the quantification capacity that it offers. Therefore, in the next pages all effective DNA-based approaches will be discussed.

Contemporary GMO detection involves three steps (Querci et al. 2010):

i. definition of the target GMOs, which can be any specific event or group of events,
ii. grouping of the target GMOs according to their genetic elements (i.e., promoters, leaders, introns, structural gene, transcript termination sequences, etc.); accurate grouping is necessary as it facilitates the next step,
iii. selection of the most appropriate target sequences that may ensure effective detection of the predefined target(s); these targets may be any part of the respective gene cassettes.

An example of a detection study that was based on this approach is the one by Paramithiotis et al. (2015), in which five screening elements, namely P-*35S*, T-*nos*, CP4-*epsps*, *bar* and *pat*, were selected to assist the detection of as many events as possible, in milk and dairy products.

Since DNA is the target of choice, then PCR is the approach that is in most of the cases employed. In the next paragraphs PCR formats that have been effectively employed are discussed.

3.1 Simplex-Multiplex PCR

PCR in either simplex or multiplex format is the simplest and straightforward detection approach. GMO presence is inferred by the amplification and concomitant detection through electrophoresis of control elements. In simplex format, only one control element may be detected per reaction; on the contrary the detection of more is possible in multiplex format. It has been employed in numerous studies for event-specific detection or screening for GMOs in a wide variety of commodities. The most important advantage is its simplicity since there is no need for sophisticated equipment or additional expertise in both laboratory work and interpretation of the results. On the other hand, accurate quantification is not possible and the throughput is rather low as only a limited amount of targets can be amplified in one reaction. In Table 1 recent characteristic studies reporting the development of multiplex PCR protocols are summarized. In these studies, multiplex PCR format has been employed for screening of products for GMO presence in general or occurrence of specific events.

3.2 Real-time PCR

Real-time PCR is the method most often employed for GMO detection purposes due to the quantification possibility that it offers. In addition,

Table 1. Recent characteristic studies reporting the development of multiplex PCR protocols for the detection of GMOs.

Target	Primers sequence	Comment	References
CaMVP-*35S*	P35S-cf-3:CCACGTCTCAAAGCAAGTGG P35S-cr-5:TCTCTCAAATGAAATGAACTC	A GM maize-specific tetraplex PCR assay targeting P-*35S*, T-*nos*, T-*35S* and *adh1* and a GM cotton-specific pentaplex PCR targeting assay targeting P-*35S*, T-*nos*, *nptII*, *pat* and *sad1* with LOD of 0.5% and 0.25%, respectively, were developed and effectively applied to screen for 94% of maize and 93% of cotton events globally approved.	Singh et al. (2016)
T-*nos*	NOS1:GAATCCTGTTGCCGGTCTTG NOS3:TTATCCTAGTTTGCGCGCTA		
T-*35S*	T35S-f:CACTGGATTTTGGTTTTAGGAA T35S-r:TTCGCTCATGTGTTGAGCAT		
pat	pat-f:GAAGGCTAGGAACGCTTACG pat-r:GCCAAAAACCAACATCATGC		
nptII	nptII-f:CGGCCGCTTGGGTGGAGA nptII-r:AGGTAGCCGGATCAAGCGTATGC		
adh1	ADH-FF3:CGTCGTTTCCCATCTCTTCCTCC ADH-RR4:CCACTCCGAGACCCTCAGTC		
sad1	S3F:CCAAAGGAGGTGCCTGTTCA S4R:TTGAGGTGAGTCAGAATGTTGTTC		
CaMVP-*35S*	P35S F:CGTGCACCATGATGTGTGATTCGAC P35S R:GTGGGATTGTGCGTCATCCCTT	Two PCR assays, a fourplex and a triplex, were developed for concurrent and efficient detection of transgenic material. The first one involves the amplification of P-*35S*, T-*nos*, *epsps* and lectin gene while the second of P-*35S*, *cry1Ab* and zein gene.	Datukishvili et al. (2015)
T-*nos*	TNOS F:GGTACCGGATCCAATTCCCGATCGTT TNOS R:CGCGCTATATTTTGTTTTCTATCGCGT		
epsps	EPSPS F:ACCGGCCTCATCCTGACGCT EPSPS R:CCGAGAGGCGGTCGCTTTCC		
cry1Ab	Cry1 F:GCACCTCCGTGGTGAAGGGC Cry1 R:AACCCACGGTGCGGAAGCTG		
Soybean lectin gene	LECT F:ACGGCACCCCAAAACCCTCG LECT R:GGAAGCGGCGAAGCTGGCAA		
Maize zein gene	ZEIN F:ACACCACCGACCATGGCAGC ZEIN R: TGGTGGCAAGTGCGCTGGAA		

Lectin gene	YjLec-F:GGGCTTGCCTTCTTTCTCGC YjLec-R:CGACGACTTGATCACCAGAC	A multiplex PCR was developed for event- specific detection of six GM soybean events (RRS, A2704-12, DP356043-5, MON89788, A5547-127 and DP305423-1). The limit of detection was 0.05%.	Kim et al. (2013)
3′-flanking region of RRS[1], GTS40-3-2	YjRRS-F:CTCGATTTCGGCAATGCCGC YjRRS-R:CAGCAGAGATCCCCAGGAAG		
5′-flanking region of A2704-12	YjA2704-F:GGGCGTTCGTAGTGACTGAG YjA2704-R:GCGTTACCCAACTTAATCGC		
3′-flanking region of DP356043-5	DP-5-F:GGCCACTAGTGAGCTCGGTA DP-5-R:GATGGTCTTTCTATTCCCCTAGCG		
5′-flanking region of MON89788	YjMON5-F:TCCCGCTCTAGCGCTTCA AT YjMON5-R:GGTTTGGAGACTCTGTACCC		
3′-flanking region of DP305423-1	JN23-1-F:CGTCAGGAATAAAGGAAGTACAGTA JN23-1-R:CAACCAAATGCCCTAAAGGATGCGT		
3′-flanking region of A5547-127	JNA5-1-F:CGAGCGCAGAAGTGGTCCTG JNA5-1-R:GTTAAATTGTGGTTATGGCGGTGC		
epsps	RR01:TGGCGCCCAAAGCTTGCATGGC RR04:CCCCAAGTTCCTAAATCTTCAAGT	A multiplex PCR assay for simultaneous amplification of *cry1Ab* and *epsps* for screening of soy and maize samples was reported.	Yoke-Kqueen et al. (2011)
cry1Ab	Cry1Ab:ACCATCAACAGCCGCTACAACGACC Cry1As:TGGGGAACAGGCTCACGATGTCCAG		

Table 1 contd. ...

...Table 1 contd.

Target	Primers sequence	Comment	References
DAS59122	DAS59122XF76:ACGTTGGATACTCACCAACATT GGTGCGCA MIR604XR24:ACGTTGGATAAACGGCTTGTCCC GCGTCATCG	An hexaplex PCR approach combined with capillary gel electrophoresis allowed concurrent targeting of five novel GM maize events (DAS59122, LY038, MON88017, MIR604 and Event 3272), as well as the maize *hmga* gene. The LOD was found at 0.1% for all events.	Holck et al. (2010)
hmga	ZMXF1:ACGTTGGATACAGGGAGGAATTTCGC AAGGA ZMXR1:ACGTTGGATGGCTACATAGGGAGCCT TGTCCT		
LY038	LY038XF78:ACGTTGGATTCAGCCTGTTACTCG TTCGCTC LY038XR99:ACGTTGGATAGTGTCCAAGGATGG CACTCG		
MON88017	MON88017XF104:ACGTTGGATTGAATTCTGCAT GCGTTTGGAC MON88017XR105:ACGTTGGATTGGTTTGTTTTC GGCAGTATGC		
MIR604	MIR604XF26:ACGTTGGATAGCTTAGCGGCCGG CACGTTG MIR604XR24:ACGTTGGATAAACGGCTTGTCCC GCGTCATCG		
event 3272	ES3272XF51:ACGTTGGATTGCTCTTGCTCTTGC GCTGC MIR604XR24:ACGTTGGATAAACGGCTTGTCCC GCGTCATCG		

the protocol is faster since detection of the amplicon takes place in real time. As in the case of end-point PCR, simplex or multiplex formats may be used; the former may be performed through the use of DNA-binding dyes whereas the latter through molecular probes. The limitations of this approach refer mainly to the rather limited amount of targets that may be amplified per reaction. In Table 2, recent representative studies reporting the development of real time PCR protocols based on TaqMan chemistry are summarized. The study by Debode et al. (2018) deserves a special reference since except for the development of a protocol for the detection of the emerging *tpin*II terminator, it attempts to address the problems raised by the use of the UGPase gene as a potato marker. More accurately, the UGPase gene is being used as a marker for quantification of GM events in potato samples. However, it occurs in many copies (more than 1000 per haploid genome) compromising quantification; in addition, non-specific amplicons may occur when other plants belonging to the Solanaceae family are analyzed using the specific PCR conditions. For that purpose, the authors developed and evaluated a protocol for the detection of the *stls* gene, as an alternative.

3.3 Digital PCR

Digital PCR is based on dilution and partitioning of the sample and the use of molecular probes within an end-point PCR, to facilitate detection of the amplicon under consideration; calculation of the amplicon copies per sample unit is achieved through Poisson statistics (Baker 2012). Compared to qPCR it offers higher sensitivity and precision without the need of calibrations and internal controls. On the other hand, qPCR is much less expensive and is characterized by higher throughput. However, being an emerging approach, factors associated with the available platforms should also be taken into consideration (Dong et al. 2015). A very comprehensive review article on the available digital PCR platforms and the factors that affect quantification has been recently provided by Demeke and Dobnik (2018). In Table 3 recent representative studies reporting the development of dPCR protocols are shown. The effectiveness of this approach as well as the ability to replace qPCR has been repeatedly exhibited (Burns et al. 2010, Corbisier et al. 2010, Morisset et al. 2013, Köppel and Bucher 2015, Gerdes et al. 2016). However, when absolute quantification is required, caution should be exercised since interlaboratory variation may occur, mostly due to differences in technical work and analysis of the results (Dobnik et al. 2018).

3.4 Loop-mediated isothermal amplification (LAMP)

LAMP involves the isothermal amplification of the target sequence using typically four primers that may hybridize to at least six different regions.

Table 2. Recent characteristic studies reporting the development of qPCR protocols for the detection of GMOs.

Target	Primers sequence	Comment	References
stls	StLS4C- F: GTACGACTTAATTATTAACGTTGTTGTG R: ATGAGTGGTCTCCAAGTGGTGAT P: FAM-AAGGAAGTTCGAGTAAATTGTCATACCTCCAACATA-BHQ	The development of two singleplex TaqMan protocols for the detection of the t*pin*II terminator and the endogenous *stls* potato gene was reported. The latter was proposed as an alternative to the *ugp* gene.	Debode et al. (2018)
t*pin*II	T-pinII- F: GATAAGAAATATGGATGATCTCTTTCTCTT R: GATGCACACATAGTGACATGCTAATC P: FAM-TAATTACACATAACACACAACTTTGATGCCCACA-TAMRA		
Endogenous gene of maize genome	zSSIIb- F: CTCCCAATCCTTTGACATCTGC R: TCGATTTCTCTCTTGGTGACAGG P: FAM-AGCAAAGTCAGAGCGCTGCAATGCA-TAMRA	The development of a protocol aiming to detect the *gat*-t*pin*II cassette that is present in three authorized events, namely 98140 maize, 356043 soybean and 73496 oilseed was reported.	Zhang et al. (2018)
Endogenous gene of oilseed rape genome	qHMG- F: GGTCGTCCTCCTAAGGCGAAAG R: CTTCTTCGGCGGTCGTCCAC P: FAM-CGGAGCCACTCGGTGCCGCAACTT-TAMRA		
Endogenous gene of soybean genome	qLec- F: GCCCTCTACTCCACCCCCA R: GCCCATCTGCAAGCCTTTTT P: FAM-AGCTTCGCCGCTTCCTTCAACTTCAC-TAMRA		
The junction DNA between the gat gene and pin II terminator	Gat-tpinII-F1: AGTTRGGMTTCAGYGAGCARGGAGA R: CCAATCCAGAAGATGGACAAGTC P2: FAM-CCKCCAGTWGGACCTCACATCCTGATGTAT-TAMRA		
E9 terminator	T-E9-F: TGAGAATGAACAAAAGGACCATATCA R: TTTTTATTCGGTTTTCGCTATCG P: FAM- TCATTAACTCTTCTCCATCCATTTCCATTTCACAGT-TAMRA	Two singleplex TaqMan methods for the detection of the E9 terminator and the endogenous lectin gene from pea were developed and their suitability for screening purposes was validated.	Debode et al. (2017)
Lectin gene of pea	lecPea1- F:TGGAATCGATGTGAACAGTATCAA R: ACAACATTAGCCTCTTCACCATTCT P: FAM- TCCGTAAACACTAAGTCGTGGAAGTTG-TAMRA		

Table 3. Recent characteristic studies reporting the development of dPCR protocols for the detection of GMOs.

Target	Primers sequence	Comment	References
MON87769	MON87769 F2: CATACTCATTGCTGATCCATGTAGA MON87769 R2: GCAAGTTGCTCGTGAAGTT MON87769 Joe: CCCGGACATGAAGCCATTTACAATTGAC	A pentaplex qPCR and four duplex droplet digital PCR were developed for the simultaneous detection of MON87769, MON87708, MON87705, FG72 and lectin. The superiority of qPCR for screening purposes and of dPCR for quantification was highlighted.	Köppel et al. (2015)
MON87708	MON87708 F: TCATACTCATTGCTGATCCATGTAG MON87708 R: AGAACAAATTAACGAAAAGACAGAACG MON87708 Fam: TCCCGGACTTTAGCTCAAAATGCATGTA		
MON87705	MON87705 F: TTCCCGGACATGAAGCCATTTAC MON87705 R: ACAACGGTGCCTTGGCCCAAAG MON87705 DY681: AAGAGACTCAGGGTGTTGTTATCACTGCGG MON87705 Fam[1]: AAGAGACTCAGGGTGTTGTTATCACTGCGG		
FG72	MAE071: AGATTTGATCGGGCTGCAGG SHA097: GCACGTATTGATGACCGCATTA TM325 Cy5: AATGTGGTTCATCCGTCTT MGB TM325 Hex[1]: AATGTGGTTCATCCGTCTT MGB		
Lectin	Lectin F: TCCACCCCCATCCACATTT Lectin R: GGCATAGAAGGTGAAGTTGAAGGA Lectin_ROX_TMP: AACCGGTAGCGTTGCCAGCTTCG Lectin_Fam[1]: AACCGGTAGCGTTGCCAGCTTCG Lectin_Joe[1]: AACCGGTAGCGTTGCCAGCTTCG		
NK603	NK603-F: ATGAATGACCTCGAGTAAGCTTGTTAA NK603-R: AAGAGATAACAGGATCCACTCAAACACT NK603-P: [6FAM]TGGTACCACGCGACACACTTCCACTC[BHQ]	Two validated qPCR protocols for quantification of NK603 and DAS1507 were transferred to dPCR platform. The use of higher probe and primer concentration than qPCR was proposed for the improvement of the separation between positive and negative signals.	Grelewska-Nowotko et al. (2018)
DAS1507	MaiY- F: TAGTCTTCGGCCAGAATGG MaiY- R: CTTTGCCAAGATCAAGCG MaiY- P: [6FAM]TAACTCAAGGCCCTCACTCCG[BHQ]		
hmg	ZM1- F: TTGGACTAGAAATCTCGTGCTGA ZM1- F: GCTACATAGGGAGCCTTGTCCT ZM1- F: [6FAM]CAATCCACACAAACGCACGCGTA[BHQ]'		

Table 3 contd. ...

...Table 3 contd.

Target	Primers sequence	Comment	References
35S	CSP-35s112bp- F: TTGGTCGTGGTGGTGGTTTCGACAGTGGTCCCAAAGA CSP-35s112bp- R: TTGGTCGTGGTGGTGGTTTAAGACGTGGTTGGAACGTCTTC CSP-35s112bp- P: FAMTGGACCCCCACCCACGAGGAGCATCBH-Q1	The development of a single universal primer-fiveplex-ddPCR approach was reported for accurate broad-spectrum screening and quantification of *35S*, *nos*, *nptII*, *pat* and *adh1*. The effectiveness was highlighted by the LOD of 0.1% and the LOQ of 0.01% (RSD < 25%) that were also reported.	Niu et al. (2018)
nos	CSP-NOS203bp- F: TTGGTCGTGGTGGTGGTTTATCGTTCAAACATTTGGCA CSP-NOS203bp- R: TTGGTCGTGGTGGTGGTTTATTGCGGGACTCTAATCATA CSP-NOS203bp-P: FAM-CATCGCAAGACCGGCAACAGG-BHQ1		
nptII	CSP-NPTII-139- QF: TTGGTCGTGGTGGTGGTTTCTATGACTGGGCACAACAGACA CSP-NPTII-139- QR: TTGGTCGTGGTGGTGGTTTCGGACAGGTCGGTCTTGACA CSP-NPTII-139- QP: FAM-CTGCTCTGATGCCGCCGTGTTCCG-BHQ1		
pat	CSP-PAT-229- QF: TTGGTCGTGGTGGTGGTTTGTCGACATGTCTCCGGAGAG CSP-PAT-229- QR: TTGGTCGTGGTGGTGGTTTGCAACCAACCAAGGGTATC CSP-PAT-229- QP: FAM-TGGCCGCGGTTTGTGATATCGTTAA-BHQ1		
adh1	CSP-adh1-173- QF: TTGGTCGTGGTGGTGGTTTCGTCGTTTCCCATCTCTTCCT CSP-adh1-173- QR: TTGGTCGTGGTGGTGGTTTCCACTCCGAGACCCTCAGT CSP-adh1-173- QP: VIC-AATCAGGGCTCATTTCTCGCTCCTCA-BHQ1		

[1] Primers used for duplex dPCR; RSD: relative standard deviation

Visualization of the reaction may be achieved either in real time through fluorescent detection or after the reaction is completed through the use of fluorescence dyes (Notomi et al. 2000, Chen et al. 2012, Randhawa et al. 2013, Huang et al. 2014). Isothermal amplification provides a competitive advantage over PCR; there is no need for expensive thermocycling equipment as amplification may be performed even in a water-bath. Although a variety of techniques based on isothermal DNA amplification have been developed, such as strand displacement amplification (Qiu et al. 2013), helicase dependent amplification (Barbieri et al. 2014), rolling circle amplification (Ali et al. 2014), etc., only LAMP has been extensively considered as a method for GMO detection. The major advantages of this approach are the efficiency of the reaction, which may be completed within 1 h (Tomita et al. 2008), and the ability to tolerate the presence of PCR inhibitors (Kaneko et al. 2007, Francois et al. 2011). The effectiveness of this approach has been repeatedly exhibited and several protocols have been developed for the detection of genetic markers, such as P-*35S*, T-*nos*, *bar*, *nptII*, *uidA*, *cry* genes, etc., or specific events, such as oilseed rape events MS8 and RF3 rice events KMD1, TT51-1 and KF7, maize T25, wheat B73-6-1, etc. (Lee et al. 2009, Rostamkhani et al. 2011, Li et al. 2013, 2014, Zhou et al. 2014, 2016, Di et al. 2014, Randhawa et al. 2013, Chen et al. 2012, Xu et al. 2013, Cheng et al. 2014). However, contradictory performance characteristics are very often reported, making method validation prior to any application necessary. In Table 4, representative recent studies reporting the development of LAMP protocols are summarized.

3.5 Next generation sequencing (NGS)

When the definition of the target GMOs is not possible then incorporation of NGS into the detection approach seems necessary. In that case detection of a control element may result in the masking of unauthorized events through the presence of authorized ones. Thus, additional information, ideally in the form of nucleotide sequence is required for effective differentiation between authorized and unauthorized events. NGS is currently the most suitable technique as it offers high throughput and allows massive DNA sequencing. Wahler et al. (2013) used NGS to sequence a GM rice variety and identify the GM event through comparison with the developer description without any prior knowledge. In addition, Kovalic et al. (2012) had already exhibited the equivalence of NGS to Southern blot-based approach for the characterization of a typical GM soybean event. Moreover, a series of studies that followed, highlighted the efficacy of this approach (Yang et al. 2013, Liang et al. 2014, Song et al. 2014, Willems et al. 2016). NGS may be used either as a detection and simultaneous characterization step, suitable for screening purposes or for sequencing of specific genomic regions previously amplified by

Table 4. Recent characteristic studies reporting the development of LAMP protocols for the detection of GMOs.

Target	Primers Sequence	Comment	References
pat	pat- F3: GGCGCAAGGTTTTAAGTCTG B3: GGTAACTGGCCTAACTGGC FIP: TACCCCGGGCTGTGTATCCCATAGGCCTTCCAAACGATCC BIP: ATTGCGCGCAGCTGGATACAAGGAGGAGCTGGCAACTCA LoopF: GCCTCATGCAACCTAACAGAT LoopR: GCATGGTGGATGGCATGAT	Two LAMP methods, one visual and one Real Time, were developed for the detection of *pat* and *pmi* marker genes. With the former, detection of up to 0.05% GM content for pat and 0.1% for pmi was possible within 40 min. Real time LAMP could detect up to 0.01% GM content within 30 min.	Singh et al. (2017)
pmi	pmi- F3: TAACCACAAGCCGGAGCT B3: GGGCCGATTTTAAAATCGCC FIP: GACCGGCTGGAGTAGGGAGACGCTGACGCCTTTCCTTG BIP: AACAGCCTGATGCCGAACGTGCGCGGGATTTTTCTTCAC LoopF: TCACGAAACGCGTTCATCG LoopR: GCGAACTGTTCGCCAGCCTG		
MON89034	MON89034- F3: CATGCTACACTGCCTACAC B3: GCACCTGAATTTGAATGGC FIP: GTACTCGTGCTCACGTCGTAGATTATTATTGGAACATGAGGCT BIP: ATCCGCCCTCTCTGTCCCTGCTGTTTCATCTGACAACA LoopF: CCGTGTGGGAGGAGAAAT LoopB: CTGCTACTCCAGCCACTG		
NK603	NK603- F3: AACAGATCAGCATCAGCG B3: ATGAATGACCTCGAGTAAGC FIP: AGGAGGAGTCTGGCCGAATTCGAAAGTTTCGTCAAAGGA BIP: TCTCTGGCATTTCCAACCCTCCGCTACAAGGCTTGC LoopF: GGCGGCTGGAAACAGT LoopB: AGAGACGTGCGTCCCT		

GA21	GA21- F3: TCGAAGCGGACAAAGC B3: TCACTTTGGGCCACCT FIP: TCTAGAGCTGCACTTCCTCTGCTGTAGTTGTTGGCTGT BIP: CCAGCTTGCATGCCTGCTTGGACTATCCCGACTCTC LoopF: GCATCCTCAACTGGGAACT LoopB: GGTCGAGGTCATTCATATGC	One visual and one Real Time LAMP assay were developed for the detection of six GM maize events, namely MON89034, NK603, GA21, Bt11, TC1507 and MON810. The limit of detection was 0.1% for NK603 and 0.01% for the remaining.	Bhoge et al. (2015)
Bt11	Bt11- F3: GCCTCGTGATACGCCTA B3: TGTACGAGCCTCTGGTC FIP: TCCCTCCATGAGCGGATACAACGTCAGGTGGCACTT BIP: TGGTGGAGACCATTTCTTGGTAGCCATGAGCGACCAT LoopF: GTTCCGCGCACATTTCC LoopB: ATCTGTAGGTGTTAGCCTCT		
TC1507	TC1507- F3: CCTCTAGAGTCGACCTGC B3: ACTGCACTGCAAATGAA FIP: GCCGAAGACTACTGAGTGTCCGCCTCTAGTTGAAGACAC BIP: CAGAATGGCCTAACTCAAGGTGCGTCAAATATCTTTGCCA LoopF: CTTACGATGAAGACATGAAC LoopB: CCTCACTCCGCTTGATCT		
MON810	MON810- F3: GGTACTGGTTCCCTCTGG B3: GGTGGTCTTACATCTAAGAAGG FIP: TCGGAAATGAAAGAAGGCTACCCACTTCTCCTTGGACATCG BIP: GCTTGCGAGCAGTCAGGTAAGCACTAGTACTGCCAA LoopF: AAGTCCTCGTTCAGGTCG LoopB: TTTGATTCATCTGAGTTTGGCT		

Table 4 contd. ...

...Table 4 contd.

Target	Primers Sequence	Comment	References
P-*35S*	p-35S- F3: CTCCTCGGATTCCATTGC B3: TCTACAGGACGGACCATG FIP: ACGATGCTCCTCGTGGGTCATCGTTGAAGATGCCTCT BIP: CGTTCCAACCACGTCTTCAAGTCTTGCGAAGGATAGTGG LoopF: ATCTTTGGGACCACTGTCG LoopB: TGATATCTCCACTGACGTAAGG	One visual and one Real Time assay were developed targeting the promoters P-*35S* and P-*fmv* and the marker genes *aadA*, *nptII* and *uidA*. Their LOD was determined at 40 and 4 copies, respectively. The flexibility, sensitivity and speed of the assays performed with the isothermal master mix on the isothermal real-time system was highlighted.	Randhawa et al. (2013)
P-*fmv*	p-FMV- F3: AACAATTCTGCACCATTCCT B3: AATTCTCAGTCCAAAGCCTC FIP: TGCATCATGGTCAGTAAGTTTCAGATGCTCGATGTTGACAAGATT BIP: TGTGCTGGAACAGTAGTTTACTTTGAAGGTCAGGGTACAGAGTC LoopF: AAGACATCCACCGAAGACTTAA LoopB: AGATTCTTCATTGATCTCCTGTAGC		
nptII	Npt-II- F3: TGATGCTCTTCGTCCAGA B3: CTCGACGTTGTCACTGAAG FIP: CCATTCGACCACCAAGCGACATCCTGATCGACAAGACC BIP: TAGCCGGATCAAGCGTATGCTCATCTCACCTTGCTCCT LoopF: CGTACTCGGATGGAAGCC LoopB: TTGCATCAGCCATGATGGATA		
aadA	aadA- F3: AGCTTTGATCAACGACCTTT B3: GATCCTGTTCAGGAACCG FIP: CAGTTCGCGCTTAGCTGGATGCTGTAGAAGTCACCATTGT BIP: GGTATCTTCGAGCCAGCCACAACGCTATGTTCTCTTGCTT LoopF: AACGCCACGGAATGATGT LoopB: TGATCTGGCTATCTTGCTGAC		
uidA	uidA- F3: GTGGACGATATCACCGTG B3: TGGTTAATCAGGAACTGTTGG FIP: CCGCTAGTGCCTTGTCCAGGGTGATGTCAGCGTTGAA BIP: GAATCCGCACCTCTGGCAAATCACACTCTGTCTGGCT LoopF: CCACCTGTTGATCCGCAT LoopB: TCTCTATGAACTGTGCGTCAC		

PCR. However, detection may only be qualitative and not quantitative and sequencing by the Sanger method is recommended for verification purposes (Fraiture et al. 2017).

3.6 Biosensors

In general, the aim of biosensors is to provide with comparable accuracy, sensitivity, speed and cost to most analytical laboratory techniques but in a more suitable for routine analysis way and ideally from non-specialized personnel. In addition, portability is a property that is also desired, as it enables on-site sample analysis. As far as GMO testing is concerned, such research is active for nearly 25 years. Biosensors consist of a biological material that acts as a receptor and a transducer that converts binding of the target analyte to the receptor into a readable signal. The biological receptor in most of the cases is a molecule, e.g., proteins, enzymes, antibodies, nucleic acids, aptamers, biomimics, etc., while the transducer may vary to piezoelectric, thermal, electrochemical, micromechanical, optical or magnetic. In the case of GMO detection, the biological receptor is in most of the cases a DNA sequence complementary to screening elements, i.e., sequences used for the detection of GMOs in general; only a limited amount of studies aimed to develop biosensors for event-specific detection (Duwensee et al. 2009, Mix et al. 2012, Liao et al. 2013, Sun et al. 2013, Barroso et al. 2015, Manzanares-Palenzuela et al. 2015, 2016a,b, 2017). Regarding transduction, only electrochemical, piezoelectric and optical has been reported so far, with the first being the strategy of choice in most of the cases due to the relatively simple and cost efficient instrumentation as well as the high sensitivity that it offers. On the other hand, optical and piezoelectric biosensors may offer the capacity for real-time monitoring of the binding process, with the latter being also label-free. However, their application is hindered by non-specific interactions (Minunni et al. 2001, Stobiecka et al. 2007).

Research so far has been focused on the development of biosensors as an effective analytical step, the final step of the downstream processing that starts from sample preparation and continues with DNA extraction and amplification. Regarding this aim, several successful attempts have been made for the detection of reference materials in synthetic samples. On the contrary, their use in the analysis of real samples is rather rare. As a result, the basic aims of a biosensor, i.e., portability and simplicity of application, are yet to be achieved. However, based on the progress made so far, it is expected that within the next decade significant steps towards this direction can be made.

4. Concluding Remarks and Future Perspectives

The use of genetically modified organisms and products thereof has been widely spread and is expected to increase over the next decade. Therefore, the strategy employed for contemporary GMO monitoring needs to be thoroughly understood and implemented. In addition, new methods capable of addressing the limitations of the current ones need to be developed and applied.

Abbreviations

Genetically modified organism	GMO
Cauliflower mosaic virus	CaMV
Deoxyribonucleic acid	DNA
Polymerase chain reaction	PCR
Limit of detection	LOD
Quantitative polymerase chain reaction	qPCR
Loop-mediated isothermal amplification	LAMP
Next generation sequencing	NGS
Relative standard deviation	RSD
Digital polymerase chain reaction	dPCR
Droplet digital PCR	ddPCR
Limit of quantification	LOQ

References

Ali, M.M., F. Li, Z. Zhang, K. Zhang, D.-K. Kang, J.A. Ankrum, X.C. Le and W. Zhao. 2014. Rolling circle amplification: a versatile tool for chemical biology, materials science and medicine. Chem. Soc. Rev. 43: 3324–3341.

Baker, M. 2012. Digital PCR hits its stride. Nat. Methods 9: 541–544.

Barbieri, D., S. Venturoli, F. Rösl and B. Rincon-Orozco. 2014. Detection of high-risk human papilloma virus type 16 and 18 using isothermal helicase-dependent amplification. Diagn. Microbiol. Infect. Dis. 79: 178–182.

Barroso, M.F., M. Freitas, M.B. Oliveira, N. de-los-Santos-Alvarez, M.J. Lobo-Castanon and C. Delerue-Matos. 2015. 3D-nanostructured Au electrodes for the event-specific detection of MON810 transgenic maize. Talanta 134: 158–164.

Bhoge, R.R.K., R. Chhabra, G. Randhawa, M. Sathiyabama and M. Singh. 2015. Event-specific analytical methods for six genetically modified maize events using visual and real-time loop-mediated isothermal amplification. Food Control 55: 18–30.

Brophy, J.A.N., T. LaRue and J.R. Dinneny. 2018. Understanding and engineering plant form. Semin. Cell Dev. Biol. 79: 68–77.

Burns, M.J., A.M. Burrell and C.A. Foy. 2010. The applicability of digital PCR for the assessment of detection limits in GMO analysis. Eur. Food Res. Technol. 231: 353–362.

Chen, X., X. Wang, N. Jin, Y. Zhou, S. Huang, Q. Miao, Q. Zhu and J. Xu. 2012. Endpoint visual detection of three genetically modified rice events by loop-mediated isothermal amplification. Int. J. Mol. Sci. 13: 14421–14433.

Cheng, Y., M. Zhang, K. Hu, F. Sun, R. Tao, X. Gao and F. Luan. 2014. Loop-mediated isothermal amplification for the event-specific detection of wheat B73-6-1. Food Anal. Methods 7: 500–505.

Corbisier, P., S. Bhat, L. Partis, V.R.D. Xie and K.R. Emslie. 2010. Absolute quantification of genetically modified MON810 maize (*Zea mays* L.) by digital polymerase chain reaction. Anal. Bioanal. Chem. 396: 2143–2150.

Datukishvili, N., T. Kutateladze, I. Gabriadze, K. Bitskinashvili and B. Vishnepolsky. 2015. New multiplex PCR methods for rapid screening of genetically modified organisms in foods. Front. Microbiol. 6: 757.

Debode, F., I. Huber, R. Macarthur, P.E. Rischitor, M. Mazzara, V. Herau, D. Sebah, D. Dobnik, S. Broeders, N.H. Roosens, U. Busch, G. Berben, D. Morisset and J. Zel. 2017. Inter-laboratory studies for the validation of two singleplex (tE9 and pea lectin) and one duplex (*pat/bar*) real time PCR methods for GMO detection. Food Control 73: 452–461.

Debode, F., K. Zdeňková, E. Janssen, A. Tizolova, P. du Jardin, G. Berben and K. Demnerova. 2018. Development of real-time PCR assays for the detection of the pin II terminator (*tpinII*) used in GM constructs and its donor organism, potato (*Solanum tuberosum*). Food Anal. Method 11: 2172–2180.

Demeke, T. and D. Dobnik. 2018. Critical assessment of digital PCR for the detection and quantification of genetically modified organisms. Anal. Bioanal. Chem. 410: 4039–4050.

Di, H., L. Shi, H. Shen, H. Yan, H. Meng, L. Li, M.J. Alam, S. Yamasaki and Y. Lei. 2014. Rapid detection of genetically modified ingredients in soybean products by real-time loop mediated isothermal amplification. J. Food Nutr. Res. 2: 363–368.

Dobnik, D., T. Demšar, I. Huber, L. Gerdes, S. Broeders, N. Roosens, F. Debode, G. Berben and J. Žel. 2018. Inter-laboratory analysis of selected genetically modified plant reference materials with digital PCR. Anal. Bioanal. Chem. 410: 211–221.

Dong, L., Y. Meng, Z. Sui, J. Wang, L. Wu and B. Fu. 2015. Comparison of four digital PCR platforms for accurate quantification of DNA copy number of a certified plasmid DNA reference material. Sci. Rep. 5: 13174.

Duwensee, H., M. Mix, I. Broer and G.-U. Flechsig. 2009. Electrochemical detection of modified maize gene sequences by multiplexed labeling with osmium tetroxide bipyridine. Electrochem. Commun. 11: 1487–1491.

Fedoroff, N.V. 2010. The past, present and future of crop genetic modification. N. Biotechnol. 27: 461–465.

Fraiture, M.-A., P. Herman, M. De Loose, F. Debode and N.H. Roosens. 2017. How can we better detect unauthorized GMOs in food and feed chains? Trends Biotechnol. 35: 508–517.

Francois, P., M. Tangomo, J. Hibbs, E.J. Bonetti, C.C. Boehme, T. Notomi, M.D. Perkins and J. Schrenzel. 2011. Robustness of a loop-mediated isothermal amplification reaction for diagnostic applications. FEMS Immunol. Med. Microbiol. 62: 41–48.

Gerdes, L., A. Iwobi, U. Busch and S. Pecoraro. 2016. Optimization of digital droplet polymerase chain reaction for quantification of genetically modified organisms. Biomol. Detect. Quantif. 7: 9–20.

Grelewska-Nowotko, K., M. Żurawska-Zajfert, E. Żmijewska and S. Sowa. 2018. Optimization and verification of droplet digital PCR event-specific methods for the quantification of GM maize DAS1507 and NK603. Appl. Biochem. Biotechnol. 185: 207–220.

Hefferon, K.L. 2015. Nutritionally enhanced food crops; progress and perspectives. Int. J. Mol. Sci. 16: 3895–3914.

Holck, A., B.O. Pedersen and E. Heir. 2010. Detection of five novel GMO maize events by qualitative, multiplex PCR and fluorescence capillary gel electrophoresis. Eur. Food Res. Technol. 231: 475–483.

Huang, X., L. Chen, J. Xu, H.-F. Ji, S. Zhu and H. Chen. 2014. Rapid visual detection of phytase gene in genetically modified maize using loop-mediated isothermal amplification method. Food Chem. 156: 184–189.

ISAAA. 2017. Global status of commercialized biotech/GM crops in 2017: biotech crop adoption surges as economic benefits accumulate in 22 years. ISAAA Brief No. 53. ISAAA: Ithaca, NY.

Kaneko, H., T. Kawana, E. Fukushima and T. Suzutani. 2007. Tolerance of loop-mediated isothermal amplification to a culture medium and biological substances. J. Biochem. Biophys. Methods 70: 499–501.

Kim, J.H., D. Jeong, Y.R. Kim, Y.K. Kwon, G.S. Rhee and D. Zhang. 2013. Development of a multiplex PCR method for testing six GM soybean events. Food Control 31: 366–371.

Köppel, R. and T. Bucher. 2015. Rapid establishment of droplet digital PCR for quantitative GMO analysis. Eur. Food Res. Technol. 241: 427–439.

Köppel, R., T. Bucher, A. Frei and H.-U. Waiblinger. 2015. Droplet digital PCR versus multiplex real-time PCR method for the detection and quantification of DNA from the four transgenic soy traits MON87769, MON87708, MON87705 and FG72, and lectin. Eur. Food Res. Technol. 241: 521–527.

Kovalic, D., C. Garnaat, L. Guo, Y. Yan, J. Groat, A., Silvanovich, L. Ralston, M. Huang, Q. Tian, A. Christian, N. Cheikh, J. Hjelle, S. Padgette and G. Bannon. 2012. The use of next generation sequencing and junction sequence analysis bioinformatics to achieve molecular characterization of crops improved through modern biotechnology. Plant Genome 5: 149–163.

Lee, D., M. La Mura, T. Allnutt and W. Powell. 2009. Detection of genetically modified organisms (GMOs) using isothermal amplification of target DNA sequences. BMC Biotechnol. 9: 7.

Li, F., W. Yan, L. Long, X. Qi, C. Li and S. Zhang. 2014. Development and application of loop mediated isothermal amplification assays for rapid visual detection of *cry2Ab* and *cry3A* genes in genetically modified crops. Int. J. Mol. Sci. 15: 15109–15121.

Li, Q., J. Fang, X. Liu, X. Xi, M. Li, Y. Gong and M. Zhang. 2013. Loop-mediated isothermal amplification (LAMP) method for rapid detection of *cry1Ab* gene in transgenic rice (*Oryza sativa* L.). Eur. Food Res. Technol. 236: 589–598.

Liang, C., J.P. van Dijk, I.M. Scholtens, M. Staats, T.W. Prins, M.M. Voorhuijzen, A.M. da Silva, A.C. Arisi, J.T. den Dunnen and E.J. Kok. 2014. Detecting authorized and unauthorized genetically modified organisms containing vip3A by real-time PCR and next-generation sequencing. Anal. Bioanal. Chem. 406: 2603–2611.

Liao, W.C., M.C. Chuang and J.A. Ho. 2013. Electrochemical sensor for multiplex screening of genetically modified DNA: identification of biotech crops by logic-based biomolecular analysis. Biosens. Bioelectron. 50: 414–420.

Manzanares-Palenzuela, C.L., N. Santos-Alvarez, M.J. Lobo-Castanon and B. Lopez-Ruiz. 2015. Multiplex electrochemical DNA platform for femtomolar-level quantification of genetically modified soybean. Biosens. Bioelectron. 68: 259–265.

Manzanares-Palenzuela, C.L., I. Mafra, J. Costa, M.F. Barroso, N. de los-Santos-Alvarez, C. Delerue-Matos and M.B. Oliveira. 2016a. Electrochemical magnetoassay coupled to PCR as a quantitative approach to detect the soybean transgenic event GTS40–3-2 in foods. Sens. Actuator B: Chem. 222: 1050–1057.

Manzanares-Palenzuela, C.L., E.G.R. Fernandes, L.M. Jobo-Castanon, B. Lopez-Ruiz and V. Zucolotto. 2016b. Impedance sensing of DNA hybridization onto nanostructured phthalocyanine-modified electrodes. Electrochim. Acta 21: 86–95.

Manzanares-Palenzuela, C.L., J.P. Martin-Clemente, M.J. Lobo-Castanon and B. Lopez-Ruiz. 2017. Electrochemical detection of magnetically-entrapped DNA sequences from complex samples by multiplexed enzymatic labelling: application to a transgenic food/feed quantitative survey. Talanta 164: 261–267.

Minunni, M., S. Tombelli, S. Pratesi, M. Mascini, P. Piati, P. Bogani and M. Buiatti. 2001. A piezoelectric affinity biosensor for genetically modified organisms (GMOs) detection. Anal. Lett. 34: 825–840.

Mix, M., J. Ruger, S. Kruger, I. Broer and G.U. Flechsig. 2012. Electrochemical detection of 0.6% genetically modified maize MON810 in real flour samples. Electrochem. Commun. 22: 137–140.

Moghissi, A.A., S. Pei and Y. Liu. 2016. Golden rice: scientific, regulatory and public information processes of a genetically modified organism. Crit. Rev. Biotechnol. 36: 535–541.

Morisset, D., D. Štebih, M. Milavec, K. Gruden and J. Žel. 2013. Quantitative analysis of food and feed samples with droplet digital PCR. PLoS One 8: e62583.

Niu, C., Y. Xu, C. Zhang, P. Zhu, K. Huang, Y. Luo and W. Xu. 2018. Ultrasensitive single fluorescence-labeled probe-mediated single universal primer–multiplex–droplet digital polymerase chain reaction for high-throughput genetically modified organism screening. Anal. Chem. 90: 5586–5593.

Notomi, T., H. Okayama, H. Masubuchi, T. Yonekawa, K. Watanabe, N. Amino and T. Hase. 2000. Loop-mediated isothermal amplification of DNA. Nucleic Acids Res. 28: e63.

Nuccio, M.L. 2018. A brief history of promoter development for use in transgenic maize applications. Methods Mol. Biol. 1676: 61–93.

Paramithiotis, S., A. Adrakta, K. Sigala, E.H. Drosinos and C. Proestos. 2015. Detection of DNA sequences originating from GMOs in milk and dairy products commercially available in Greece. Eur. J. Nutr. Funct. Foods 26: 52–56.

Qiu, L., H. Zhou, W. Zhu, L. Qiu, J. Jiang, G. Shen and R. Yu. 2013. A novel label-free fluorescence aptamer-based sensor method for cocaine detection based on isothermal

circular strand-displacement amplification and graphene oxide absorption. New J. Chem. 37: 3998–4003.
Querci, M., M. Van den Bulcke, J. Žel, G. Van den Eede and H. Broll. 2010. New approaches in GMO detection. Anal. Bioanal. Chem. 396: 1991–2002.
Raman, R. 2017. The impact of Genetically Modified (GM) crops in modern agriculture: a review. GM Crops Food. 8: 195–208.
Randhawa, G.J., M. Singh, D. Morisset, P. Sood and J. Zel. 2013. Loop-mediated isothermal amplification: rapid visual and real-time methods for detection of genetically modified crops. J. Agric. Food Chem. 61: 11338–11346.
Rostamkhani, N., A. Haghnazari, M. Tohidfar and A. Moradi. 2011. Rapid identification of transgenic cotton (*Gossypium hirsutum* L.) plants by loop-mediated isothermal amplification. Czech J. Genet. Plant. 47: 140–148.
Singh, M., R.K. Bhoge and G. Randhawa. 2016. Crop-specific GMO matrix-multiplex PCR: A cost-efficient screening strategy for genetically modified maize and cotton events approved globally Food Control 70: 271–280.
Singh, M., R.K. Bhoge and G. Randhawa. 2017. Real-time and visual loop-mediated isothermal amplification: Efficient GMO screening targeting *pat* and *pmi* marker genes. Food Control 71: 248–254.
Song, Q., G. Wei and G. Zhou. 2014. Analysis of genetically modified organisms by pyrosequencing on a portable photodiode-based bioluminescence sequencer. Food Chem. 154: 78–83.
Stobiecka, M., J.M. Cieśla, B. Janowska, B. Tudek and H. Radecka. 2007. Piezoelectric sensor for determination of genetically modified soybean roundup ready® in samples not amplified by PCR. Sensors 7: 1462–1479.
Sun, W., Y. Zhang, A. Hu, Y. Lu, F. Shi, B. Lei and Z. Sun. 2013. Electrochemical DNA biosensor based on partially reduced graphene oxide modified carbon ionic liquid electrode for the detection of transgenic soybean A2704–12 gene sequence. Electroanal. 25: 1417–1424.
Tomita, N., Y. Mori, H. Kanda and T. Notomi. 2008. Loop-mediated isothermal amplification (LAMP) of gene sequences and simple visual detection of products. Nat. Protocols 3: 877–882.
Wahler, D., L. Schauser, J. Bendiek and L. Grohmann. 2013. Next-generation sequencing as a tool for detailed molecular characterisation of genomic insertions and flanking regions in genetically modified plants: a pilot study using a rice event unauthorised in the EU. Food Anal. Methods 6: 1718–1727.
Willems, S., M.-A. Fraiture, D. Deforce, S.C. De Keersmaecker, M. De Loose, T. Ruttink, P. Herman, F. Van Nieuwerburgh and N. Roosens. 2016. Statistical framework for detection of genetically modified organisms based on next generation sequencing. Food Chem. 192: 788–798.
Xu, J., Q. Zheng, L. Yu, R. Liu, X. Zhao, G. Wang, Q. Wang and J. Cao. 2013. Loop-mediated isothermal amplification (LAMP) method for detection of genetically modified maize T25. Food Sci. Nutr. 1: 432–438.
Yang, L., C. Wang, A. Holst-Jensen, D. Morisset, Y. Lin and D. Zhang. 2013. Characterization of GM events by insert knowledge adapted re-sequencing approaches. Sci. Rep. 3: 2839.
Yoke-Kqueen, C., C. Yee-Tyan, K. Siew-Ping and R. Son. 2011. Development of multiplex-PCR for genetically modified organism (GMO) detection targeting EPSPS and Cry1Ab genes in soy and maize samples. Int. Food Res. J. 18: 515–522.
Zhang, F.-L., B. Niu, L.-J. Chang, Q. Yin, P. Luo, D. Wang, S.-R. Lei, L.-A. Guo, J. Song, W.-J. Liu and Q. Deng. 2018. A new construct specific real-time PCR method for screening GMO ingredients with *gat-tpin*II cassette in foods, feeds and seeds. Food Control 86: 266–274.
Zhou, D., J. Guo, L. Xu, S. Gao, Q. Lin, Q. Wu, L. Wu and Y. Que. 2014. Establishment and application of a loop-mediated isothermal amplification (LAMP) system for detection of *cry1Ac* transgenic sugarcane. Sci. Rep. 4: 4912.
Zhou, D., C. Wang, Z. Li, Y. Chen, S. Gao, J. Guo, W. Lu, Y. Su, L. Xu and Y. Que. 2016. Detection of bar transgenic sugarcane with a rapid and visual loop-mediated isothermal amplification assay. Front. Plant Sci. 7: 279.

7

Authenticity of Maize/Corn

Konstantinos V Kotsanopoulos

1. Introduction

Zea mays belongs to the grass family and is amongst the most diverse plant species. Maize plants are equipped with both male and female reproduction systems and reproduce via cross-pollination and self pollination. Wind carries pollen from the tassel to other plants and fertilization takes place. The female reproductive structure of maize (the ear) can be from 2.5 to > 45 cm long. Even though there is a very high number of maize "races", most of hybrids used by the food industry have been obtained from very few races (Eckhoff et al. 2003). The production of maize worldwide is the highest among cereals, as the cereal is used widely for human food, animal feed and to produce biofuel. A number of technologies are currently employed across the globe using maize-derived flour and corn meals as raw materials for the production of precooked refined maize flour, dehydrated nixtamalized flour, fermented flours, etc. These products contain various levels of vitamins and minerals and are processed following a number of methods, which can significantly affect their nutritional composition (Gwirtz and Garcia-Casal 2014). Research around maize has historically focused on creating plants resistant to diseases or pests and capable of producing high yields in different climates. Genetically modified (GMO) maize is very important for the markets around the globe, while extensive research has been carried out on the genetic modification of the plant, its nutritional characteristics or modifications to render it more suitable to be

School of Agricultural Sciences, Department of Ichthyology and Aquatic Environment, University of Thessaly, Fytoko St, 38446 Nea Ionia Magnesias, Volos, Hellas, Greece.

used as animal feed and monocrop cultivation (Staller 2010). Matsuoka et al. (2002) noted that maize races developed in pre-Columbian eras are extraordinarily diverse both morphologically and genetically. These researchers came to the conclusion that maize arose from a single domestication in south Mexico around 9,000 years ago. Their analyses also indicated that the most ancient race of maize existing nowadays grows in Mexican highlands.

Maize is also of particularly high importance due to its huge variety of uses in processed foods where it can be both a principal and minor ingredient. The authenticity of maize is associated not only with the identification of cultivars and the geographical origin of the crop, but also with whether it is a GMO as well as its content in fatty acids, phenolics, pigments and heavy metal levels present in the grains together with data obtained via sensory analysis. As a result, it is to be expected that multivariate analysis plays a significant role in the analysis of data that allow conclusions to be reached regarding the authenticity of the product (Arvanitoyannis and Vlachos 2009). The levels of a number of nutrients in maize were determined by Ullah et al. (2010) who reported that the samples they examined contained moisture levels ranging from 9.201 to 10.908%, ash ranging from 0.7 to 1.3%, fats ranging from 3.21 to 7.71%, protein ranging from 7.71 to 14.60%, crude fibre ranging from 0.80 to 2.32% and carbohydrates ranging from 70 to 75%. It was therefore made clear that different varieties can contain very different amounts of protein, fats, and crude fibre. Protein is particularly important as it can define the overall quality of the crop. In another study, Zarkadas et al. (2000) determined the amino acid levels in a number of cultivars of quality protein maize (QPM) that contained opaque-2 modifier genes. The total protein content of the cultivars examined was in the range of 8.0 to 10.2%. For comparison, the protein levels in the commercial varieties Dekalb DK435 and Pioneer 3925 were determined as 7.9% and 10.3%, respectively. High lysine levels (4.43–4.58 g of lysine/100 g of protein) were found in 4 of those varieties, with the rest containing 3.43 to 4.21 g of lysine/100 g of protein, in comparison to Dekalb DK435 and Pioneer 3925, that have been found to contain 2.9 and 3.1 g of lysine/100 g of protein, respectively. It was also noted that the lysine content of these varieties could increase further by either increasing the levels of high-lysine-containing nonzein proteins or/and by reducing the 19 and 22 kDa alpha-zein in endosperms, which could lead to the production of maize of higher quality. Furthermore, Zarkadas (1997) also made an effort to measure the quantity of protein and amino acids of a common variety of white floury maize and evaluate the quality of its protein based on the FAO/WHO protein digestibility corrected amino acid scoring pattern. It was shown that the proteins of the examined variety were very similar to the amino acid content of floury-2 mutant maize. It was found that this variety contained higher levels of lysine, tryptophan, histidine, arginine and aspartic acid than the normal flint CO255 maize, and lower levels of glutamic acid, leucine, alanine, tyrosine, and phenylalanine. The methionine content was the highest among all other maize analysed. It was therefore concluded that this variety was rich in essential amino acids, with the exception of lysine, and had a protein digestibility-corrected amino acid score of 37.0 to 40.0% in comparison to 27.5 for flint CO255, 50.6% for floury-2 and 67% for QPM-C13. Also, the study of Zarkadas et al. (1995) was designed for the quantitative determination and comparison of the total protein content and the amino acid composition of three northern adapted (latitude > 45° N) maize cultivars (Dent C0251, Flint C0255 inbred line and Pioneer 3953) with a protein maize inbred (QPM-C13). An assessment of their nutritional value using their FAO/WHO amino acid scoring pattern was also made. The total protein content varied among the samples ranging from 7.95% in QPM to 8.2% (Pioneer), 10.5% (Dent), and 11.79% (Flint). Nevertheless, the QPM maize protein was of better quality than the common maize protein, containing twice the amount of lysine and arginine, while it was also richer in tryptophan and cysteine. Lysine was the only amino acid found at lower levels, but its amino acid score was 67% versus 28.5, 31.0, and 33.0% found for Pioneer, Dent, and Flint, respectively. It was therefore proved that the quality of a product containing maize can be improved by improving the protein quality of breeding maize but also that the protein quality of cereals can be evaluated by assessing their amino acid composition.

It is thus apparent that different types of maize can significantly differ in both nutritional and qualitative characteristics, and it is therefore essential that robust, accurate and cost-effective methods are available to authenticate the different varieties and their contaminants or evaluate the quality of the raw material or foodstuffs containing maize. Furthermore, EU legislation and legislation in a number of other countries mandates that approval and labelling is required for any GMOs contained in food. This has driven the development of methods capable of detecting such materials, while for screening purposes it is preferable that these methods can detect many different GMOs (Heide et al. 2008). An overview of some of the most important studies that have been conducted to verify authentication of maize is given in Table 1.

2. Authenticity Analysis of Maize and Major Contaminants

2.1 Genetic markers

DNA markers are used to identify variations in the nucleotide sequence of a genome thus highlighting the presence and degree of inter and intra-

Table 1. Overview of key techniques used for the authentication of maize.

Parameter tested	Extraction method	Test method	Markers/Test method characteristics	Results/LOD	References
Genetic structure of Brazilian sweet corn	RAPD amplification products resolved on 1.4% agarose gel; SSR products resolved in 6% polyacrylamide gel	RAPD, SSR	RAPD primers: TGCGCCCTTC, GTCCACACGG, ACGGATCCCT, GAGGATCCCT, CCTGATCACC, GGTGATCAGG, CCGAATTCCC, GGGAATTCGG, CCGATATCCC, GGGATATCGG, GTAGCACTCC; fragment sizes: 300-1700 bp; (mean number of bands per primer: 6.5) SSR loci (5′→3′): Umc 1620: CCACCGAGTGACTAGTTGTGAGAG/ CCTTTCAATGTTCATGTTCTCTTCC, 1023: CTTGTGCCACCACATGCAGTA/ CAGTTTGGAACAGGGAAAAGTACG, 1097: CTCGTCAACGTCAACCCAAGTAAG/ CTGTTAGATGTGCGACAACAGAGC, 1034: GTGTTTCCGTTTCGCTGATTTTAC/ TCATCCATGTGACAGAGACGACTT, 1331: TTATGAACGTGGTCGTGACTATGG/ ATATCTGTCCCTCTCCCACCATC, 1648: CTGCAGTACGTGAGCCTGTACG/ GCTTGAGCTGTGAGGAAGTTTTG	Higher similarity found among populations than within; or dominant markers, RAPD dominant markers presented lower polymorphism information contents	Bered et al. 2005

Table 1 contd. ...

...Table 1 contd.

Parameter tested	Extraction method	Test method	Markers/Test method characteristics	Results/LOD	References
Genetic diversity among 30 maize hybrids	RAPD amplification products resolved on 1.5% agarose gel; SSR products resolved in 3% agarose in 1 x TBE buffer	RAPD, SSR	RAPD primers: CACACTCCAG, CAGCTCACGA, GGCTCATGTG, CCGGTACCGA, TAGGACAGTC, CCTTTCCCGA; (16 polymorphic bands) SSR loci (49 polymorphic bands) (5′→3′): Phi053: CTGCCTCTCAGATTCAGAGATTGAC AACCCAACGTACTCCGG Umc1600: CGATCAGTGCGTGGAGAGTA TAGGCATGCATTGTCCATTG Bnlg1043: TTTGCTCTAAGGTCCCCATG CATACCCACATCCCGGATAA Umc1659: CAAGCTTGCTACTGTGATTTCTCG AACTTCTCGGTGATCTTGTCCATC Bnlg1867: CCACCACCATCGTAGGAGTT CAGTACACAGCAGGCAGCTC Umc2258: AAGATTGTATAAATGGCAGCCACG GAATAAGACCAGACAGCACCGAAC Phi126: TCCTGCTTATTGCTTTCGTCAT GAGCTTGCATATTTCTTGTGGACA Bnlg1257: CGGACGATCTTAGCAAACA ACGGTCTGCGACAGGATATT Umc2071: ATACACGCAGTTACCCGAAGGTT ACTGATGGTGTTCTTGGGTGTTTT Bnlg1647: CGTCGTCTGTGGACGTACTG AGAAGCTCACAAGCCTGCTC Umc2101: CTAGCTAGTTTGGTGCGTGGTGAT CCCGGCTAGAGCTATAAAGCAAGT Umc2265: CTGGACGTGGACTCAGACACC AAGACGGTCCCGAAGAAAGC	SSR: average polymorphism information content: 0.71 RAPD: average polymorphism information content: 0.61;	Cholastova et al. 2011

Genetic diversity (12 elite lines of sweetcorn)	amplification products resolved on 10% (w/v) denaturing polyacrylamide gel	SSR and artificial neural network with the self-organizing map algorithm	markers obtained from the Maize Genetic Data Bank: bnlg1367, bnlg1371, bnlg1927, bnlg2190, bnlg2191, mmc0001, mmc0111, mmc0181, umc1069, umc1071, umc1137, umc1152, umc1549, umc1636, umc1757, umc2047, umc2165, umc2214, umc2292, umc2308	3.85 alleles per locus; heterozygosity > 0.5 for 90% and polymorphism information content > 0.6 for 40% of loci, thus capable of detecting genetic differences	Ferreira et al. 2018
Detection of *Zea mays* L. in processed foods	Genomic DNA: agarose gel electrophoresis using 1.0% SeaKem LE agarose (Cambrex) gel; PCR products: agarose gel electrophoresis using 2.0% SeaKem LE agarose (Cambrex) gel	Three uniplex and one triplex PCR systems targeting invertase and zein genes	PCR primers: plant1/plant2 (500–600 bp), IVTAS1/IVTAS2 (226 bp), IVRf/IVRr (140 bp), ZEINf/ZEINr (102 bp) and new primers ZEINf: 5′-ACACCACCGACCATGGCAGC-3′ and ZEINr: 5′- TGGTGGCAAGTGCGCTGGAA-3′	Multiplex PCR was the most reliable and rapid	Gabriadze et al. 2014

Table 1 contd. ...

...Table 1 contd.

Parameter tested	Extraction method	Test method	Markers/Test method characteristics	Results/LOD	References
Detection of maize T25, GA21, TC1507, MON863, MON810, NK603, BT176, BT11 and *hmga* maize reference gene	fluorescence capillary electrophoresis (separately to RT-PCR)	Nineplex PCR	Primer names and amplicon sizes: T25ES2FMH2VIC/T25ES2RMH (126 bp) GA21ESFF86MH2NED/GA21ESRR84MH2 (135 bp) BT176PEPCMH2-FAM/BT176CRY04MH2 (186 bp) TC1507PRI1FMH2NE/TC1507PRI2RMH2 (210 bp) MON863ESFMH2-FAM/MON863ESRR110MH2 (265 bp) BT11AHJ2MH2VIC/BT11RBMHB2 (273 bp) ZM2FFMH2-NED/ZM3RRMH2 (373 bp) MON810FF26MH2VIC/MON810RR26MH2 (395 bp) NK603ESBFF90MH2N/NK603ESBRR98MH2 (449 bp)	detection limit of 0.1% for each GMO	Heide et al. 2008
Detection of genetic modification in thermally processed maize	1.5% agarose gels, staining with ethidium bromide	PCR (amplification of maize-specific invertase gene ivr1)	amplicon 211 bp; Primers: Cry01: 59-ATG GAC AAC AAC CCC AAC ATC-39 Cry02: 59-GGT CTT CAG GCC GAT CTG GTT-39 Cry03: 59-CTC TCG CCG TTC ATG TCC GT-39 Cry04: 59-GGT CAG GCT CAG GCT GAT GT-39 ivr1-F: 59-CCG CTG TAT CAC AAG GGC TGG TAC C-39 ivr1-R: 59-GGA GCC CGT GTA GAG CAT GAC GAT C-39	Limit of detection: 0.01%	Hupfer et al. 1998
Detection of aflatoxins (B1, B2, G1, and G2)	Extraction of aflatoxins from samples using tri-portions of acetonitrile/water (80:20, v/v) (10 + 7 + 7 ml)	Liquid chromatography (LC)/electrospray ionization tandem mass spectrometry (MS)	Analytes and (retention times - min): AFM_1 (10.8), AFG_2 (12), AFG_1 (12.8), AFB_2 (12.9), AFB_1 (13.9)	Sample recoveries: 81 to 101%; quantification limit: 0.1 to 0.6 μg/kg	Cavaliere et al. (2007)

Detection of fumosins	Extraction of fumosins with acetonitrile/water (75:25)	LC/tandem MS	Analytes and (retention times - min): FB_1 (8.2), FB_2 (9.6), FB_3 (8.9), FB_4 (10.2), Diclofenac (12.0)	Sample recoveries: 91–105%; detection limits: ≤ 2 ng/g (FB_1) and ≤ 1 ng/g (FB_2)	Faberi et al. 2005
Simultaneous detection of aflatoxins, ochratoxin A and Fusarium toxins	Double extraction process, using water followed by methanol	LC/tandem MS following multi-toxin immunoaffinity clean up	Extraction of the mycotoxins within 4 min; clean up of extract by multi-toxin immunoaffinity column; raw maize, corn flakes and maize snacks used for process validation	Sample recoveries: > 70%; quantification limits: 1–30 μg/kg	Lattanzio et al. 2014
Detection of hydrolysed and partially hydrolysed fumosins	Shaking for 1 h with methanol-acetonitrile-citrate/phosphate buffer (25:25:50, v/v/v) or heat (55°C) extraction and subsequent shaking at room temperature for 30 min with methanol-acetonitrile-citrate/phosphate buffer	LC-high-resolution MS	Fumosin and partially hydrolyzed fumonisins ranged from 60–5700 and 30 to 200 μg/kg, respectively	Quantification limit 10 μg/kg	De Girolamo et al. 2014
Simultaneous quantitative determination of deoxynivalenol, fumonisin B1 and zearalenone	Extraction of toxins from initial samples with acetonitrile/water (75:25 v/v) mixture (40°C, 2000 psi)	LC-atmospheric pressure chemical ionization tandem MS	Use of stable isotopically labelled and structural analogues internal standards	Limit of detection: 3–10 μg/kg, limit of quantification: 10–50 μg/kg	Royer et al. 2004

Table 1 contd. ...

...Table 1 contd.

Parameter tested	Extraction method	Test method	Markers/Test method characteristics	Results/LOD	References
Geographical origin authentication	Petroleum ether and shaking used for oil extraction	FTIR Spectroscopy	Development of classification models (PLS-DA and SIMCA) based on authentic grain maize samples from three different countries (Ukraine, USA, Peru)	100% sensitivity for solid samples in both classification models; oil extraction led to 100% and 70% sensitivity for PLS-DA and SIMCA, respectively	Achten et al. 2019
	ASE 300 accelerated solvent extractor with petroleum benzine as solvent		First evaluation of spectra with principal component analysis (PCA); construction of models to determine geographical origin based on LDA and SIMCA	80% correct classification in most cases	Nietner et al. 2013
Geographical origin authentication	Grinding of grains and homogenisation only	NIR/PCA+ PLS-DA and Competitive adaptive reweighted sampling for variable selection NIR/PCA+ OPLS-DA and Competitive adaptive reweighted sampling for variable selection	Use of corn distillers dried grains with solubles originating from Jilin (China), Heilongjiang (China), USA and Europe	100% sensitivity in most cases	Zhou et al. 2015

	Maize kernels: water rinsed and dried; Maize kernels: grounded, sieved	ICP-MS and composition analysis with one-way ANOVA, PCA, k-nearest neighbor and stepwise linear discriminant analysis	Mineral element fingerprinting; samples from 3 provinces in China	One-way ANOVA indicated significant differences between locations in 19 elements; Preliminary visual classification groups via PCA; K-nearest neighbour analysis 83.9% correct classification on training set, and 82.2% on prediction set; SLDA - 92.2% total correct classification rate based on Na, Cr, Rb, Sr, Mo, Cs, Ba, and Pb	Wang et al. 2020
Detection of irradiation	Crushing, transfer PBS solution, mixing, sieving	DNA Comet Assay	Lysis time: 45 min, Staining time: 80 min	Successful identification of irradiated and non-irradiated samples; length of comets increased with irradiation dose	Khan et al. 2005
	Oils extracted using hexane and anhydrous Na_2SO_4.	Gas chromatography	Study of hydrocarbons patterns to determine irradiation	Detection of alkenes 17:1, 16:2, 17:2, and 16:3 in irradiated but not in non-irradiated samples	Hwang et al. 2001

species diversity. They can be highly informative as a result of their large numbers and stability. Polymorphic DNA markers have found numerous uses, including the assessment of genetic variation and the development of genetic molecular maps and marker-assisted selections (Scarano and Rao 2014). Several studies have been published on the use of genetic (DNA-based) markers for the detection of adulteration in maize and some of the most important ones are discussed below.

Bered et al. (2005) aimed at characterizing the genetic structure of a total of three Brazilian sweet corn populations and investigating the genetic diversity within and among them. Random amplification of polymorphic DNA (RAPD) and simple-sequence repeats (SSR) were employed for the detection of inter and intra-population variation. Polymerase chain reaction (PCR) was carried out using 11 RAPD and six microsatellite primer pairs. When estimating any genetic similarity, the markers were considered both combined and independently. Measurements of each locus variability were taken using the polymorphism information content (PIC) and the degree of correlation between pairwise distance matrices was determined. Based on the genetic similarity and the molecular variance determined, it was proved that the degree of similarity was higher among populations than within them, while RAPD showed lower PIC values compared to SSR.

Cholastova et al. (2011) also made an effort to assess molecular genetic diversity among 30 hybrids using RAPD and SSRs. Computing a number of factors such as the number of polymorphic bands, average number of alleles per locus, effective alleles per locus, heterozygosity, effectiveness index of analysis and PIC was carried out to allow a comparison to be made between the two methods. SSRs gave better results providing an average PIC of 0.71, while 0.61 was the average value in the case of RAPD. An estimation of genetic similarities was made using Nei and Li's coefficients for SSR and Jaccard's coefficient for RAPD. In the case of SSR, the average genetic similarity was 58.8%, while for RAPD, an average value of 49.5% was given. Clustering of hybrids was achieved using an unweighted pair group method with arithmetic mean (UPGMA) and the correlation between similarity matrices for RAPD and SSR was 0.11.

Yu et al. (2001) noted that heterotic groups of *Zea mays* L. have not been defined to a great extent. SSRs have therefore been used to assign inbreds. In their study these researchers aimed at determining whether SSRs can clearly group white corn inbreds. An analysis was carried out on 18 white corn inbreds belonging to the populations Pride of Saline, Jellicorse, Neal's Paymaster, Johnson County White, and Boone County White, four yellow corn inbreds belonging to Iowa Stiff Stalk Synthetic (BSSS), and six non-BSSS yellow corn inbreds using a total of 82 SSR marker loci. There were five alleles per locus as an average. The diversity of single crosses between inbreds belonging to different populations was very close to that of the BSSS × non-BSSS yellow corn single crosses. Cluster analysis allowed the inbreds to be categorized into 5 groups with the four BSSS inbreds (i.e., B14, B37, B73, and B84) assigned to one single group. On the other hand, the other four groups had yellow and white corn inbreds from different populations. The first three principal components comprised 22% of the SSR variation among inbreds and therefore the molecular marker data underscored the lack of clear heterotic groupings among inbreds obtained from open-pollinated cultivars of white corn. It was assumed that white-endosperm conversions of BSSS inbreds can be assigned to one broad heterotic group, while non-BSSS could be grouped to an opposite heterotic pack of inbreds (Yu et al. 2001). Furthermore, Ferreira et al. (2018) used 20 microsatellites to evaluate the genetic diversity of a number of elite lines of sweet corn. An artificial neural network with the self-organizing map algorithm was employed to genetically differentiate the lines. A total of three groups were formed using this algorithm, while the expected heterozygosity was high (He > 0.5) for 90% of the loci and the PIC was high (PIC > 0.6) for 40% of the loci.

The study of Gabriadze et al. (2014) aimed at developing a method based on PCR to rapidly screen food materials for corn. A PCR-based DNA corn zein gene marker was designed and a comparison was made between three uniplex PCR and one triplex PCR method for invertase and zein genes. Flour, chips, flakes and snacks were used for this experiment. The PCR products were analysed using electrophoresis on agarose gel, demonstrating that multiplex PCR is highly accurate and can rapidly identify maize contained in processed foodstuffs.

Moreover, GMO maize detection can be achieved using genetic markers that allow for very low detection limits. Heide et al. (2008) for example, reported the development of a highly effective nineplex PCR method that could be used for event-specific detection of maize T25, GA21, TC1507, MON863, MON810, NK603 and specific detection of BT176, BT11 as well as for detecting the endogenous *hmga* maize reference gene. The primers for the PCR were labelled with fluorescent groups and fluorescence capillary electrophoresis was used for detecting the amplicons. The limit of detection of the method was about 0.1% for each GMO and the results were in agreement with the results obtained via using a quantitative real-time 5′-nuclease PCR in 85 of the 88 analyses. The method was therefore found to be effective for detecting GMO maize in food and feed materials.

Also, the development of a method was reported by Hupfer et al. (1998) that was used to detect GMO maize with resistance to insects in thermally treated products of maize. The maize was genetically modified to express a synthetic gene that encodes a truncated version of the CryIA(b) protein. The transgene had more possibilities to be detected when a 211 bp PCR amplicon was chosen. The method was rendered highly specific due to the use of a promoter region that was adjacent to the *cryIA(b)* gene in the target sequence. Polenta was also used as a model for cereal-derived foodstuffs to examine the effects of pH on the detection capability of the method. Heat treatment of polenta at a pH of 9.5–9.5 did not lead to further decomposition when compared to treatment at neutral pH. The *cryIA(b)* gene could be fully recovered after 1 hour and the 211 bp fragment could still be found.

2.2 Mycotoxins detection

Mycotoxins such as aflatoxins, deoxynivalenol (DON), fumonisins, zearalenone (ZEN) and ochratoxin are amongst the most dangerous contaminants found in cereals. The diseases caused by these toxins can vary widely and the majority of them occur following consumption of contaminated grains or products thereof. Diagnosing mycotoxicosis can be hard as the symptoms can be similar to diseases caused by other factors. As a result, diseases caused by mycotoxins can be diagnosed following testing for these toxins that is performed following sampling, sample preparation and analysis (Richard 2007). Maize has been found to support the growth of microorganisms that produce mycotoxins while fumonisins are also commonly found in it. Mycotoxin contamination can increase as a result of exposure to stress factors during plant growth, late harvesting of crops, incomplete drying, storage under improper conditions and lack of knowledge about the subject. The risk can be controlled via education on the risks of mycotoxins, earlier harvesting, quick drying, sorting, proper sanitation, use of appropriate storage facilities and others (Bankole and Adebanjo 2003). It is therefore clear that a number of preventive steps need to be taken to ensure maize is not contaminated with mycotoxins which in certain cases can increase the costs. Furthermore, crops contaminated with mycotoxins should not be made available for food or feed, the profit loss that this entails as well as the costs of destroying/disposing of the material may lead to its illegal release by certain individuals. According to Food and Drug Administration (FDA) (21 U.S. Code § 342), a food is considered to be adulterated when it contains poisonous substances which make its consumption harmful (FDA, 21 USC 342). There is therefore a need for robust methods that can detect these contaminants in maize and a large number of studies have examined the potential of spectroscopic, chromatographic methods and their combinations for the effective determination of mycotoxins in this product.

Cavaliere et al. (2007) reported the use of a liquid chromatography/electrospray ionization tandem mass spectrometry (LC/ESI-MS/MS) method that can be used to measure the aflatoxins B1, B2, G1, and G2 in maize. The extraction of aflatoxins from the samples was carried out using tri-portions of acetonitrile/water (80:20, v/v) and part of the extract was diluted with water and cleaned up with a 100 mg Carbograph-4 cartridge. After the internal standard AFM1 was added, the extract was subjected to analysis by LC/ESI-MS/MS in positive ion mode with the use of multiple reaction monitoring with a triple-quadrupole instrument. A C18 column (at 45°C) employed with a mobile phase gradient of acetonitrile/water with 2 mmol/L ammonium formate was used. Even though the matrix suppression effect was minimal, quantitation was allowed following an external calibration procedure using matrix-matched standard solutions as a means of improving the accuracy level. Eighty-one to one hundred and one percent of samples were recovered at four spiking levels, and the quantification limit ranged from 0.1 to 0.6 µg/kg. This method is characterised by high sensitivity, accuracy and selectivity and can be used as a confirmatory procedure to verify conformity with regulations on maximum levels of mycotoxins.

Similarly, a sensitive liquid chromatography/tandem mass spectrometry (LC/MS/MS) method was used by Faberi et al. (2005) to determine the levels of type B fumonisin in food made from maize. The extraction of FB1 and FB2 toxins was achieved via homogenising the samples using acetonitrile/water (75:25, v/v, 50 mmol/L formic acid, 25 mL final volume) and defatting the extract on C18 phase. Carbograph-4 cartridges were used to clean up the extracts. High Performance Liquid Chromatography (HPLC) with ESI-MS was then used for the analysis. Recoveries ranging from 91 to 105% (RSD% $\leq$ 8%) were achieved, while the limit of detection of the method was $\leq$ 2 ng/g for FB1 and $\leq$ 1 ng/g for FB2. Testing of two different spiking levels (5000 and 100 ng/g for FB_1 and 1000 and 20 ng/g for FB_2) was carried out and the quantities of toxins were determined using matrix-matched standards and diclofenac. Identification of the fumonisins, FB3 and FB4 was also carried out in corn meal using an information-dependent acquisition protocol incorporating neutral loss scan, enhanced resolution scan, and enhanced production scan. An estimated quantitation of FB3 and FB4 was given as peak area ratios relative to the FB2 response since no standards were available.

Another method was described by Royer et al. (2004) who attempted to simultaneously quantify DON, fumonisin B1 and ZEN in maize using liquid chromatography-atmospheric pressure chemical ionization tandem mass spectrometry (LC-APCIMS/MS) supported by isotopically targeted and structural analogues internal standards. In this process, an accelerated solvent extraction and a two solid-phase clean-up using strong anion exchange resin and a Mycosep® column were used. A calculation of recoveries was given by spiking blank maize at DON levels of 200, 400 and 1000 µg/kg, at 70%, for fumonisin B1 (100, 200 and 1000 µg/kg) at 90%, and for ZEN (50, 100 and 200 µg/kg) at 40%. LC-APCIMS/MS analyses were conducted in collision-induced dissociation on an ion-trap instrument rendering the method highly selective and sensitive. Ions were extracted from two transition reactions, monitored by LC-APCIMS/MS for each analyte, allowing limits of detection of 10, 20 and 3 μg/kg to be achieved for DON, fumonisin B1 and ZEN, respectively, as well as limits of quantification of 50, 50 and 10 μg/kg, respectively.

More recently, aflatoxins (B1, B2, G1, G2), ochratoxin A, fumonisins (B1, B2), ZEN, DON, nivalenol, T-2 and HT-2 toxins were quantitatively determined in cereals and derived foodstuffs using wet chemistry. The toxins were extracted within 4 minutes using extraction with water and methanol. The extract was cleaned up using a new multi-toxin immunoaffinity column. An evaluation of the performance of the method was made via single laboratory validation. Raw wheat and maize, corn flakes and maize snacks were tested for validation purposes with the validation assay being performed at 50, 100 and 150% of EU maximum allowed levels per toxin. Analysis of variance (ANOVA) was used to assess the reproducibility of the method as well as the error contributions from repeatability, between day effects, and the effect of the use of different matrices. Recoveries were in most cases > 70% for all mycotoxins tested, with a relative standard deviation lower than 37%. The quantification limits ranged from 1 μg/kg to 30 μg/kg. The performance of the system was evaluated by analysing reference materials. It was finally proved that the method can be effectively used as a single-analysis method to confirm compliance with maximum levels dictated by regulations (Lattanzio et al. 2014).

Finally, a Liquid Chromatography-High Resolution Mass Spectrometry (LC–HRMS) method was used in the study of De Girolamo et al. (2014) to simultaneously determine fumonisins in 24 naturally contaminated samples of maize and maize-based products. Most samples (18 out of 24) contained fumonisins B1 and B2. These toxins were found together with both partially and fully hydrolyzed toxins in a variety of samples (masa flours, tortilla chips, kernels, maize meal, cous-cous, corn-cakes and cornflakes). The levels of toxins ranged from 60 to 5700 μg/kg in the case of fumonisins (sum of FB1 and FB2), 10 to 210 μg/kg in the case of partially hydrolyzed fumonisins (sum of PHFB1 and PHFB2) and 30 to 200 μg/kg in the case of hydrolyzed fumonisins (sum of HFB1 and HFB2). This was the first study that reported an attempt to isolate PHFB2 co-existing with FB1, FB2, PHFB1, PHFB2, HFB1 and HFB2 in maize products. Taking into account the increasing use of nixtamalized and maize-based food, monitoring the presence of fumonisins and their partially and fully hydrolyzed forms in them is essential in establishing a level of food safety risk.

2.3 Detection of irradiation

Irradiation is used in countries around the world to inhibit the growth of microorganisms. However, consumer trends towards more "natural" food have driven the implementation of regulations that control the use of irradiation and enable consumers to make informed decisions on whether they wish to purchase irradiated food, mainly via mandatory labelling. As irradiation can be very effective, there is a risk that it can be used to prolong the shelf life of food and its use is not communicated across the supply chain. It is therefore necessary that methods are developed to determine whether maize has been irradiated. Some of the most important studies that report the development and use of such methods are detailed below. Whole grains of maize were examined in the study of Khan et al. (2005) to determine whether they had been irradiated using microgel electrophoresis of single cells. The use of this DNA Comet Assay was suggested as a quick screening method for detecting irradiation, while it is not radiation specific. Maize was treated with radiation doses of 0.5, 1 and 3 or 5 kGy, which cover the radiation range commonly used to disinfect food. Following electrophoresis, maize cells subjected to irradiation showed comets which indicated that the fragmented DNA had stretched towards the anode. The size and shape of the comets depended on the radiation dose used. The cells of the product that had not been irradiated appeared to have intact cells with round nuclei or nuclei with short faint tails. Thus, both samples that had been subjected to treatment and samples that had not, were effectively determined using this DNA Comet Assay. The length of comets increased with increasing the doses, an observation that could be useful for determining the level of radiation used.

Also, Hwang et al. (2001) indicated that alkanes and alkenes can be formed from lipids in maize, brown rice, and wheat when irradiated. They therefore made an effort to examine whether patterns of hydrocarbons in cereals can act as markers for the identification of irradiation treatment. Irradiation of maize, brown rice, and wheat with doses of 0.5, 1, and 3 kGy was applied. Hexane and Na_2SO_4 were used for the extraction of oils from the samples. An isolate containing hydrocarbons was obtained using florisil column chromatography and gas chromatography (GC) was used for the analysis. The first six Florisil column chromatography fractions (10 ml each) were used and the alkenes 17:1, 16:2, 17:2, and 16:3 were detected principally in the third, fourth, and fifth fractions. These fractions were then combined and following evaporation to a volume of 4 mL, they were analysed using GC. The alkenes 17:1, 16:2, 17:2, and 16:3 were found in samples subjected to irradiation but not in reference samples. No significant differences in the amounts of hydrocarbons were observed though by roasting either irradiated or non-treated samples.

Moreover, another study was carried out to develop a method for detecting irradiated cereals. Following grounding and irradiation of samples at 2.5, 5.0, 7.5, 10, 15 kGy using a Co60 irradiator, a decrease in viscosity was seen with increasing the irradiation doses. Specifically, a decrease in the viscosities of Panicum miliaceum (Korean) and Andropogon sorghum (Korean) pastes was observed from 143.38 ± 0.44 and 35.92 ± 1.90 in control samples to 6.60 ± 1.16 and 3.86 ± 0.32, respectively, following irradiation at 15 kGy. Similar observations were made in the case of Chinese samples. Regression equation and coefficients of viscosity of Panicum miliaceum from Korean and Chinese samples were 0.80 ($y = -27.789x + 150.17$) and 0.98 ($y = -3.367x + 88.93$), respectively, while in the case of Andropogon sorghum, the same values were 0.84 ($y = -6.0466x + 35.49$) and 0.84 ($y = -13.346x + 101.67$), for Korean and Chinese samples, respectively at 50 rpm. All samples showed a decrease in specific parameters with increasing rpm. Parameter values were found to depend on radiation doses, indicating that all values of irradiated samples were lower than those of non-irradiatedones. It therefore became apparent that the method is suitable for detecting irradiated cereals and could be expected to be suitable for maize (Kim et al. 2001).

2.4 Determination of geographical origin

The determination of the geographical origin of maize using Fourier transform infrared spectroscopy (FT-IR) was attempted in the study of Achten et al. (2019). The method was rapid and non-destructive. A classification model was developed on the basis of the use of attenuated total reflexion (ATR)-FTIR spectroscopy data to verify the geographical origin of the unprocessed feed material grain maize. Two different sample preparation processes were followed and Partial Least Squares Discriminant Analysis (PLS-DA) and Soft Independent Modelling of Class Analogy (SIMCA) were used for the classification. Optimization of preprocessing data was done to reveal the most effective preprocessing method to achieve accurate classification. ATR-FT-IR and multivariate analyses were used for differentiating Ukrainian, American and Peruvian maize samples. The models were validated following separate test sets and permutation tests. Data preprocessing rendered the classification models 100% sensitive in the case of solid samples, while oil extraction led to 100% and 70% sensitivity, for PLS-DA and SIMCA, respectively. It was therefore suggested that FTIR spectroscopy in combination with multivariate classification can be suitable for verifying the geographical origin of grain maize. It was also noted that although the method can also be suitable for the classification of oil extracts, analysing solid material has certain benefits. Solid samples can more easily and quickly be prepared, often without a need for specialised laboratory equipment. Oil extraction, on the other hand, requires the use of organic solvents and laboratory equipment.

FT-IR was also used in the study of Nietner et al. (2013) to examine distillers dried grains and solubles (DDGS). ATR-FT-IR spectroscopy was used to examine the samples in their solid form and as extracted oils. An evaluation of the spectra was made initially with principal component analysis (PCA) in accordance with their botanical (corn, rice, wheat) and geographical origin (Canada, China, European Union, India, United States). Statistical models were subsequently developed to characterise the botanical and geographical origin of the samples using linear discriminant analysis (LDA) and SIMCA. A randomized batchwise process was used to validate the models showing more than 80% correct classification in the majority of cases. It was therefore demonstrated that ATR-FT-IR in combination with multivariate data analysis can be effectively used to verify the origin of DDGS.

Promising results were also provided in the study of Zhou et al. (2015) in which 137 corn DDGS samples originating from the Jilin and Heilongjiang Provinces of China, USA and Europe were examined. Near infrared spectrometry (NIR) in combination with different chemometric techniques were employed by two independent laboratories to assess whether these methods can successfully classify the samples according to their geographical origins. Using the same data, one of the laboratories constructed a PLS-DA model, while the other laboratory designed an orthogonal PLS-DA model. It was demonstrated that samples from different countries have distinct chemical characteristics. NIR spectroscopy in combination with chemometrics could successfully classify DDGS according to their geographical origins. Also, using the same sample dataset, two independent laboratories obtained spectral data and analysed them using different chemometrics software packages, ultimately ending up however, with very similar results. A complete determination of DDGS samples from all regions was made using PLS-DA or orthogonal PLS-DA models. Competitive adaptive reweighted sampling-selected variables were also used, leading to a simplification of the classification model.

Another important study was that of Vermeulen et al. (2015), in which DDGS were used in order to compare analytical strategies for the determination of the origin of samples. A number of techniques were examined for their potential and performance in authenticating DDGS, among which spectroscopic and spectrometric techniques in combination with multivariate data analysis, but also traditional methods such as DNA and stable isotope ratio analyses. An external validation procedure was employed for analysing sample sets (blind analysis) and comparing the analytical techniques. All techniques allowed the botanical origin (corn and wheat) of samples to be determined and some of them also differentiated the samples as per their geographical origin. Specifically, NIR and mid-infrared spectroscopy (MIR) were found to be quite fast and cost-effective. With regards to the methodology followed, the study indicated that the botanical origin should be determined before the geographical origin. Consideration should be given to the overlap between geographical and processing origin when the data are being evaluated.

Therefore, these studies prove that IR technologies, in combination with statistical models and chemometrics, are very effective in identifying the geographical origin of maize. The construction of databases though with reference samples could allow the methods to be more easily and effectively applied by the industry and regulators for routinely controlling the authenticity of maize.

Finally, in the study of Wang et al. (2020), the mineral element composition of samples was used to develop a fingerprinting technique that was used for determining the geographical origin of maize. Ninety maize samples originating from Jilin, Gansu, and Shandong provinces in China were obtained. Measurements of the levels of elements were taken using inductively coupled plasma mass spectrometry (ICP-MS). The data were subjected to multivariate statistical analysis, including one-way ANOVA, PCA, *k*-nearest neighbour analysis, and stepwise LDA. One-way ANOVA indicated that the levels of 19 elements in the samples differed significantly among the different geographical areas. PCA considering these 19 elements could allow a preliminary classification to be made but *k*-nearest neighbour analysis correctly classified 83.9% of samples in the training set and 82.2% in the prediction set. The stepwise LDA model was based on the levels of Na, Cr, Rb, Sr, Mo, Cs, Ba, and Pb, succeeding in correctly classifying 92.2% of samples with cross-validation. It was therefore concluded that this element fingerprinting technique in combination with multivariate statistical analysis could be quite effective in classifying maize samples according to their geographical origin.

3. Conclusion

Maize is the cereal with the highest production around the world. In the food industry, the cereal is used for the production of various raw materials and ingredients. Although a very limited number of maize races are used for the production of food, the nutritional and qualitative characteristics of maize have been found to differ considerably between varieties. GMO maize offers a number of advantages that have allowed the production of the cereal to significantly increase. Legislation in several parts of the world, however, dictates that consumers should be informed of the presence of GMO in food and methods are therefore needed to verify its presence or absence. The development of techniques for the authenticity of maize though is also of paramount importance to verify the nutritional characteristics, quality and geographical origin of maize, as well as its use as an adulterant to replace more expensive cereals. A number of techniques have been used to detect adulteration of maize. Genetic markers based on DNA have been used to detect genetic diversity and characterise the genetic structure of maize. These methods are highly accurate, being able to categorise maize into different origin grains and identify maize in processed foods. They can also be very effective in identifying GMO material in maize with very low limits of detection. Moreover, as one of the most important contaminants in maize is mycotoxins, combinations of chromatographic and spectrometric techniques have been used to quantify a number of these toxins, allowing levels as low as 0.1 μg/kg (Gavaliere et al. 2007) to be detected, thus providing a high level of trust in ensuring food safety. Similarly to GMO, the use of irradiation must also be communicated to consumers via mandatory labelling in big parts of the world. It is therefore important that methods are developed to detect irradiation in maize. Current studies have attempted to detect the results of irradiation use via observing changes in the size and shape of commets in cells (Khan et al. 2005), using molecular markers such as alkanes and alkenes (lipidomics) (Hwang et al. 2001) and monitoring changes in viscosity of cereal pastes (Kim et al. 2001). Depending on the markers used, not only the use of irradiation, but also the irradiation doses can be estimated. Finally, geographical origin claims are commonly used as marketing tools around the globe and often as indicators of quality. It is therefore important to ensure that any such claims are genuine and do not mislead industry stakeholders and consumers. Methods such as FT-IR have therefore been used, usually in combination with chemometrics to differentiate samples of different geographical origins with sensitivities that often reach 100%.

Abbreviations

Analysis of variance	ANOVA
Attenuated total reflexion	ATR
Deoxynivalenol	DON
Distillers dried grains and solubles	DDGS
Food and Drug Administration	FDA
Fourier transform infrared spectroscopy	FT-IR
Gas chromatography	GC
Genetically Modified Organisms	GMO
High Performance Liquid Chromatography	HPLC
Inductively coupled plasma mass spectrometry	ICP-MS
Linear discriminant analysis	LDA
Liquid chromatography-atmospheric pressure chemical ionization tandem mass spectrometry	LC-APCIMS/MS

Liquid Chromatography-High Resolution Mass Spectrometry	LC–HRMS
Liquid chromatography/electrospray ionization tandem mass spectrometry	LC/ESI-MS/MS
Liquid chromatography/tandem mass spectrometry	LC/MS/MS
Mid-infrared spectroscopy	MIR
Near infrared spectrometry	NIR
Partial Least Squares Discriminant Analysis	PLS-DA
Polymerase chain reaction	PCR
Polymorphism information content	PIC
Principal component analysis	PCA
Random amplification of polymorphic DNA	RAPD
Simple-sequence repeats	SSR
Soft Independent Modelling of Class Analogy	SIMCA
Unweighted pair group method with arithmetic mean	UPGMA

References

Achten, E., D. Schütz, M. Fischer, C. Fauhl-Hassek, J. Riedl and B. Horn. 2019. Classification of Grain Maize (*Zea mays* L.) from different geographical origins with FTIR spectroscopy—A suitable analytical tool for feed authentication? Food Analytical Methods 12: 2172–2184.

Arvanitoyannis, I.S. and A. Vlachos. 2009. Maize Authentication: Quality control methods and multivariate analysis (Chemometrics). Crit. Rev. Food Sci. Nutr. 49: 501–537.

Bankole, S.A. and A. Adebanjo. 2003. Mycotoxins in food in West Africa: current situation and possibilities of controlling it. African Journal of Biotechnology 2(9): 254–263.

Bered, F., TdF. Terra, M. Spellmeier and J.F.B. Neto. 2005. Genetic variation among and within sweet corn populations detected by RAPD and SSR markers. Crop. Breed. Appl. Biotechnol. 5: 418–425.

Cavaliere, C., P. Foglia, C. Guarino, M. Nazzari, R. Samperi and A. Laganà. 2007. A sensitive confirmatory method for aflatoxins in maize based on liquid chromatography/ electrospray ionization tandem mass spectrometry. Rapid Communications in Mass Spectrometry 21: 550–556.

Cholastova, T., M. Soldanova and R. Pokorny. 2011. Random amplified polymorphic DNA (RAPD) and simple sequence repeat (SSR) marker efficacy for maize hybrid identification. African Journal of Biotechnology 10(24): 4794–4801.

De Girolamo, A., V.M.T. Lattanzio, R. Schena, A. Visconti and M. Pascale. 2014. Use of liquid chromatography–high-resolution mass spectrometry for isolation and characterization of hydrolyzed fumonisins and relevant analysis in maize-based products. Journal of Mass Spectrometry 49: 297–305.

Eckhoff, S.R., M.R. Paulsen and S.C. Yang. 2003. Maize. pp. 3647–3653. *In*: Caballero, B. [ed.]. Encyclopedia of Food Sciences and Nutrition. Academic Press, IL, USA.

Faberi, A., P. Foglia, E. Pastorini, R. Samperi and A. Laganà. 2005. Determination of type B fumonisin mycotoxins in maize and maize-based products by liquid chromatography/ tandem mass spectrometry using a QqQlinear ion trap mass spectrometer. Rapid Communications in Mass Spectrometry 19(2): 275–282.

FDA – Food and Drugs Administration. Title 21-FOOD AND DRUGS CHAPTER 9-FEDERAL FOOD, DRUG, AND COSMETIC ACT SUBCHAPTER IV-FOOD. Available at: https://uscode.house.gov/browse/prelim@title21/chapter9/subchapter4&edition=prelim (Accessed 30.12.2020).

Ferreira, F., C.A. Scapim, C. Maldonado and F. Mora. 2018. SSR-based genetic analysis of sweet corn inbred lines using artificial neural networks. Crop. Breed. Appl. Biotechnol. 18: 309–313.
Gabriadze, I., T. Kutateladze, B. Vishnepolsky, M. Karseladze and N. Datukishvili. 2014. Application of PCR-based methods for rapid detection of corn ingredients in processed foods. International Journal of Nutrition and Food Sciences 3(3): 199–202.
Gwirtz, J.A. and M.N. Garcia-Casal. 2014. Processing maize flour and corn meal food products. Ann. N Y Acad. Sci. 1312(1): 66–75.
Heide, B.R., E. Heir and A. Holck. 2008. Detection of eight GMO maize events by qualitative, multiplex PCR and fluorescence capillary gel electrophoresis. European Food Research and Technology 227: 527–535.
Hupfer, C., H. Hotzel, K. Sachse and K.-H. Engel. 1998. Detection of the genetic modification in heat-treated products of Bt maize by polymerase chain reaction. Z Lebensm Unters Forsch A 206: 203–207.
Hwang, K.T., J.S. Hong, J.S. Yang, H.S. Sohn and C.L. Weller. 2001. Detection of Alkanes and Alkenes for Identifying Irradiated Cereals JAOS 78(11): 1145–1149.
Khan, A.A., H.M. Khan and H. Delincée. 2005. DNA Comet Assay—a rapid screening method for detection of irradiated cereals and tree nuts. Food Control 16(2): 141–146
Kim, H.-K., D.-S. Kang, M.-g. Choi and J.-H. Kwon. 2001. Detection of irradiated dried cereals from korea and china by viscometric method. Korean Journal of Food Science and Technology 33(6): 645–650.
Lattanzio, V.M.T., B. Ciasca, S. Powers and A. Visconti. 2014. Improved method for the simultaneous determination of aflatoxins, ochratoxin A and *Fusarium* toxins in cereals and derived products by liquid chromatography–tandem mass spectrometry after multi-toxin immunoaffinity clean up. Journal of Chromatography A 1354: 139–143.
Matsuoka, Y., Y. Vigouroux, M.M. Goodman, J. Sanchez, E. Buckler and J. Doebley. 2002. A single domestication for maize shown by multilocus microsatellite genotyping. PNAS 99(9): 6080–6084.
Nietner, T., M. Pfister, M.A. Glomb and C. Fauhl-Hassek. 2013. Authentication of the botanical and geographical origin of distillers dried grains and solubles (DDGS) by FT-IR spectroscopy. Journal of Agricultural and Food Chemistry 61: 7225–7233.
Richard, J.L. 2007. Some major mycotoxins and their mycotoxicosis—An overview. International Journal of Food Microbiology 119: 3–10.
Royer, D., H.U. Humpf and P.A. Guy. 2004. Quantitative analysis of *Fusarium* mycotoxins in maize using accelerated solvent extraction before liquid chromatography/atmospheric pressure chemical ionization tandem mass spectrometry. Food Additives and Contaminants 21(7): 678–692
Scarano, D. and R. Rao. 2014. DNA markers for food products authentication. Diversity 6: 579–596.
Staller, J.E. 2010. An introduction to maize cobs and cultures. pp. 1–6. *In*: Maize Cobs and Cultures: History of *Zea mays* L. Springer-Verlag, Heidelberg, Germany.
Ullah, I., M. Ali and A. Farooqi. 2010. Chemical and nutritional properties of some maize (*Zea mays* L.) varieties grown in NWFP, Pakistan. Pak. J. Nutr. 9(11): 1113–1117.
Vermeulen, P., T. Nietner, S.A. Haughey, Z. Yang, N. Tena, H. Chmelarova, S. van Ruth, M. Tomaniova, A. Boix, L. Han, C.T. Elliott, V. Baeten and C. Fauhl-Hassek. 2015. Origin authentication of distillers' dried grains and solubles (DDGS)—application and comparison of different analytical strategies. Analytical and Bioanalytical Chemistry 407: 6447–6461.
Wang, F., H. Zhao, C. Yu, J. Tang, W. Wu and Q. Yang. 2020. Determination of the geographical origin of maize (*Zea mays* L.) using mineral element fingerprints. Journal of the Science of Food and Agriculture 100(3): 1294–1300.
Yu, J., H. Lu and R. Bernardo. 2001. Inconsistency between SSR groupings and genetic backgrounds of white corn inbreds. Maydica. 46(2): 133–139.

Zarkadas, C.G., Z. Yu, R.I. Hamilton, P.L. Pattison and N.G.W. Rose. 1995. Comparison between the protein quality of northern adapted cultivars of common maize and quality protein maize. J. Agric. Food Chem. 43: 84–93.

Zarkadas, C.G. 1997. Assessment of the protein quality of native white floury maize designated IAPO-13, by amino acid analysis. J. Agric. Food Chem. 45: 1062–1069.

Zarkadas, C.G., R.I. Hamilton, Z.R. Yu, V.K. Choi, S. Khanizadeh, N.G. Rose and P.L. Pattison. 2000. Assessment of the protein quality of 15 new northern adapted cultivars of quality protein maize using amino acid analysis. J. Agric. Food Chem. 48(11): 5351–5361.

Zhou, X., Z. Yang, S.A. Haughey, P. Galvin-King, L. Han and C.T. Elliott. 2015. Classification the geographical origin of corn distillers dried grains with solubles by near infrared reflectance spectroscopy combined with chemometrics: A feasibility study. Food Chemistry 189: 13–18.

8

Authenticity of Rice

Konstantinos V Kotsanopoulos

1. Introduction

Rice is one of the most important food staples around the world. Its demand is expected to increase as the world's population increases. As more and more fraud attempts are reported in many supply chains, it is of paramount importance that efficient and rapid techniques are established to monitor the authenticity of rice (Teye et al. 2019). Rice is grown around the world and has a high genetic diversity. As differentiation between seeds of different rice varieties is very hard when based only on visual observation, both the seeds and the processed products can be easily substituted with lower quality varieties. Figure 1 shows 4 different commercial types of rice as sold in the market in Greece and the UK. It is apparent that a clear differentiation of different types requires experience and expertise, while it can be impossible when claims on the geographical origin of products are investigated.

In an effort to protect consumers and the reliability of the market, various methods have been used over the years to unambiguously differentiate between cultivars and accurately quantify adulterants, as well as to discriminate between products of different geographical origins. Table 1 synoptically presents the techniques used in various studies investigating rice authenticity and their effectiveness.

Recent innovations in biotechnology and the progress made in DNA based techniques have allowed the development of rapid and efficient methods that can detect adulteration at commercial levels (Vemireddy

---

School of Agricultural Sciences, Department of Ichthyology and Aquatic Environment, University of Thessaly, Fytoko St, 38446 Nea Ionia Magnesias, Volos, Hellas, Greece.

Figure 1. Common rice types obtained from markets in Greece and United Kingdom.

et al. 2015). Rice's high genetic diversity is a result of a very high range of geographic and ecologic adaptations. A type of grass, that was the early progenitor of rice, had been differentiated into distinctly different forms in a variety of regions in an area now called the Gondwana supercontinent, more than 130 million years ago. As this continent was splitting up and drifting apart, the continents that exist today were formed. Two distinct cultivars, the African rice (*Oryza glaberrima* Steud.) and Asian (*O. sativa* L.), evolved separately in West Africa and South and Southeast Asia, respectively. A long cultivation, domestication and diversification period that lasted approximately 12,000 years followed the rising of human civilization in Africa and Asia. Therefore, a number of changes in rice cultivation are reflected upon human manipulation of rice genes and the rice ecosystem as a whole (Chang 2003). It is also clear that the history of rice is closely linked to the emergence of the world economy and the networks of industrial capitalism that have been developed around the world. Rice has played a very important role in connecting the histories of Africa, America, Europe and most of all Asia, having often been the reason for the development of certain countries and currently being food to over half the world's population (Bray 2015). As a food, rice can be very easily adapted in the majority of environments, but not always prolifically. Half of the amount of cultivated rice is irrigated (irrigated cultivars produced 75% of almost 700 million tons of rice harvested in 2010, of which an amount ranging from 30 to 35% is discarded due to husking, milling and polishing). Wetland rice cultivation starts with the

Table 1. Overview of key techniques used for the authentication of rice.

Technique	Scope	Effectiveness	References
SSLPs	Basmati vs. other rice	100% differentiation using several primer sets	Bligh 2000
SSR	Differentiation between Taiwan rice varieties and foreign rice cultivars	Sensitivity as low as 3 to 7.5%	Chuang et al. 2011
	Investigation of genetic diversity of cultivars	Genetic relationships effectively determined	Giarrocco et al. 2007
	Assessment of genetic relationships (Indian rice)	Indian genotypes can be distinguished from indica and japonica, while crossbred varieties can be distinguished from traditional Basmati. Genetic relationship effectively determined	Jain et al. 2004
	Rice varieties identification (24 markers across 2 chromosomes)	Maternal lines data reliably verified paternal lines data modified, successful construction and analysis of genotype data of 103 rice varieties	Jie-zheng et al. 2007
Microsatellite and fragrance typing + HRM	Discrimination between Basmati and non-Basmati varieties	Very sensitive detecting a ratio of 1:100 of non-fragrance in fragrance rice	Ganopoulos et al. 2011
Amplification of single copy gene and microsatellites	Discrimination of cereal products, detection of GMO	Extracted DNA useful for gene fragment amplification and microsatellite analysis. Two thirds of microsatellites were not polymorphic and could not be used for discrimination	Ren et al. 2006
ICP-MS	Organic vs non-organic rice (comparison of concentration of 20 elements)	Effective differentiation using chemometrics	Borges et al. 2015
	Thai jasmine rice vs. French, Indian, Italian, Japanese and Pakistani (comparison of concentration of 21 elements)		Cheajesadagul et al. 2013
	Geographical authentication of Basmati rice (determination of $^{87}Sr/^{86}Sr$ ratios and rare earth elements composition of rice, soil and water)	$^{87}Sr/^{86}Sr$ ratios can be effectively used as a tracer for the differentiation of Basmati rice originating from India and other countries	Lagad et al. 2017
	Geographical origin ($^{87}Sr/^{86}Sr$)—testing of soil, water and plants	Strong link between the fraction in plants and in water & soil	Song et al. 2014

Table 1 contd. ...

...Table 1 contd.

Technique	Scope	Effectiveness	References
Laser ablation ICP-MS	Classification according to origin and type (distribution of (Ca, Cu, Fe, K, Mg, Mn, P and Zn)	Overall accurate classification and cross-validation of 93.8% and 91.1% for origin and 100% and 97.9% for type.	Promchan et al. 2016
IRMS & ICP-MS and chemometrics (PCA and OPLS-DA)	Geographic authentication (light element (C, N, O, S) isotope ratios and compositions of 25 elements)	PCA could distinguish samples from most countries, but not Philippines and Japan. OPLS-DA discriminated between Korea and other countries	Chung et al. 2018
IRMS and ICP-MS + canonical discriminant analysis	Geographical discrimination (different continents), C-13, O-18, B, Ho, Gd, Mg, Rb, Se and W	Effective differentiation	Kelly et al. 2002
IRMS	Determination of geographical origin (C and N, $d^{13}C$, $d^{15}N$, and $d^{18}O$)	Effective for rapid discrimination	Suzuki et al. 2008
Mixed-mode cation exchange-based solid-phase extraction and direct infusion-MS/MS-based metabolomics	Geographical discrimination (countries)	Quick and reliable discrimination with combination of mixed-mode cation exchange-based solid-phase extraction with 100% methanol and direct infusion-MS in positive ion mode or combination of mixed-mode cation exchange-based solid-phase extraction with 20% methanol and targeted direct infusion -MS/MS in positive ion mode	Lim et al. 2017
ICP-AES	Determination of geographical origin, Cu, Ag, Zn, Cr, Ca, Ba, Cd, Bi, K, Pb, and In and chemometrics	Clear discrimination using PLS-DA	Chung et al. 2015
ICP-OES and chemometrics	Geographical origin classification (32 elements)	91.30% correct classification	Gonzálvez et al. 2011
IRMS	Discrimination of geographical origin (C, N, O and S stable isotope ratios and chemometrics)	Effective discrimination of geographical origins but not between rice genotypes	Chung et al. 2016
	Organic vs conventional rice ($d^{13}C$ and $d^{15}N$)	Can assist in discrimination where pesticide residue analysis alone is not sufficient	Chung et al. 2017

Atomic fluorescence spectrometry	Determination of Se concentration	Detection limit 0.5 ng/ml	Chen et al. 2002
NIR and SVDD	Differentiation of black rice varieties	Specificity 100%, sensitivity 94.2%	Chen et al. 2018
NIR and PCA, BP-ANN	Discrimination between varieties	100% successful discrimination	Liang et al. 2009
NIR hyperspectral imaging and chemometrics	Rice cultivar identification (Spectra 1,039 nm to 1,612 nm used to construct classification models)	80–100% correct classification depending on models	Kong et al. 2013
NIR	Discrimination between Basmati and other long-grain rice	Misclassification in 8% of Basmatis and 14% of other rice.	Osborne et al. 1993
NIR and chemometrics	Geographical and quality grades discrimination	> 90% correct classification, rapid discrimination	Teye et al. 2019
Aroma extract dilution analyses [following sample preparation and High-Resolution GC/ Olfactometry (HRGC/O) and High-Resolution GC/ Mass Spectrometry (GC/ MS)]	Aroma compounds identification	41 odor-active compounds identified; differences were found in compounds between rice varieties	Jezussek et al. 2002
Capillary GC-MS	Analysis of volatile components (identification of 70 volatiles)	Highly sensitive	Mahatheeranont et al. 2001
GC-MS-Selected Ion Monitoring (GC-MS-SIM	Formation of 2-acetyl-1-pyrroline in aromatic rice at different temperatures	Determination of N source and temperature required for formation	Yoshihashi et al. 2015
NMR & chemometrics	Discrimination of rice varieties and geographical discrimination	Effective for official screening control	Monakhova et al. 2014
Laser induced breakdown spectroscopy	Verification of protected designation of origin of Argentine brown rice (emission lines of C, Ca, Fe, Mg and Na were selected)	84% accuracy, 100% sensitivity, 78% of specificity	Pérez-Rodríguez et al. 2019

nursery seedlings that are ultimately transplanted to paddies. This is a particularly labour-intensive process that can lead to the production of one to three crops per year (depending on rice type, geographical area and environmental conditions). Wet rice agriculture takes place mostly in tropical latitudes, covering southern China and Southeast Asia, India, Africa, the United States, Mexico and South America. Rain-fed lowland rice accounts for 20% and upland rice for 5% of the total amount available. Upland rice agriculture, also known as dry cultivation, is prevalent in South America and Africa. Finally, deep-water rice grows in water with a depth of 50 cm (30 in) or more, mostly in Bangladesh or other regions with deeply flooded river valleys. Rice plants grow very quickly above water, to a height of up to 4 meters (Marton 2014). Rice is consumed across the world, but the highest level of consumption is seen in Asia and Africa. Numerous genetically modified (GM) and non-GM cultivars grow and are sold in several regions depending primarily on environmental and soil conditions. Numerous analytical-, enzymatic-, and genomic-based methods, in combination with sensory analysis have been employed (often unsuccessfully) for the detection of fraudulent or accidental rice adulteration. It was proved that the use of multivariate analysis can offer a series of benefits, when data is analysed, as it allows the effective differentiation between different origin, and/or rice cultivars. Even though sensory analysis is performed by a trained panel whose results could be subjective, any conclusions produced can still be as accurate as analysis-derived data if the panel is properly trained (Vlachos and Arvanitoyannis 2008). Starch, protein and lipids are the fundamental components of rice that have an effect on its quality. High-quality rice for direct consumption is characterised by low amylose and protein levels and a large breakdown, based on amylograph measurements. Nevertheless, there are important differences amongst cultures as to quality preferences, with the most significant factors affecting its acceptance in Asia and the United States being its appearance and post-cooking aroma (Zhou et al. 2002).

2. Botanical and Geographical Origin of Rice

Genetic studies employing a range of different methodologies have revealed the existence of a deep population structure within domesticated rice. The two main varieties that have been determined (the Indica and Japonica subspecies) can be subcategorized into numerous subpopulations that exist within each variety. The link between these two varieties dates at least 100,000 years back, far preceding rice domestication. The products of crosses between subspecies are characterized by sterility and distinct domestication traits, thus suggesting that different populations are fixed for certain groups of alleles that dictate these traits. A high number of domestication quantitative trait loci have been found in crosses between the subspecies as well as wild and domesticated rice were crossed. Several quantitative trait loci are detected in the same genomic regions, indicating that either a single gene with pleiotropic effects or closely related clusters of genes underlie these quantitative trait loci. In the past few decades, numerous domestication loci have been cloned from rice, covering also the gene that controls pericarp colour and two loci for shattering (Sweeney and McCouch 2007). Data obtained using genetics and archaeology show evidence of domestication of *O. sativa japonica* in the Yangtze River valley of China. It is important to note that even though there are indications that rice has been cultivated in the area since 8000 BP, the key domestication trait of nonshattering had not been fixed for at least 1,000 years. Rice cultivation started in India as early as 5000 BP, although the domesticated indica subspecies appears at present to be a product of the introgression of selected alleles from japonica (Gross and Zhao 2014).

As the agro-food market becomes more and more globalised and products are transported across countries and continents, there is a much higher potential for cases of accidental or incidental mislabelling. It is therefore imperative that appropriate measures are put in place to trace food and identify its geographical origin. Food of high market price with identified geographical origin is vulnerable to adulteration, an issue that is of interest not only to consumers but also to farmers, retailers and national authorities. Nowadays, stable isotope ratio analysis combined with other chemical methods becomes more and more important for the verification of the authenticity of food (Zhao et al. 2014).

3. Authenticity of Rice

3.1 DNA-based techniques

Bligh (2000) reported that the price of Basmati rice is 2–3 times higher than the price of other types of rice across the world. There is therefore a need for the development of methods that could be used to detect non-Basmati long-grain rice within Basmati samples. Fluorescent simple sequence length polymorphisms (SSLPs) among known Basmati rice cultivars and potential non-Basmati samples were therefore employed in an effort to detect adulteration. Fluorescent detection also offers the benefit of allowing crude estimations of the level of adulteration present in blind test samples. Several rice cultivars originating from India, Pakistan and the US were first screened with selected SSLP primers. Among the commercial primer sets and the sets already available, a selection was made of 12 sets that were all able to distinguish between Basmati and non-Basmati rice. A very low number of primer sets were able to distinguish all Basmati cultivars from non-Basmati cultivars, but the use of a range of primer sets allowed the differentiation of all Basmati cultivars from the non-Basmati cultivars. It is expected that the conduction of further analyses using this methodology and larger samples of grain could provide more accurate results in the detection of adulteration. It is important to note also that even though this method can be effective for detecting the presence of an adulterant, the adulterant could not be accurately quantified. Furthermore, as the method relies on genetics, it could also be potentially used to test adulteration of other cultivars of high value, such as Thai Fragrant rice, and generally, in cases that morphological tests are not reliable.

In the study Chuang et al. (2011), microsatellite marker [Simple Sequence Repeats (SSR)] analysis was employed to discriminate between Taiwan and other rice cultivars in an effort to authenticate and trace the products. A total of 32 microsatellite DNA markers and 36 rice varieties originating from different countries were employed to conduct a comparative polymorphism analysis. A total of 306 alleles were detected in 32 loci. A number of 3 to 21 alleles were present per locus, with an average of 9.6. Polymorphic information content (PIC) values ranged from 0.205 (RM7023) to 0.926 (RM333), with an average of 0.75. The markers of most importance that could allow domestic Taiwan rice identities to be efficiently differentiated were RM21, RM22, RM101, RM333, RM475, RM1387, RM5704, and RM7545. Principle component analysis (PCA) score plot and clustering analysis allowed the sufficient discrimination between two distinct groups of rice varieties according to their geographic origins. These results could therefore be used to detect and prevent fraud by providing SSR marker sets and workflow to regularly verify genotypes and premium rice variety purity. The technology of DNA fingerprinting can find practical use in rice adulteration detection, as it is sufficiently sensitive (3 to 7.5%) when different ratios of rice varieties are mixed.

The genetic diversity of *O. sativa* cultivars that are traditionally significant for rice breeding and production in Argentina were assessed at a DNA level in the study of Giarrocco et al. (2007). A survey of sixty-nine accessions was carried out with 26 SSR markers showing the genomic relationships among cultivars. Two hundred and nineteen polymorphic bands were found. Cluster analysis relied on comparing the similarities of cultivar genetic characteristics pair-wise, categorising *O. sativa* into indica and japonica, and further categorising the japonica group into the subgroups of tropical and temperate. This categorisation was found to be in agreement with pedigree information and nearly the total number of Argentina-released cultivars were assigned to the japonica cluster. The use of DNA polymorphism analysis indicated that there were genomic links in Argentine rice germplasm, thus allowing a database to be constructed that could be valuable for identifying cultivars, conserving local germplasm as well as for breeding programs.

Also, the aim of the work of Ganopoulos et al. (2011) was to create a DNA-based method coupled with High Resolution Melting (HRM) analysis that could be employed for the traceability of rice via the use of five microsatellite markers, genotyping Basmati and non-Basmati rice. The extracted data were also used for the detection of the presence of Basmati rice in commercial rice products. Furthermore, the 8 bp deletion in badh2 gene combined with HRM were used for both DNA-typing of Basmati and non-Basmati varieties as well as for an accurate quantitative determination of the adulteration of Basmati rice with non-Basmati. HRM was characterised by high sensitivity in genotyping rice varieties and detecting admixtures, while was also effective in the detection of a ratio of 1:100 of non-fragrance in fragrance rice. It was concluded that HRM analysis can offer higher resolution, lower the cost and can be used as an alternative to other techniques. Its use could also be extended to the quantification of adulteration in rice varieties and commercial food products containing rice.

In another study, the genetic links between Indian aromatic and quality rice (*O. sativa*) germplasms were evaluated with the use of 30 fluorescently labelled rice microsatellite markers. A total of 69 rice genotypes were employed, including 52 Basmati and other scented/quality rice varieties originating from different parts of India, while 17 indica and japonica varieties were used as controls. The detection of 235 alleles at the 30 SSR loci was achieved. Sixty-two (26.4%) of these alleles were detected exclusively in Basmati and other scented/quality rice germplasm accessions. Several alleles were found per locus, with the number ranging from 3 to 22, with an average of 7.8. PIC values were between 0.2 and 0.9, with an average of 0.6, while the size between the smallest and the largest allele for a given microsatellite locus ranged between 3 bp and 68 bp. Out of the 30 SSR markers, a total of 20 could be used to differentiate between traditional Basmati varieties, while a panel of eight markers was employed to distinguish premium traditional Basmati, cross-bred Basmati, and non-Basmati rice varieties that all have different prices in the market. Estimates of inferred ancestry or similarity coefficients were employed for clustering varieties, differentiating between high-quality Indian aromatic and quality rice genotypes and both indica and japonica cultivars. Differentiation between crossbred varieties and traditional Basmati rice could be achieved. This study demonstrated that Indian aromatic and quality germplasm is genetically distinct from other groups within *O. sativa* and has followed a long, independent evolutionary path. It was also indicated that further studying of the genetic diversity of Indian aromatic/quality rice germplasm is needed (Jain et al. 2004). Moreover, in the study of Jie-zheng et al. (2007), an assay was conducted of 63 major inbred varieties and parental lines of major F1 hybrids commonly used in Chinese rice production with rice microsatellites that had been subjected to screening in a previous study as well as additional microsatellites from four chromosomes. A set of 24 markers was chosen to be used for identifying rice varieties. The markers were distributed on all 12 chromosomes of rice with 2 markers on each chromosome. Genotyping with the markers was performed with the 63 major varieties and parental lines, as well as 41 major F1 hybrids, and verification of alleles detected in each line at each marker locus followed. A reliable verification of maternal lines was achieved via matching marker genotypes of corresponding F1, maternal and paternal lines of hybrid rice, while information on the paternal lines was modified, and the removal of a false hybrid was allowed. A database that contained genotype data of a total of 103 major rice varieties and parental lines at the 24 marker loci was compiled and subjected to analysis. It was demonstrated that SSRs had a greater discrimination potential for hybrids than for their parental lines. The results extracted by analysing genetic similarity were similar. Any pairs of these rice lines differed at more than one locus, with pairwise polymorphic loci detected in the hybrid group as 16.0 ± 3.8, while in the maternal and paternal groups averaged 12.0 ± 4.7 and 12.7 ± 3.4, respectively.

In another study, extraction of DNA from commercial rice cereal foodstuffs was performed using modified conventional methods (CTAB, SDS and a commercial kit) with yields that ranged from 1.4 to 10.7 lg DNA per g of sample. This DNA was used as a template to amplify a single copy rice gene (i.e., MIPS) fragment (of approximately 850 bp) as well as microsatellite DNAs (120–400 bp). Further discrimination of the products was possible using six microsatellite markers. The importance of DNA analysis for quality control purposes and authenticity verification of rice in foods as well as to detect the presence of GMO was examined. This study showed that the extracted DNA can be effectively used to both amplify rice gene fragments and for microsatellite analysis. Approximately 75% of microsatellites in eight samples were found to be polymorphic and could therefore not be employed to discriminate varieties. It was though possible to differentiate between 8 product samples using 6 rice microsatellite markers. It is important to note that two cereal samples produced by the same manufacturer on different calendar years were characterized by different microsatellite profiles, thus indicating that different rice varieties were used to produce the two cereal products (Ren et al. 2006).

3.2 Spectrometric analysis

3.2.1 Inductively coupled plasma spectroscopy

As the marketplace of rice expands around the world, a number of concerns are raised around the authenticity control of the product. The development of numerous analytical methods has been reported that their purpose is to differentiate the geographical origin of products containing rice seeds and products thereof (Lim et al. 2017). ICP (Inductively Coupled Plasma) spectroscopy is an analytical method that is employed for the detection and measurement of elements. The technique relies on ionization of samples using a highly hot plasma, commonly formed of argon gas. In an ICP spectroscopic instrument, argon gas passes through a torch apparatus. The gas is charged with an electromagnetic coil and lit with a Tesla unit. Once the plasma is lit, the spectroscopic unit is switched off the Tesla device. The acceleration of the electrons by the torch leads to collisions with argon atoms, and eventually a collision results in an argon atom shedding one of its electrons. Acceleration of released electrons by the electromagnetic unit's magnetic field follows and the process continues until the rate of electron release is balanced by the rate of electrons that recombine with argon ions that have shed an electron. A gas flow is allowed to pass through the middle of the plasma, leading to the creation of a channel that has a lower temperature than the adjacent plasma but a much higher temperature than a chemical flame. The samples to be analysed are introduced into this channel, commonly in the form of a mist of liquid (https://www.xrfscientific.com/what-is-icp-spectroscopy/).

In the study of Borges et al. (2015), 50 rice samples (18 organic and 32 non-organic) originating from Brazil were subjected to analysis using ICP-MS. Twenty elements (As, B, Ba, Ca, Cd, Ce, Co, Cr, Cu, Fe, K, La, Mg, Mn, Mo, P, Pb, Rb, Se and Zn) were analysed to detect differences in their levels between the two rice types. The levels of As, B, Ba, Co, Cr, Cu, Mn, P and Zn were significantly higher in non-organic rice in comparison to organic, while the K, Ca, Mo, Rb and Se levels were lower in non-organic in comparison to organic samples. The levels of the rest of the elements that were tested (Cd, Ce, Fe, La, Mg and Pb) were found to be statistically equivalent between the two types of rice. The use of PCA, Soft Independent Modeling of Class Analogy (SIMCA), Hierarchical Cluster Analysis (HCA) and K-nearest neighbors (KNN) rendered the discrimination of the two types of rice feasible based on elemental fingerprints, while they can be employed as alternative methods to verify the authenticity of samples. ICP-MS was also used in the study of Cheajesadagul et al. (2013), who analysed 31 Thai jasmine and 5 foreign (France, India, Italy, Japan and Pakistan) rice samples using high resolution ICP-MS following acid digestion. Rice flour standard reference samples were used to assess how accurate the whole process was. The levels of 21 elements were assessed and employed as chemical markers in order to differentiate the origin of the samples. The rice samples were classified in accordance with their composition in elements following a radar plot and multivariate data analysis that included PCA and discriminant analysis (DA). Differentiation between Thai jasmine rice and foreign rice could be achieved using radar plots and multivariate data analysis. Moreover, the use of DA allowed the differentiation of Thai jasmine samples in accordance with their geographical origin (northern, north-eastern or central areas of Thailand). It was concluded that multi-element fingerprinting in combination with the use of multivariate statistical techniques can be successfully employed for authenticating rice samples.

The study of Chung et al. (2018) also aimed at discriminating between rice samples that originated from six Asian countries using combinations of elemental/isotopic composition analysis and chemometrics. PCA was capable of distinguishing samples originating from the majority of countries, apart from those that originated from Philippines and Japan. Moreover, orthogonal projection to latent structure-DA (OPLS-DA) could clearly discriminate between rice originating from Korea and other countries. The major common variables in these models that allowed the samples to be differentiated were $\delta^{34}S$, Mn, and Mg. More specifically, the isotope ratios of C, N, O and S and the levels of 25 elements were analysed using Isotope ratio-MS (IRMS) and ICP-MS, respectively. An evaluation of the discriminative power of the elements was made using chemometric techniques, such as PCA and OPLS-DA. It was demonstrated in this study that by combining multielement and light-element isotope ratios, samples originating from different geographical regions of Asia could be effectively discriminated.

Similarly, Kelly et al. (2002) examined rice samples originating from the USA, Europe and Asia using IRMS and ICP-MS. C-13, O-18, B, Ho, Gd, Mg, Rb, Se and W were used as chemical markers. Canonical DA was used for their identification allowing the samples to be clearly differentiated in accordance with their regions of origin. Elevated concentrations of boron (> 2500 ppb) were found in American samples, while high concentrations of holmium were detected in samples of rice cultivated in Arkansas. Samples of rice cultivated in Europe were found to contain high concentrations of Mg, while Indian/Pakistani samples were found to have low levels of oxygen-18 (^{18}O). Furthermore, in the study of Lagad et al. (2017) samples of Basmati rice, soil and water were tested to determine their $^{87}Sr/^{86}Sr$ ratios and their rare earth elements levels in order to evaluate the potential of these measurements for use in the determination of the geographical origin if rice. An effort was also made to determine the source of Sr in rice plants. The levels of Sr, Rb and rare earth elements in rice and soil were determined using Q-ICP-MS. A total of 82 Basmati rice samples were tested and found to contain $^{87}Sr/^{86}Sr$ ratios ranging from 0.71143 to 0.73448 and rare earth elements (La, Ce, Pr, Nd, Sm, Eu, Gd, Dy, Er, Yb) at ppb levels. Statistical analysis indicated that strong correlation existed between $^{87}Sr/^{86}Sr$ ratios of rice, silicate and carbonate fractions of soil. The good correlation and chosen values of $^{87}Sr/^{86}Sr$ found in rice and water led to the conclusion that the water was the source of the element. Rice cultivated in southern Uttar Pradesh was found to contain higher $^{87}Sr/^{86}Sr$ ratios in comparison to samples originating from other regions of Indo-Gangetic Plain. This was attributed to the higher $^{87}Sr/^{86}Sr$ ratio content of Ganga river in comparison to other rivers. $^{87}Sr/^{86}Sr$ ratios can therefore be employed as a marker for the differentiation of Indian Basmati rice and other types of rice originating from the same country. In the study of Song et al. (2014), an effort was made to establish the geographical origin of rice, but no strong links were found between the products and the environments in the cultivation areas. Soil, water (stream water and groundwater), and *O. sativa* samples obtained from different regions of South Korea were subjected to analysis in order to determine their Sr isotope ratios ($^{87}Sr/^{86}Sr$). Measurements of Sr levels were taken using an X7 ICP-MS. Sequential leaching of soil indicated that Sr in exchangeable and carbonate fractions (bioavailable) presented a lower $^{87}Sr/^{86}Sr$ ratio in comparison to that found in the silicate fraction. This was in agreement with the low $^{87}Sr/^{86}Sr$ ratio found in the plants. Although the bedrock–soil–water–plant system was found to be strongly linked, statistical analysis showed that $^{87}Sr/^{86}Sr$ ratios in plants were mostly in agreement with those found in water and the exchangeable fraction of soil. This was the first study to examine the potential use of $^{87}Sr/^{86}Sr$ as a marker in South Korea and showed that the geographical origin of the product could be determined by examining the exchangeable fraction of soil and water.

In the study of Lim et al. (2017), a methodology was suggested that made use of direct infusion-mass spectrometry (DI-MS) to rapidly differentiate rice products originating from different countries. In this study, a polytetrafluoroethylene (PTFE) filter was used to eliminate the matrix effect and a mixed-mode cation exchange (MCX) solid-phase extraction (SPE) with 20% methanol, and an MCX SPE with 100% methanol were measured. Following that, PLS-DA and random forests were used to determine the method with the highest discrimination potential. It was demonstrated that the combined use of MCX SPE with 100% methanol and DI-MS in positive ion mode (accuracy = 1.000, R2 = 0.916, Q2 = 0.720, B/W-based p-value = 0.015) or the combined use of MCX SPE with 20% methanol and targeted DI-MS/MS in positive ion mode (accuracy = 1.000, R2 = 0.931, Q2 = 0.849, B/W-based p-value = 0.002) offered optimum discrimination ability. Moreover, the differential expression of metabolites such as sodiated lysophosphatidylcholine, lysophosphatidylcholine, lysophosphatidylethanolamine and lysophosphatidylglycerol was determined providing an additional differentiation route.

The study of Promchan et al. (2016) on the other hand, investigated the elemental imaging in a longitudinal section of a single rice grain, making use of laser ablation ICP-MS (LA-ICP-MS). In this study it was also attempted to differentiate rice based on its geographical origin and type using LA-ICP-MS with LDA. Visualization of the distributions of Ca, Cu, Fe, K, Mg, Mn, P and Zn in rice grain were given in the form of elemental images. The imaging indicated that high levels of essential elements were present in the embryo and endosperm. No uniform elemental distributions were found in rice grain. Furthermore, the levels of 20 elements detected in core endosperm were assessed and employed as chemical markers for the discrimination of the origin and type of samples. The LDA can be used for the effective differentiation of rice samples in accordance with their geographical origin within Thailand (Northeast or South regions of Thailand) as well as their types. The samples were classified correctly in the majority of cases with overall correct classification and cross-validation of 93.8% and 91.1% for classification according to geographical origin and 100% and 97.9% for classification according to type.

In the study of Gonzálvez et al. (2011), rice samples were examined for trace elements using ICP optical emission spectroscopy (ICP-OES) in an effort to classify them according to their geographical origin. A total of 107 samples claimed to be originating from different Spanish regions such as Valencia, Tarragona, Murcia and Extremadura, as well as Japan, Brazil and India were initially tested, while the results were validated using 46 samples that were derived from the regions mentioned above. The data were analysed using LDA, and the levels of 32 elements (Al, As, Ba, Bi, Cd, Ca, Cr, Co, Cu, Fe, Pb, Li, Mg, Mn, Mo, Ni, K, Se, Na, Sr, Tl, Ti, Zn, La, Ce, Pr, Nd, Sm, Eu, Ho, Er and Yb) were used as analytical variables. The method provided satisfactory results, allowing 91.30% of the samples to be correctly classified.

Finally, in the study of Chung et al. (2015) an effort was made to verify the geographical origin of rice samples using ICP atomic emission spectroscopy (ICP-AES) and chemometric analysis. Measurements of the profiles of 25 elements in brown rice were taken using ICP-AES and the data were processed using data-mining processes, such as PCA and PLS-DA. Clear discrimination in accordance with their geographical origin was provided using PLS-DA for samples originating from different countries. A total of 11 elements (Cu, Ag, Zn, Cr, Ca, Ba, Cd, Bi, K, Pb, and In) were of high importance for a successful discrimination.

3.2.2 IRMS

IRMS is a technique capable of providing information about the geographical, chemical and biological origin of samples. The origin of organic samples is determined by taking advantage of the relative isotopic abundances of elements comprising the samples. Due to the fact that the local level of isotope ratios of elements (e.g., C, H, O, S, N) can change via various kinetic and thermodynamic routes, determination of these ratios can be used to differentiate samples that are otherwise identical in nature. Although numerous different sample introduction methods can nowadays be used, combustion is frequently employed in the case of bulk isotopic analysis, while gas and liquid chromatography are primarily employed for real-time isotopic analysis of certain substances within mixtures (Muccio and Jackson 2009).

The study of Chung et al. (2016) aimed at discriminating between the different geographical origins of rice products cultivated in countries of Asia, by analysing the C, N, O and S stable isotope ratios and processing the data using chemometrics. The $\delta^{15}N_{AIR}$, $\delta^{18}O_{VSMOW}$ and $\delta^{34}{}_{SVCDT}$ values of brown rice were much more significantly affected by the geographical origin of the product than the $\delta^{13}{}_{CVPDB}$ value. Specifically, the combined use of $\delta^{18}{}_{OVSMOW}$ and $\delta^{34}{}_{SVCDT}$ allowed for a more efficient discrimination of rice according to its geographical origin compared to other combinations. The use of PCA allowed the geographical origins to be clearly discriminated but was not effective in the case of different rice genotypes. Particularly, PCA enabled the clear discrimination of rice samples originating from Philippines versus China and Korea. The results of this study indicated that analysis of light element isotope composition together with the use of chemometrics could be effective for discriminating the geographical origin of rice and can therefore be a useful tool against rice products' mislabelling.

Also, the study of Chung et al. (2017) investigated the use of $d^{13}C$ and $d^{15}N$ analyses as potential tools for authenticating environmentally friendly-produced rice sold in Korea. The use of $d^{13}C$ and $d^{15}N$ in different rice grains proved that the rice samples produced using methods friendly to the environment, could be effectively differentiated from conventional rice. No multi-residue pesticides were found in any of the rice samples tested, thus indicating that $d^{13}C$ and $d^{15}N$ analyses could be used as complementary tools to verify the authenticity of environmentally friendly produced rice sold in Korea whenever discrimination cannot be achieved simply by testing for pesticide residues. It was therefore suggested that complementary analyses covering compound-specific isotope ratio analysis could allow more reliable conclusions to be made around the authenticity of rice samples.

Finally, in another study, Suzuki et al. (2008) determined the C and N concentrations as well as the stable C, N, and O isotopic levels ($d^{13}C$, $d^{15}N$, and $d^{18}O$) of polished rice, aiming at developing a simple technique for differentiating rice according to its geographical origin. The first effort to apply the method focused on samples of the cultivar "Koshihikari", that originated from 14 different areas in Australia (n = 1), Japan (n = 12), and USA (n = 1). In the case of all samples, the C and

N concentrations and the isotopic compositions were in accordance with those found in general plant materials [37.2–40.0% (C content), 0.8–1.4% (N content), –27.1 to –25.4% (d^{13}C), +0.4 to +9.0% (d^{15}N), and +18.8 to +22.9% (d^{18}O)]. It was found though that the cultivation area was well distinguished by a pentagonal radar plot on the basis of the elemental and isotopic compositions. Therefore, useful information can be provided by comparing the C and N levels and d^{13}C, d^{15}N, and d^{18}O values that could enable the quick differentiation of rice samples as per their geographical origin.

3.2.3 Atomic fluorescence spectrometry (AFS)

AFS is based on trace element analysis. In this technique, atomization of a sample solution in a flame or non-flame cell is followed by illumination of the resulting atoms. Measurements of the atomic fluorescence that is generated when a portion of the excited atoms undergo radiational deactivation emitting radiation toward the detection device are taken (Winefordner and Elser 1971). Chen et al. (2012) used hydride generation AFS in order to determine the selenium content of regular polished rice in China and selenium-enriched polished rice cultivated using fertilizer rich in selenium in the forms of selenite and selenate. The average selenium concentration of regular rice was measured as 0.011 µg/g. Based on a daily dietary rice intake of 300–500 g, the total selenium intake from regular rice for adults was estimated at 7.5–12.5 µg per person per day. A significant increase in the selenium concentration of rice with levels ranging from 0.471 to 0.640 µg/g was found following foliar application of fertilizer rich in selenium at a rate of 20 g of Se ha^{-1}. It is important to note that when a selenate-based fertilizer was used, the selenium content of rice was 35.9% higher than the level found after the use of a selenite-based fertilizer, which indicates that selenate enriched fertilizer can more efficiently increase the Se content of rice products.

Yu et al. (2016) reported the development of a new digestion method that used a UV-assisted Fe^0 Fenton reaction to determine trace Cd in rice by hydride generation AFS. The method was simple, cost-efficient, required low levels of reagents and moderate reaction conditions. In this study it was used for the analysis of a reference material and real rice samples. The sample and reagents (1 ml) (0.0500 g rice powder, 0.2% (m/v) Fe, 0.75% (v/v) HNO_3 and 18% (v/v) H_2O_2) were subjected to irradiation by UV-light for a period of 50 min. The excess Fe^0 was subsequently separated using a magnet and spectral analysis followed. The limit of detection (LOD) in the case of Cd was 0.02 mg kg^{-1}. The recovery obtained in the case of the reference material was 103% and spiked recoveries with 0.40 mg kg^{-1} Cd in rice samples were at 93% and 101%. The t-test indicated that any differences between the reference material value and the determined value were insignificant, while there was also no significant difference between the proposed method and microwave-assisted digestion coupled with ICP-MS (MWD-ICP-MS) at 95% confidence level (Yu et al. 2016).

Finally, in the study of Jianbo et al. (2002), trace amounts of germanium were determined using hydride generation AFS (HG-AFS) associated on-line with flow injection coprecipitation preconcentration. The samples were spiked with Ni^{2+} (500 μg ml^{-1}). They were inserted into the flow injection system using time-based injection and were mixed with a NaOH solution (50 g l^{-1}). Preconcentration of germanium was done via coprecipitating with nickel hydroxide. Elution of the precipitate followed with 20% (v/v) phosphate acid and the sample was directed into the HG-AFS system. The technique was characterized by both speed and simplicity. The LOD (3σ) was determined at 0.11 μg l^{-1}.

3.2.4 Near Infrared (NIR) spectroscopy

The use of NIR has also been reported in studies examining the authenticity of rice. Chen et al. (2018) mentioned that black rice is a rice variety that is of high significance for the Southeastern Asian market. However, a lot of incidents have been reported according to which low-priced black rice is sold as a higher quality product. It is therefore apparent that rapid, reliable, inexpensive and easy-to-use methods are needed to detect adulteration attempts and verify the authenticity of the product. The potential of NIR spectroscopy and support vector data description (SVDD) for use in this area was investigated. PCA was employed for exploratory analysis and feature extraction. Two data description methods [KNN data description (KNNDD) and GAUSS] were employed as reference. A collection of 142 samples from three brands were subjected to spectral analysis. The analysis of the data led to the construction of three types of data descriptions. As an average, the optimized SVDD model performed well, being 100% specific and 94.2% sensitive on the independent test set with tight boundaries. It was therefore suggested that SVDD in combination with NIR can be effectively used for authenticity verification of black rice. In another study, identification of five varieties of rice obtained from the market was performed using visual/NIR spectroscopy. The collection of the spectra was conducted using a spectrometer. Thirty-five samples per rice variety were used. The samples were randomly assigned to two groups. The first group with 150 samples was used as a calibrated set, while the second one with 25 samples was the validated set. Analysis of the samples followed with the whole wave band and the characteristic wave band (400 ~ 500 nm, 910 ~ 1400 nm and 1940 ~ 2300 nm) models, respectively. Following pretreatment of samples data, the spectra data obtained were processed with PCA. The anterior 9 principal components analyzed by PCA acted as input variables of back-propagation artificial neural network (BP-ANN) which also covered one hidden layer. The values of the rice varieties acted as output variables of BP-ANN, allowing a three-layers BP-ANN discrimination model to be constructed. The ANN-BP model allowed the prediction of the 25 unknown samples, indicating that 100% discrimination can be achieved using both models, while more accurate results can be obtained by using the characteristic wave band model than that of the whole wave band model. It was therefore demonstrated that it is possible to differentiate between rice varieties and verify the authenticity of rice using visible/NIR reflectance spectra as a quick and non-contact solution (Liang et al. 2009).

Kong et al. (2013) reported the development of a NIR hyperspectral imaging system. Specifically, NIR hyperspectral imaging was used in combination with multivariate data analysis for identifying different rice seed cultivars. Apart from PLS-DA, SIMCA, KNN Algorithm and Support Vector Machine (SVM), an innovative learning algorithm tool called Random Forest (RF) was also employed. The use of 1,039 nm to 1,612 nm spectra as full spectra allowed classification models to be built. PLS-DA and KNN models allowed 80% of samples to be accurately classified, while SIMCA, SVM and RF models were all 100% accurate in both the calibration and prediction sets. A selection of the best twelve wavelengths were made by weighted regression coefficients of the PLS-DA model. The PLS-DA, KNN, SVM and RF models were constructed on the basis of these wavelengths. These wavelength-based models (apart from PLS-DA) allowed more than 80% of the samples to be accurately classified. Full spectra-based models performed better than the optimal wavelengths-based ones. It was therefore suggested that hyperspectral imaging can be employed for identifying rice seed cultivars and RF can be used to effectively classify them. NIR was also used in the study of Osborne et al. (1993). In this study NIR transmission spectroscopy was employed for the discrimination between Basmati and other long-grain rice varieties. The basis of the discrimination was the Fisher linear discriminant function obtained by calculating the first few principal component scores of the NIR spectra. The discriminant rule was cross validated. Classification of nine Basmati and 53 samples of other rice varieties was accurately obtained from NIR spectra measured on 200 g bulk samples. However, 8% of the Basmati samples and 14% of the other samples were not classified correctly on the basis of spectra of individual grains. The use of average spectra obtained from 23 grains allowed 16 Basmatis to be correctly classified but with a 20% error rate on 100 other samples. Additionally, the study of Teye et al. (2019) reported the use of an innovative hand-held NIR spectrometry used in combination with chemometrics. The method aimed at examining rice authenticity and quality in real time. A total of 520 rice samples were used covering various qualities (high quality, mid quality and low quality) and originating from Ghana, Thailand, and Vietnam. Several methods were used to pre-process the data with multiplicative scatter correction (MSC) showing the optimum performance. PCA was employed for extracting data from the spectral data set showing that rice samples of different categories can be effectively categorised under the first three PCs using MSC. KNN performed well when used for classifying the samples according to their quality grades, with 91.62% and 91.81% of samples being classified correctly in the case of the training set and the prediction set, respectively. Categorization per country of origin was also achieved in 90.84% and 90.64% of cases in the training and the prediction set, respectively. KNN and SVM were used to differentiate local and imported rice products with 100% success in both the training and prediction sets. It was therefore proved that hand-held spectrometry coupled with MSC-PCA-KNN could be employed to rapidly, effectively and non-destructively classify rice samples in accordance with different quality grades, countries of origin as well as to distinguish between local and imported rice.

3.2.5 Nuclear magnetic resonance (NMR) spectroscopy

NMR spectroscopy is a robust method that allows rapid analysis of mixtures at a molecular level while eliminating the need for samples' separation and purification and could therefore be very useful for the analysis of food samples. Although from a research perspective it is becoming more and more popular, the method is not extensively used, mainly due to the fact that it is expensive, its use requires expertise and has relatively low sensitivity (Hatzakis 2019). A method was reported though in which 400-MHz ^{1}H NMR spectroscopy and multivariate data analysis were employed for discriminating 46 authentic rice samples in accordance with their variety. The most effective pre-treatment method for the samples involved preparation of aqueous rice extracts at pH 1.9. Independent component analysis (ICA) was used for differentiating clusters of Basmati, non-Basmati long-grain rice, and round-grain rice, as well as to cluster them as per their geographical origin. ICA performed better than PCA when used to verify the authenticity of rice. The chemical shifts of the principal saccharides and acetic acid were the main factors used to cluster the samples. The methods used to classify the samples were: LDA, factorial DA, PLS-DA, SIMCA, and ICA, with PLS-DA and ICA being the most specific (0.96 in both cases). The sensitivity of PLS-DA and ICA was 0.94 and 1.0, respectively. NMR spectroscopy in combination with chemometric analysis could therefore be useful as a screening method in regulatory enforcement and official controls (Monakhova et al. 2014).

3.2.6 Other spectrometric techniques

Other spectrometric techniques based on gas chromatography (GC)-MS (GCMS) or laser induced breakdown spectroscopy (LIBS) have also been successfully used to authenticate rice.

The aroma compounds present in 4 types of cooked brown rice, namely Improved Malagkit Sungsong, Basmati 370 (B 370), and Khaskhani, and Indica (German supermarket sample), were studied by Jezussek et al. (2002). The methodology that followed the preparation of the aroma extracts involved the use of High-Resolution GC/Olfactometry (HRGC/O) and High-Resolution GC/MS. The identification of the aromas was done with extract dilution analysis. Among the 41 odor-active compounds that were found, eleven were reported for the first time as rice components. 2-Amino acetophenone, which was recorded for the first time as a constituent of rice, presented the highest flavor dilution factor among the 30 to 39 odor-active substances found in all rice types tested. 2-Acetyl-1-pyrroline, was characterised by a distinct popcorn-like aroma-note and was found to be an important aroma substance in Improved Malagkit Sungsong, B 370, and Khaskhani, but was not of high importance in the case of Indica. Flavor dilution factors that differed amongst the different rice types, were detected in the case of the aroma substance 3-hydroxy-4,5-dimethyl-2(5H)-furanone (Sotolon; seasoning-like), which was lower in Improved Malagkit Sungsong and Khaskhani than in B370. In Improved Malagkit Sungsong, a spicy smelling aroma was detected that was not determined with a very high flavor dilution factor. The same aroma contributed with lower flavor dilution factors to the overall aromas of B 370 and Khaskhani, while was not detected in Indica rice. The Indica samples were obtained from the market in Germany and presented the highest differences in their overall aroma, in comparison to the three Asian brown rices. In another study, extraction of volatile components of Khao Dawk Mali 105 brown rice (not cooked) was conducted via indirect steam distillation under reduced pressure and controlled temperature conditions in order to avoid unintentional cooking. The fresh extract was analysed using capillary GC/MS and more than 140 volatile components were detected. Identification of a total of 70 volatiles was also achieved, including 2-acetyl-1-pyrroline, an important aroma compound of cooked rice. An effort was also made to develop an improved method that could be used to quantify 2-acetyl-1-pyrroline in uncooked brown rice. A solvent extraction process was used to simplify the method. A capillary GC system was employed to quantitatively analyse the samples. The system was equipped with a flame ionization detector using a more selective column, CP-Wax 51, for amines. This improved system increased considerably the capacity of the method to detect 2-acetyl-1-pyrroline in rice extracts so that detection of 2-acetyl-1-pyrroline was possible even in an extract of just 0.5 g of uncooked Khao Dawk Mali (Mahatheeranont et al. 2001).

In another study, the development of the flavour substance 2-acetyl-1-pyrroline in seedlings and callus of aromatic rice (Khao Dawk Mali 105) was examined. A determination of the levels of 2-acetyl-1-pyrroline was given using GC-MS-SIM, which employed an isotope dilution method. The levels detected increased in the presence of proline, ornithine, and glutamate in solution, with proline leading to an increase of more than 3-fold in comparison to the level found in the control. Tracer experiments that were conducted using ^{15}N-proline, ^{15}N-glycine, and proline-1-^{13}C offered evidence that the nitrogen source of 2-acetyl-1-pyrroline was proline, while the carbon source of the acetyl group was not the carboxyl group of proline. The formation of 2-acetyl-1-pyrroline in aromatic rice occurred at temperatures below that of thermal generation in bread baking and the aroma was formed in the aerial part of aromatic rice from proline as the nitrogen precursor (Yoshihashi et al. 2015).

It is also important to note that the protection of Designation of Origin (DOO) claims is very important in the case of rice due to the extensive consumption of the product worldwide. Therefore, LIBS was examined as an analytical technique for PDO certification, due to the fact that its results are provided quickly, and the analysis of multiple elements does not require extensive sample preparation. LIBS spectra data were therefore assessed for PDO certification of Argentine brown rice. Samples from two PDOs were subjected to analysis using LIBS coupled to spark discharge. The spectral data were selected by extreme gradient boosting (XGBoost), an algorithm that has been employed in machine learning, but is not frequently used in chemical studies. A selection of emission lines of C, Ca, Fe, Mg and Na was made, and optimum classification was achieved using a KNN algorithm. This method was proved to be 84% accurate, 100% sensitive and 78% specific in classifying the samples. It is a method that is not characterised by high complexity and could therefore be used for rice certification testing (Pérez-Rodríguez et al. 2019).

3.3 HPLC

HPLC was used in the study of Tian et al. (2004), in which isolation of two hydroxycinnamate sucrose esters, 6′-O-(E)-feruloylsucrose and 6-O-(E)-sinapoylsucrose, from methanol extracts of rice bran was reported. HPLC was then used to analyse soluble and insoluble phenolic compounds as well as 6′-O-(E)-feruloylsucrose and 6′-O-(E)-sinapoylsucrose from white, brown, and germinated brown rice. It was indicated that considerably high levels of insoluble phenolic compounds were present in the samples in comparison to soluble phenolics, while almost all substances found in germinated brown rice and brown rice were present in higher concentrations compared to those found in white rice. 6′-O-(E)-Feruloylsucrose (1.09 mg/100 g of flour) and 6′-O-(E)-sinapoylsucrose (0.41 mg/100 g of flour) were the most abundant soluble phenolic compounds in brown rice. The level of the two hydroxycinnamate sucrose esters was reduced by about 70% during germination, while there was a significant increase in the level of free phenolic acid, with the level of ferulic acid increasing in brown rice from 0.32 mg/100 g to 0.48 mg/100 g of flour and becoming the major phenolic ingredient in brown rice following germination. The level of sinapinic acid increased to 0.21 mg/100 g of flour, a ten-fold higher level than that found in brown rice (0.02 mg/100 g of flour). Furthermore, an increase in the total level of insoluble phenolic compounds was seen from 18.47 mg/100 g of flour in brown rice to 24.78 mg/100 g of flour in germinated brown rice. HPLC was also used in the study of Sabir et al. (2017). In this study HPLC fingerprint analysis in combination with chemometrics were employed for discriminating between red and the white rice bran of Indonesian origin. The principal component in rice bran is γ-oryzanol, which is composed of cycloartenol ferulate, cyclobranol ferulate, campesterol ferulate and b-sitosterol ferulate. These and other compounds were separated using C18 and methanol-acetonitrile with a gradient elution system. PCA and DA were then employed to enable the discrimination of the samples. The discrimination between red and white rice bran was effectively carried our using DA and the predictive ability of the model indicated that the test samples could be satisfactorily classified. It was finally suggested that this method could be effectively employed for rice bran quality control purposes, to identify and discriminate between red and white rice bran.

4. Conclusions

Rice is a widely grown crop and one of the most important food sources for the world's population. It is characterised by high genetic diversity and its cultivation around the world has allowed the development of a very high number of varieties and crop qualities. However, as visual discrimination of seeds of different origins or grades is particularly difficult, the crop and processed foods containing it are prone to adulteration attempts. Those fraudulent activities can take many forms, from substitution of higher-quality and more expensive seeds with lower quality ones to geographical origin mislabelling. A number of methods have therefore been developed to detect adulteration and protect the interests of the markets and consumers. DNA-based techniques such as SSLP and SSRs have been successfully used as they are sensitive and can be employed when different ratios of rice varieties are mixed. Although very effective in detecting adulterants though, they cannot be used to reliably quantify them. Spectrometric analyses based on ICP spectroscopy have been used in numerous studies to detect and quantify elements in samples using the differences in these levels as markers. The technique is particularly effective in differentiating samples based on their geographical origin. Another spectrometric technique, IRMS, can be used to differentiate organic samples by measuring the relative abundance in isotopes. This technique, usually combined with chemometrics, has been successfully used to differentiate samples based not only on their geographical origin but also on cultivation methods (e.g., conventional versus environmentally friendly). Differences in cultivation methods can also be detected using AFS, while NIR spectroscopy has been successfully used to differentiate between cultivars and is particularly useful in detecting adulteration of high-quality rice with cheaper alternatives. NMR, although expensive, is rapid and requires limited samples' preparation, but it is only recently that its use has started being seen in food materials. Furthermore, techniques based on GC/MS can offer valuable insights when used to detect specific components in rice, such as specific aroma compounds. Finally, the use of HPLC has recently been explored in few studies, in combination with chemometrics, as a tool for analysing phenolics or other compounds. It is therefore apparent that due to the nature of the product and its extensive consumption, there is an urgent need to improve the current methods that are used to detect adulteration, but also develop new techniques that will allow the product to be tested rapidly, accurately and cost efficiently.

Abbreviations

Back-propagation artificial neural network	BP-ANN
Discriminant analysis	DA
Direct infusion-mass spectrometry	DI-MS
Designation of Origin	DOO
Gas Chromatography	GC
Genetically modified	GM
Hierarchical Cluster Analysis	HCA
Hydride generation atomic fluorescence spectrometry	HG-AFS
High-Resolution GC/Olfactometry	HRGC/O
High Resolution Melting	HRM
Independent component analysis	ICA
Inductively Coupled Plasma	ICP
Inductively Coupled Plasma atomic emission spectroscopy	ICP-AES
Inductively Coupled Plasma optical emission spectroscopy	ICP-OES
Isotope ratio mass spectrometry	IRMS
K-nearest neighbors	KNN
K-nearest neighbor data description	KNNDD
Laser ablation	LA
Linear discriminant analysis	LDA

Laser induced breakdown spectroscopy	LIBS
Limit of detection	LOD
Mixed-mode cation exchange	MCX
Mass Spectrometry	MS
Multiplicative scatter correction	MSC
Microwave-assisted digestion coupled with Inductively Coupled Plasma-Mass Spectrometry	MWD-ICP-MS
Near Infrared	NIR
Nuclear Magnetic Resonance	NMR
Orthogonal projection to latent structure-discriminant analysis	OPLS-DA
Principle component analysis	PCA
Polymorphic information content	PIC
Partial least squares	PLS
Polytetrafluoroethylene	PTFE
Random Forest	RF
Soft Independent Modeling of Class Analogy	SIMCA
Solid-phase extraction	SPE
Simple sequence length polymorphisms	SSLPs
Single Sequence Repeats	SSR
Support vector data description	SVDD
Support Vector Machine	SVM

References

Bligh, H.F.J. 2000. Detection of adulteration of Basmati rice with non-premium long-grain rice. Int. J. Food. Sci. Technol. 35: 257–265.

Borges, E.M., J.M.L.N. Gelinski, V.C.O. Souza, F. Barbosa, Jr. and B.L. Batista. 2015. Monitoring the authenticity of organic rice via chemometric analysis of elemental data. Food Res. Int. 77: 299–309.

Bray, F. 2015. Introduction—global networks and new histories of rice. pp. 1–35. *In*: Bray, F., P.A. Coclanis, E.L. Fields-Black and D. Schäfer [eds.]. Rice Global Networks and New Histories. Cambridge University Press, New York, USA.

Chang, T.-T. 2003. Origin, domestication, and diverseification. pp. 3–26. *In*: Smith, C.W. and R.H. Dilday [eds.]. Rice—Origin, History, Technology, and Production. John Wiley & Sons, Inc. New Jersey, USA.

Cheajesadagul, P., C. Arnaudguilhem, J. Shiowatana, A. Siripinyanond and J. Szpunar. 2013. Discrimination of geographical origin of rice based on multi-element fingerprinting by high resolution inductively coupled plasma mass spectrometry. Food Chemistry 141: 3504–3509.

Chen, H., C. Tan and Z. Lin. 2018. Authenticity detection of black rice by near-infrared spectroscopy and support vector data description. Int. J. Anal. Chem., 2018: 1–9.

Chen, L., F. Yang, J. Xu, Y. Hu, Q. Hu, Y. Zhang and G. Pan. 2002. Determination of selenium concentration of rice in china and effect of fertilization of selenite and selenate on selenium content of rice. J. Agric. Food Chem. 50: 5128–5130.

Chuang, H.-Y., H.-S. Lur, K.-K. Hwu and M.-C. Chang. 2011. Authentication of domestic Taiwan rice varieties based on fingerprinting analysis of microsatellite DNA markers. Bot Stud. 52: 393–405.

Chung, I.-M., J.-K. Kim, J.-K. Lee and S.-H. Kim. 2015. Discrimination of geographical origin of rice (*Oryza sativa* L.) by multielement analysis using inductively coupled plasma atomic emission spectroscopy and multivariate analysis. J. Cereal Sci. 65: 252–259.

Chung, I.-M., J.-K. Kim, M. Prabakaran, J.-H. Yang and S.-H. Kim. 2016. Authenticity of rice (*Oryza sativa* L.) geographical origin based on analysis of C, N, O and S stable isotope ratios: a preliminary case report in Korea, China and Philippine. J. Sci. Food Agric. 96: 2433–2439.
Chung, I.-M., S.-K. Park, K.-J. Lee, M.-J. An, J.-H. Lee, Y.-T. Oh and S.-H. Kim. 2017. Authenticity testing of environment-friendly Korean rice (*Oryza sativa* L.) using carbon and nitrogen stable isotope ratio analysis. Food Chemistry 234: 425–430.
Chung, I.-M., J.-K. Kim, J.-K. Lee, S.-K. Park, J.-H., Lee, N.-Y. Son, Y.-I. Jin and S.-H. Kim. 2018. Geographic authentication of Asian rice (*Oryza sativa* L.) using multielemental and stable isotopic data combined with multivariate analysis. Food Chemistry 240: 840–849.
Ganopoulos, I., A. Argiriou and A. Tsaftaris. 2011. Adulterations in Basmati rice detected quantitatively by combined use of microsatellite and fragrance typing with High Resolution Melting (HRM) analysis. Food Chemistry 129: 652–659.
Giarrocco, L.E., M.A. Marassi and G.L. Salerno. 2007. Assessment of the genetic diversity in argentine rice cultivars with SSR markers. Plant Genetic Resource 47(2): 853–858.
Gonzálvez, A., S. Armenta and M. de la Guardia. 2011. Geographical traceability of "Arròs de Valencia" rice grain based on mineral element composition. Food Chemistry 126: 1254–1260.
Gross, B.L. and Z. Zhao. 2014. Archaeological and genetic insights into the origins of domesticated rice. PNAS 111(17): 6190–6197.
Hatzakis, E. 2019. Nuclear Magnetic Resonance (NMR) spectroscopy in food science: A comprehensive review. Compr. Rev. Food Sci. Food Saf. 18(1): 189–220.
https://www.xrfscientific.com/what-is-icp-spectroscopy/ (Accessed on 13.09.2020)
Jain, S., R.K. Jain and S.R. McCouch. 2004. Genetic analysis of Indian aromatic and quality rice (*Oryza sativa* L.) germplasm using panels of fluorescently-labeled microsatellite markers. Theor. Appl. Genet. 109: 965–977.
Jezussek, M., B.O. Juliano and P. Schieberle. 2002. Comparison of key aroma compounds in cooked brown rice varieties based on aroma extract dilution analyses. J. Agric. Food Chem. 50: 1101–1105.
Jianbo, S., T. Zhiyong, T. Chunhua, C. Quan and J. Zexiang. 2002. Determination of trace amounts of germanium by flow injection hydride generation atomic fluorescence spectrometry with on-line coprecipitation. Talanta. 56: 711–716.
Jie-zheng, Y., S. Yong-feng, E. Zhi-guo, Z. Rui-zhen, C. Jie, Z. Zhi-wei and Z. Jie-yun. 2007. Construction and testing of a primary microsatellite database of major rice varieties in China. Rice Sci. 14(4): 247–255.
Kelly, S., M. Baxter, S. Chapman, C. Rhodes, J. Dennis and P. Brereton. 2002. The application of isotopic and elemental analysis to determine the geographical origin of premium long grain rice. Eur. Food Res. Technol. 214: 72–78.
Kong, W., C. Zhang, F. Liu, P. Nie and Y. He. 2013. Rice seed cultivar identification using near-infrared hyperspectral imaging and multivariate data analysis. Sensors 13: 8916–8927.
Lagad, R.A., S.K. Singh and V.K. Rai. 2017. Rare earth elements and $^{87}Sr/^{86}Sr$ isotopic characterization of Indian Basmati rice as potential tool for its geographical authenticity. Food Chemistry 217: 254–265.
Liang, L., Z.X. Liu, M.H. Yang, Y.X. Zhang and C.H. Wang. 2009. Discrimination of variety and authenticity for rice based on visual/near infrared reflection spectra. J. Infrared Millim. W. 28(5): 353–356.
Lim, D.K., C. Mo, N.P. Long, J. Lim and S.W. Kwon. 2017. A rapid and reliable method for discriminating rice products from different regions using MCX-based solid-phase extraction and DI-MS/MS-based metabolomics approach. J. Chromatogr. B Biomed. Appl. 1061-1062: 185–192.
Mahatheeranont, S., S. Keawsa-ard and K. Dumri. 2001. Quantification of the rice aroma compound, 2-Acetyl-1-pyrroline, in uncooked khao dawk mali 105 brown rice. J. Agric. Food Chem. 49: 773–779.

Marton, E. 2014. Rice A Global History. Reaktion Books Ltd, London, UK.
Monakhova, Y.B., D.N. Rutledge, A. Roßmann, H.-U. Waiblinger, M. Mahler, M. Ilse, T. Kuballa and D.W. Lachenmeier. 2013. Determination of rice type by 1 H NMR spectroscopy in combination with different chemometric tools. J. Chemom. 28: 83–92.
Muccio, Z. and G.P. Jackson. 2009. Isotope ratio mass spectrometry. Analyst. 134: 213–222.
Osborne, B.G., B. Mertens, M. Thompson and T. Fearn. 1993. The authentication of basmati rice using near infrared spectroscopy. J. Near Infrared Spec. 1(2): 77–83.
Pérez-Rodríguez, M., P.M. Dirchwolf, T.V. Silva, R.N. Villafañe, J.A.G. Neto, R.G. Pellerano and E.C. Ferreira. 2019. Brown rice authenticity evaluation by spark discharge-laser-induced breakdown spectroscopy. Food Chemistry 297: 124960.
Promchan, J., D. Günther, A. Siripinyanond and J. Shiowatana. 2016. Elemental imaging and classifying rice grains by using laser ablation inductively coupled plasma mass spectrometry and linear discriminant analysis. J. Cereal Sci. 71: 198–203.
Ren, X., X. Zhu, M. Warndorff, P. Bucheli and Q. Shu. 2006. DNA extraction and fingerprinting of commercial rice cereal products. Food Res. Int. 39: 433–439.
Sabir, A., M. Rafi and L.K. Darusman. 2017. Discrimination of red and white rice bran from Indonesia using HPLC fingerprint analysis combined with chemometrics. Food Chemistry 221: 1717–1722.
Song, B.-Y., J.-S. Ryu, H.S. Shin and K.-S. Lee. 2014. Determination of the source of bioavailable Sr using $^{87}Sr/^{86}Sr$ Tracers: A case study of hot pepper and rice. J. Agr. Food Chem. 62(38): 9232–9238.
Suzuki, Y., Y. Chikaraishi, N.O. Ogawa, N. Ohkouchi and T. Korenaga. 2008. Geographical origin of polished rice based on multiple element and stable isotope analyses. Food Chemistry 109: 470–475.
Sweeney, M. and S. McCouch. 2007. The Complex History of the Domestication of Rice 100(5): 951–957.
Teye, E., C.L.Y., Amuah, T. McGrath and C. Elliott. 2019. Innovative and rapid analysis for rice authenticity using hand-held NIR spectrometry and chemometrics. Spectrochim. Acta A Mol. Biomol. Spectrosc. 217: 147–154.
Tian, S., K. Nakamura and H. Kayahara. 2004. Analysis of phenolic compounds in white rice, brown rice, and germinated brown rice. J. Agr. Food Chem. 52: 4808–4813.
Vemireddy, L.R., V.V. Satyavathi, E.A. Siddiq and J. Nagaraju. 2015. Review of methods for the detection and quantification of adulteration of rice: Basmati as a case study. Int. J. Food Sci. 52: 3187–3202.
Vlachos, A. and I.S. Arvanitoyannis. 2008. A review of rice authenticity/adulteration methods and results. Crit. Rev. Food Sci. Nutr. 48: 553–598.
Winefordner, J.D. and R.C. Elser. 1971. Atomic fluorescence spectrometry. Anal. Chem. 43(4): 24–42.
Yoshihashi, T., N.T.T. Huong and H. Inatomi. 2002. Precursors of 2-Acetyl-1-pyrroline, a potent flavor compound of an aromatic rice variety. J. Agr. Food Chem. 50: 2001–2004.
Yu, H., X. Ai, K. Xu, C. Zheng and X. Hou. 2016. UV-assisted Fenton digestion of rice for the determination of trace cadmium by hydride generation atomic fluorescence spectrometry. Analyst. 141: 1512–1518.
Zhao, Y., B. Zhang, G. Chen, A. Chen, S. Yang and Z. Ye. 2014. Recent developments in application of stable isotope analysis on agro-product authenticity and traceability. Food Chemistry 145: 300–305.
Zhou, Z., K. Robards, S. Helliwell and C. Blanchard. 2002. Composition and functional properties of rice. Int. J. Food Sci. 37: 849–868.

9

Authenticity of Wheat

Konstantinos V Kotsanopoulos,[1,*] *Neethu Asokan,*[2] *P Saranraj,*[2] *K Amala*[2] and *R Bright*[2]

1. Introduction

Cereals are amongst the most widely used ingredients in food and feed production (Hernández et al. 2005). Their composition is very important for assessing the quality and safety of a final product while the accurate identification of cereal species and cultivars is of paramount importance for ensuring the grains are correctly handled, processed and marketed as well as to protect the rights of plant breeders (Terzi et al. 2005). Wheat is commonly cultivated in temperate climates to be used for food and animal feed. It is highly successful as it easily adapts and can produce high yields. It is very important from a commercial point of view as it contains gluten that is responsible for the viscoelastic properties of dough which allows it to be processed into bread, pastas and other foodstuffs. Gluten however can also cause food allergies and coeliac disease. From a nutritional point of view, it contains essential amino acids, minerals, and certain vitamins, as well as phytochemicals and fibres, which are found in high concentrations especially in whole grains (Shewry 2009).

Food analysis for authenticity purposes is an important subject in the food and beverage industries (Bönick et al. 2017). Analysis for authenticity verification, whether conducted on ingredients, additives or foodstuffs can prevent incorrect labelling or product description, substitution of

[1] School of Agricultural Sciences, Department of Ichthyology and Aquatic Environment, University of Thessaly, Fytoko St, 38446 Nea Ionia Magnesias, Volos, Hellas, Greece.
[2] Department of Microbiology, Sacred Heart College (Autonomous), Tirupattur – 635 601, Tamil Nadu, India.
* Corresponding author: kostaskot@yahoo.gr

materials with cheaper alternatives, other forms of adulteration and loss of consumer trust and satisfaction (Primrose et al. 2010). Furthermore, food manufacturers and the industry as a whole should ensure that products marketed are authentic and develop systems that ensure the products are traceable. In Europe traceability is a legal requirement (EU regulation 178/2002), while most countries around the world have established similar regulatory systems (Bönick et al. 2017).

2. Wheat Varieties

According to Joshi et al. (2007a) a total of 154 wheat varieties (131 in bread wheat or *Triticum aestivum*, 22 in durum or *Triticum durum* and 1 in dicoccum or *Triticum dicoccum*) have been reported. Different crop varieties can be used to allow farmers and processors to produce suitable products in accordance with consumer demands. When the quality of products depends on the use of a specific wheat variety, it is of paramount importance that suitable methods are used that can accurately identify that variety. Variety identification is thus important for farmers and an important study subject for geneticists and biotechnologists. It is known that the morphological characteristics of wheat grains are inheritable (Harper et al. 1970) and they play a very significant part in species and variety identification. Visual analysis depending on the color, size, shape and texture of samples is simple, but requires experience and expertise (Shouche et al. 2001). Thousands of wheat species exist but the species used in the bakery and paste industries are *T. aestivum* (soft wheat) and *T. durum* (hard wheat) (Raba et al. 2007). A third species of wheat, *T. dicoccum* is cultivated in India. *T. aestivum* accounts for nearly 95% of wheat cultivated, whereas 4% is durum wheat or *T. durum* and only 1% is dicoccum wheat or *T. dicoccum* (Gupta 2004). Soft wheat is characterised by lower gluten levels, which makes it a better choice for bakery products. Hard wheat on the other hand has a less elastical gluten, which is not ideal for bakeries but is very important for pastes. These two species together with a number of by-products such as cus-cus and several assortments, are obtained from durum wheat. In France, Italy and Greece, the production of pastes involves the use of durum wheat flour only. The use of hard wheat flours allows the pastes to develop fine surfaces of white-yellow colour that can be modelled and maintain their shapes. Once dried, the pastes develop a glassy structure. Their volume increases significantly with boiling, while they maintain their shape (Raba et al. 2007). According to Gupta (2004), heat stress, diseases such as wheat rusts (yellow, brown, and black) and leaf blight are amongst the biggest challenges in wheat cultivation.

2.1 Wheat varieties' improvements

Fischer and Byerlee (1991) stated that South Asia's rice-wheat zones are heat stressed. Researchers have also reported that wheat varieties in India are often exposed to high temperatures (exceeding 35°C) (Tandon 1994). High temperature stress is considered to be a major factor in reducing wheat yield and deteriorating the quality of wheat (Stone and Nicolas 1995). Wardlaw (1994) previously reported that temperatures higher than 15–18°C can result in a reduction in grain weight at maturity.

Heat stress has been considered as a major factor in India's reduction of expected production volumes (Nagarajan 2005). There is therefore an urgent need to develop genotypes that are tolerant to heat stress or that mature early, before the rise in temperature. Resistant cultivars' farming was reported to be the most financially sustainable and ecologically safe means of developing stable wheat varieties. Also, the practice of farming rust-resistant varieties of wheat has protected Indian wheat cultivation from severe rust disease epidemics for several years (Singh et al. 2005). Wheat (*T. aestivum*) cultivars have shown durable or race-nonspecific resistance (Johnson 1988, Singh et al. 2005). Wheat can also be susceptible to a severe spot blotch disease that causes heavy yield loss in late sown wheat crops (Joshi et al. 2007b, Joshi et al. 2007c, Pandey et al. 2005).

3. Authenticity Analysis of Wheat

The importance of the development of analytical methods that are able to verify food authenticity is apparent and extensive research has been conducted towards this direction. Certain analytical strategies, for example, that have been developed to verify the authenticity of products such as wheat, spelt and rye have relied on *in silico* digestion to confirm the specificity of peptides and liquid chromatography-tandem mass spectrometry analysis (LC-TMS) for quantification (Bönick et al. 2017). The composition of foods that contain cereals is of paramount importance for the quality and safety of products. It is therefore necessary to develop highly specific and effective methods to ensure the absence of contaminants in foods, especially those intended or advertised as suitable for consumers with coeliac disease, where the different species need to be accurately determined. Furthermore, pasta made from durum wheat is considered to be of higher quality than that made of bread wheat or a mixture of the two and therefore methods need to be developed to qualitatively and quantitatively detect specific cereal species in foodstuffs (Terzi et al. 2003).

Some researchers have stated that digital image analysis could be an unbiased and accurate quantitative method for estimating morphological features of grains. Zayas et al. (1986), for example, used digital image analysis to distinguish and differentiate between wheat varieties. In recent developments advancements in food authenticity analysis using genetics and MS techniques have also been reported (Woolfe et al. 2013). The existence of adulterated food and mislabelled products in the markets across the world is a major economic and reputational risk for the food industry, manufacturers and traders. It is therefore crucial in the case of wheat to establish methodologies for checking the shape, size and colour of grains and the different varieties, while the maintenance of traceability records is essential to ensure compliance throughout the supply chain (Shouche et al. 2001). It is also important to note that EU legislation mandates that ingredients containing allergens or intolerances are labelled. Wheat (including common wheat, durum wheat, spelt wheat and khorasan wheat), rye, barley, oats or their varieties contain gluten which is an allergen but also part of the world's population have an intolerance to it (OJEU 2011, OJEU 2014).

Several studies have been published on classification and authenticity analysis of wheat using machine algorithms and image processing (Saini et al. 2014). Proteomics using proteins, peptides and their fragment patterns have also been used to determine wheat quality and authenticity (Michaletti et al. 2018, Zörb et al. 2009a, Zörb et al. 2009b, Zörb et al. 2006), while polymerase chain reaction (PCR) can be used to detect foreign DNA.

Other molecular markers have also been employed to allow authenticity analysis of specific rust resistance wheat genes (Datta et al. 2008), while crystallization with additives is a fingerprint method that has allowed the differentiation between organic and non-organic samples (Kahl et al. 2014). Table 1 synoptically presents a number of techniques used in studies investigating wheat authenticity and their effectiveness.

3.1 DNA-based methods

Allmann et al. (1993) described a rapid, highly sensitive and specific analysis based on PCR that can be used to determine wheat contamination. Initially, primers specific for highly conserved eukaryote DNA sequences were employed to isolate nucleic acid substrates for the PCR amplification. Following that, PCR amplification of a highly repetitive and specific genomic wheat DNA segment was carried out to detect wheat. A total of 35 different samples were used to test the assay, that ranged from bakery additives to heated and processed foodstuffs. Furthermore, a comparison was made between the PCR method and an immunochemical assay that can be used for the detection of the wheat protein gliadin. When the two assays were combined, wheat contamination was characterised in high detail. PCR was also used in the study of Bryan et al. (1998). This study aimed at detecting adulteration of pasta described as durum wheat-derived by hexaploid bread wheat. The use of PCR allowed the detection of a sequence amplified in the D-genome of wheat, even when present

Table 1. Techniques used for the authentication of wheat

Technique	Scope/Details	Markers	Effectiveness	References
Metabolomics and proteomics	Drought stress effect on proteome and metabolome of Bahar and Kavir	Total protein content, amino acids, organic acids, sugar	Bahar-significant reduction in protein content; Kavir- minor effect	Michaletti et al. 2018
Proteomics	Authentication of organic *T. aestivum* L. vs. conventional	Total protein content, storage proteins, enzymes of carbohydrate metabolism, peroxidase, and other proteins	Total of 16 proteins identified as suitable for differentiation	Zörb et al. 2009a
Metabolomics	Authentication of organic wheat vs conventional	Metabolite profiles and lipids, cations, starch and protein concentrations	Differences found between organic and conventional, but tend to disappear as the grains mature	Zörb et al. 2009b
Metabolomics/ GC–MS	Authentication of organic *T. aestivum* L. vs. conventional	Amino acids, organic acids, sugars, sugar alcohols, sugar phosphates, and nucleotides	No significant differences in metabolites examined between organic and conventional	Zörb et al. 2006
Lipidomics	Differentiation between wholemeal of spelt and winter wheat	Linoleic, palmitic, oleic acids a-linolenic acids expressed as oleate/palmitate ratio	Higher levels of oleic acid in spelt (16–17.5%) in comparison to winter wheat (10.4–10.9%) while winter wheat was richer in palmitic, linoleic, and a-linolenic acids. The oleate/palmitate ratio clearly differentiated the species	Ruibal-Mendieta et al. 2004
	Differentiation between hole ground spelt (*T. aestivum* ssp. *spelta* L.) and hexaploid winter wheat (*T. aestivum*)	Free, bound and total lipids	1·3 times higher free lipids and 1·2 times higher total lipids in spelt in comparison to wheat	Ruibal-Mendieta et al. 2002

Table 1 contd. ...

...Table 1 contd.

Technique	Scope/Details	Markers	Effectiveness	References
DNA fingerprinting, TBP light assay	Authentication of wheat species	Tubulin-based polymorphism: 5′-6-FAM-AACTGGGCBAARGGNCAYTA-YAC-3′ and 5′-ACCATRCAYTCRTCDGCRTTYTC-3′; TBP: 2-5Fw:5′-CTGAGAACTGTGACTGYCTKCAAGGTA-3′ 2–5Rev: 5′-GTGGCATACTTGGAAWCCTGRWAACRT-3′ 4Fw: 5′-GCCTCCAGGGTAATTAATTATACCATC-3′ 4Rev: 5′-CAAACTTGGAATCCTGTAAAGGAAGC-3′ 3Fw: 5′-TGCAAGGTATACCCTCCTCC-3′ 3Rev: 5′-GCARACTTGRAATCCTGTGRTA-3′	Sensitivity of 0.5–1% w/w in binary mixtures of durum wheat in einkorn and emmer flour	Silletti et al. 2019
DNA microsatellites, RT-PCR	Detection of *T. aestivum* L. in semolina and durum wheat bread (*T. turgidum* L. var. durum).	Microsatellites: 10 primer pairs; RT-PCR: forward and reverse primers of microsatellite GDM111, 50 ng DNA template	Threshold: 3% common wheat in semolina and 5% common wheat in breads; 2.5% using RT-PCR	Pasqualone et al. 2007
Microsatellites, PCR	Verification of authenticity of Saratovskaya 29/Yanetzkis Probat inter-varietal wheat chromosome substitution lines	Ninety-five markers revealing polymorphism of the amplified loci between vars. S29 and YP	55% of markers were polymorphic between S29 and YP, available for all chromosome arms except 4DS, 6DS and 7DS, 17 out of 21 lines showed the microsatellite pattern of the donor variety	Pestsova et al. 2000
RT-PCR	Quantitative detection of wheat, distinguishment of bread from durum wheat	acetyl-CoA carboxylase sequences: forward - TGCCCATTGTCGGCCTTA reverse - GCATTCCAACCATCTGCCC probe - TGCCTCGACAACACCATCGCTATCC RALyase: forward - GGAGCGGTTGCTGTTGGA reverse - GCCTCGGTGCATTAATAATCTTG	Limit of detection: 1 genome copy; limit of quantification: 10 genome copies	Hernández et al. 2005
	Detection of adulteration of durum wheat semolina with soft wheat (quantification of soft wheat in semolina, pasta and breads)	Forward and reverse primers and a dual-labelled probe	Reliable quantifications obtained	Sonnante et al. 2009

RT-PCR and Padlock probes	detection of wheat *T. aestivum* (D-genome)	RT-PCR: Forward primer: 5′ AAGGAGCTCGCCAACGG 3′; reverse primer: 5′ AACCGAGGGTCCAGAAGAGAC 3′ and probe: 5′ FAM- GCTAGATGGCTGGCTTCTATTTCATGATCCA-DQ 3′ Padlock probe: 5′ target - ATGGCTGGCTTCTATTTCATGATC, cZIP sequence – ATTGACCAAACTGCGGTGCG, 3′ target - TTTTCAAATATCCAGCTAG	Detection limit: 2.5% bread wheat in Farro	Prins et al. 2010
PCR	Wheat varieties Arina, CWRS, Eiger, Forno, and Zenith	wheat PCR primers: TR01 5′-GCGGCGTGTGCCACGTACGTGGTTT and TR02 5′-GAACGGGCGTTACGTGGACACGGGA	Rapid, sensitive, specific	Allmann et al. 1993
	Durum what adulteration	DNA sequence amplified in the D-genome of wheat. Optimal primers: a)GTCCTATATCTTGAGGCCGCAAGN b)CTTCTGACGGGTCAGGGGCACO c)AACCCACTGTACCTGAGTATATATCP d)CTGAATGCCCCTGCGGCTTAAG	Rapid, cost effective	Bryan et al. 1998
ICP-OES, ICP-MS and chemometrics	Differentiation between organic and conventional winter wheat and geographical origin	C, N, K, Mg, P, S, Ca, Fe, Mn, B, Zn, Cu, Mo, Ba, Sr, Na	If used with chemometrics can allow differentiation	Laursen et al. 2011
HR-ICP-MS and X-ray fluorescence	Verification of geographical origin	Be, Na, Mg, Ca, Sc, Ti, V, Cr, Mn, Fe, Co, Cu, Zn, Ga, Se, Rb, Sr, Y, Zr, Cd, Cs, and Pb	Significant correlations found between wheat grain and soil content in Cr, Mn, Ga, Rb, Sr, Zr, and Cd	Zhao et al. 2013
Q-ICP-MS	Tracing back to grinding machine and contamination from machine use	Al, Cd, Co, Cr, Cu, Fe, Mn, Mo, Ni, Pb	Tracing was possible in most cases, no contamination free devices were identified	Cubadda et al. 2001

Table 1 contd. ...

...Table 1 contd.

Technique	Scope/Details	Markers	Effectiveness	References
GC-MS	Differentiation between organic and conventional wheat	Metabolites: amino acids, organic acids, sugars, sugar alcohols, sugar phosphates and nucleotides	No significant differences in metabolite composition and quality	Zörb et al. 2006
UHPLC-HRAM-MS & chemometrics	Differentiation between breads made from whole and refined wheat flour	Alk(en)ylresorcinols, diglycerides, and phosphatidylethanolamine	Effective differentiation but bread crumb and crust did not differ in chemical profiles	Geng et al. 2016
(UPLC-ESI)-MS/MS	Detection of durum wheat adulteration with common wheat	tryptic peptide of puroindoline and a purothionin A-1 peptide	Limit of detection 0.01%, limit of quantification 0.03%	Russo et al. 2014
NIR and chemometrics	Identification of hard red winter and hard red spring wheat	Protein content, NIR hardness, protein & hardness, multiple linear regression, PCA/Mehalanobis distance classifier, PLS, ANN	95–98% accuracy with ANN model	Delwiche et al .1995
	Discrimination of wheat varieties (Sarina, Bolero, Berdún, Soisson, Chamorro, Artur Nick, Berdun, Marius, Anza, Kalifa, Galibier and Quality)	Kernel colour, kernel hardness, moisture content, protein content, wet gluten, deformation energy	99.5% correct identification	Miralbés et al. 2008

at very low levels, while being rapid and cost-effective. It offered the additional advantage that in contrast to the analysis of proteins, it can be used even when highly processed products are examined. In another study, DNA microsatellites were evaluated in an effort to establish an effective method for detecting soft wheat (*T. aestivum* L.) in semolina and durum wheat bread (made of *T. turgidum* L. var. durum). A D-genome-specific repetitive DNA sequence was selected for detecting common wheat in semolina (threshold 3%) and breads (threshold 5%) using qualitative PCR or real-time PCR (threshold 2.5%). The development of such a method is very important in order to check the authenticity of the product during production. It was also shown that real-time PCR can be used to quantify common wheat adulteration in semolina (Pasqualone et al. 2007). Sonnante et al. (2009) also noted that as in Italian pasta and durum wheat, only durum wheat semolina can be used, it is important to develop a method that can detect different types of wheat in these products. These authors described a quantitative real-time PCR method that used a dual-labelled probe. The construction of standard curves based on 100% dilutions of soft wheat flour, pasta or bread was completed. For the purposes of the experiment, increased levels of soft wheat were used for the preparation of durum wheat semolina, pasta and bread to test how accurate the method was. It was revealed that quantification could be reliably obtained in particular in the case of samples that contained lower amounts of soft wheat DNA.

Taqman qPCR assays were tested by Hernández et al. (2005) for their potential in identifying wheat. These researchers reported that wheat (*T. aestivum*) is commonly found as an ingredient in a wide range of processed foodstuffs and thus specific assays are required to detect its presence and quantify it. Additionally, the increasing use of genetically modified varieties enhances the need for accurate detection in order to safeguard consumers' rights but also for environmental risk assessment purposes. Four independent real-time PCR assays were therefore developed and proved to be capable of identifying and quantifying barley, rice, sunflower and wheat, targeting γ-hordein, gos9, helianthin, and acetyl-CoA carboxylase sequences, respectively. It was proved that the assays were effective in detecting and quantifying DNA from the target species. Furthermore, the distinguishment of durum wheat from bread was achieved by simultaneously amplifying RALyase. The limits of detection were 1 genome copy for barley, sunflower, and wheat and 3.3 copies in the case of rice real-time PCR systems, while the quantification limits were 10 genome copies for barley, sunflower, or wheat and about 100 haploid genomes for rice real-time PCR systems. However, it was noted that the wheat assay detects rye simultaneously and with similar sensitivity in an agarose gel.

In another study, Moon et al. (2016) reported the use of multiplex end-point PCR that relies on the chloroplast gene *rpo*, for identifying wheat. The method was developed using species-specific primers to simultaneously detect Poaceae species such as adlay, barley, maize, rice, and wheat, looking for sequence polymorphism in DNA-directed RNA polymerase beta-prime chain genes (rpoC2). Species-specific primer pairs were developed using a common forward primer and reverse primers that were characterised by different sequence polymorphisms and were at different gene locations. The wheat primer pair was constructed to allow the amplification of products of five chloroplast genes of 221 bp. Multiplex PCR (DNA template 0.01–10 ng/μL) was employed for the optimization of the amplification. Moreover, the use of species-specific primers allowed the detection of components of seven commercial flour-mixed products.

The use of DNA fingerprinting based on tubulin-based polymorphism and the development of a new assay (tubulin-based polymorphism *light*), were also reported by researchers to authenticate wheat species. A number of commercial foodstuffs were tested using this assay to prove its effectiveness. The assay was sensitive enough to detect amounts as low as 0.5–1% w/w in binary mixtures of durum wheat in einkorn or emmer flour and was capable of authenticating the composition of food samples, effectively detecting adulteration.

A padlock probe targeting the D genome in conjunction with microarray hybridization was used for the detection of wheat DNA in bread wheat admixtures. In this study a method was developed to examine the purity of Farro della Garfagnana DNA in complex cereal mixtures. A combination of padlock probe ligation and multiplex microarray detection effectively identified admixtures. One of the most common adulterants of Farro della Garfagnana is common bread wheat (*T. aestivum*), containing the BBA^uA^uDD genome. As Farro harbours the BBA^uA^u genome, when the D-genome is absent, it can be safely assumed that no bread wheat is present. The limit of detection of this method is at least 2.5% bread wheat in Farro (Prins et al. 2010). Furthermore, Voorhuijzen et al. (2012) reported the development of a DNA-based multiplex detection method using a padlock probe ligation and microarray detection tool. The specificity level of 14 plant-related padlock probes was assessed and validated in mixtures that contained seven or nine plant species or varieties. One nucleotide difference in target sequence sufficed to determine whether a specific target was present. At least 5% FdG was detected in mixtures containing less expensive bread wheat, but it was suggested that even lower levels of adulterant could be detected.

In the study of Pestsova et al. (2000) a total of ninety-five wheat microsatellite markers were employed for the verification of the authenticity of the set of Saratovskaya 29/Yanetzkis Probat inter-varietal wheat chromosome substitution lines developed using Saratovskaya 29 as the recipient variety. Polymorphic markers were present for all chromosome arms apart from 4DS, 6DS and 7DS. Testing of each chromosome substitution line was conducted using 2–8 microsatellite markers. It was proved that the majority of lines were correct. Out of 21 lines, 17 presented the expected microsatellite pattern of the donor variety. No substitution with Yanetzkis Probat took place in the case of two entire chromosomes, 1B and 7A, and two chromosome arms, 3AL and 6DS. The researchers concluded that microsatellite markers can be effectively used to test the authenticity of genetic stocks of wheat.

Finally, according to Berke et al. (1992), one of the most important limitations to the genetic improvement of wheat (*T. aestivum* L.) was that there was limited information about quantitative trait loci. In their study, these researchers determined the chromosomal locations of Quantitative trait loci that control grain yield and yield components, grain test weight, the height of the plant and anthesis date. Reciprocal sets of chromosome substitution lines in duplicate among two wheat cultivars, namely Cheyenne (CNN) and Wichita (WI), were employed for the identification of the chromosomes that bear quantitative trait loci. Following a number of field trials, it was proved that Wichita had major quantitative trait loci on chromosomes 3A and 6A increasing the grain yield in Cheyenne and major quantitative trait loci on chromosome 36 increasing the grain yield in Cheyenne, while in the case of Cheyenne, major quantitative trait loci were found on chromosomes 3A and 6A decreasing the grain yield in Wichita. The increase in grain yield by WI 3A and WI 6A was a result of the increase in seed weight, while the decrease in grain yield by CNN 3A and CNN 6A was a result of a decrease in culms per square meter. The decrease in grain yield by WI 3B was the result of a decrease in winter hardiness, which led to a decrease in seed weight and culms per square meter. One Wichita and seven Cheyenne chromosomes bearing quantitative trait loci were identified and found to affect the seeds per culm. Also, seven Wichita and five Cheyenne chromosomes bearing quantitative trait loci that affect seed weight, one Wichita and three Cheyenne chromosomes bearing quantitative trait loci that affect culms per square meter, and finally at least one chromosome bearing Quantitative trait loci that affect each of the other agronomic traits were found.

3.2 Proteomics

Proteomics is the study of proteins and the way they function. The generation of an enormous amount of large-scale data sets for protein–protein interactions, organelle composition, protein activity patterns and protein profiles that has been made in the past few years, especially in medicine, has enabled significant technological improvements, the initiation of international proteomics projects and access to results required in order for scientists to take full advantage of proteomics. Proteomics is the study of the proteome and includes the study of not only the proteins in a cell, but also the set of protein isoforms and modifications, the way they interact, their structural characteristics and their higher-order complexes (Tyers and Mann 2003). Some of the most important studies involving the use of proteomics to authenticate wheat are discussed below.

Michaletti et al. (2018) aimed at examining the biochemical networks of wheat leaves under water deficiency conditions. Proteomics and metabolomics were used to examine two spring-wheat cultivars (Bahar, drought-susceptible; Kavir, drought-tolerant). Drought stress-induced effects on Bahar leaf proteome severely decreased the total protein content, particularly affecting photosynthetic proteins, enzymes related to the metabolism of sugar and nitrogen and their potential of harmful molecules' detoxification. On the other hand, very limited impact was seen at the protein level in Kavir stressed leaves. Metabolome analyses suggested that the quantities of amino acids, organic acids, and sugars are most heavily affected by water deficiency. In Bahar, the levels of proline, methionine, arginine, lysine, aromatic and branched chain amino acids increased. Tryptophan's levels increased via the shikimate pathway and seemed to sustain auxin production (indoleacrylic acid), while glutamate's levels reduced probably as a result of polyamine (spermine) synthesis. An effect on Kavir metabolome was observed as a result of water deficiency but the effect was relatively limited with only purine metabolism and one more pathway changing to a great extent. These results provided a better insight into the mechanisms that control plant cell response to drought stress, providing also further information on molecules that can be useful in crop improvement trials.

Zörb et al. (2009a) followed a protein profiling approach to differentiate between organic and conventional wheat. The levels of 1049 proteins present in wheat grains (*T. aestivum* L., cv. Titlis) of two growing seasons that involved both organic and conventional plots were determined. Differences between organic and conventional wheat were observed in the levels of 25 proteins. Storage proteins, enzymes of carbohydrates' metabolism, a peroxidase and other proteins were examined in both the organic and conventional crops. The total protein content was higher in conventional wheat. Although important from an authenticity perspective, these differences would not significantly affect the nutritional value of the product. The use of a number of filters and calculations, among which one considering also seasonal changes allowed for 16 of the 25 proteins with different levels in organic and conventional wheat to be retained. These proteins could be used as signatures for the distinctiont between the two cultivation methods.

Finally, Zörb et al. (2009b) analysed organic and conventionally produced wheat. Their study aimed at detecting whether organic or conventional farming systems could have an effect on the levels of metabolites and substances in developing ears as well as in matured grain. Thus, broadband metabolite profiles in combination with lipids, cations, starch and protein levels of wheat ears in the last phase of grain development and of matured grain from both types of agriculture were examined. It was shown that seed metabolism and supply of developing ears are different in organic and non-organic samples. It was noted though that any differences in 62 metabolite levels become insignificant when the grains mature. This indicates that there is a high degree of homeostasis in the final seed regardless of the method of cultivation.

3.3 Lipidomics

Spelt has been found to be prone to adulteration with wheat. This is believed to be due to the fact that food grade spelt is significantly more expensive than wheat, as the yields produced are generally lower and also due to the fact that the crop requires dehulling with its consequential losses (Ruibal-Mendieta et al. 2004). In the study of Ruibal-Mendieta (2002) a determination of free lipids, bound lipids and total lipids was made in whole ground spelt (*T. aestivum* ssp. *spelta* L.) and winter wheat (*T. aestivum*) samples using Soxhlet. A total of ninety-one samples of spelt and 78 samples of winter wheat were subjected to analysis. It was proved that spelt has different levels of lipids compared to winter wheat, with a higher level of free lipids and total lipids (1:3 times and 1:2 times as an average, respectively). Furthermore, it was shown that free lipids comprised a bigger part of total lipids in spelt compared to winter wheat. The lipid content could therefore be used to differentiate between spelt and winter wheat. Ruibal-Mendieta et al. (2004) also used the oleate/palmitate ratio to clearly differentiate between wholemeal spelt (*T. spelta* L.) and winter wheat (*T. aestivum* L.). To do that, these researchers studied the fatty acid profiles of a number of spelt and winter wheat samples of plants that grew in Belgium in two different harvest periods. It was shown that spelt contained much higher levels of free lipids, total lipids and bound lipids than winter wheat and a clear differentiation was possible when these differences were expressed as oleate/palmitate ratio.

3.4 Spectroscopic techniques

Elemental analysis has also been used for wheat authentication. Differences in the composition of elements can offer valuable information for the identification of wheat types.

The presence and concentration of elements in organic and conventional winter wheat, spring barley, faba bean, and potato was analysed in the study of Laursen et al. (2011). Inductively coupled plasma-optical emission spectrometry (ICP-OES) and ICP-MS were employed for these analyses. The cultivation of the crops was carried out in a period of two years at three geographically different locations, each used to accommodate one conventional and two organic systems. It was demonstrated that the conventional system produced the highest quantities of crops with the only exception being the nitrogen-fixing faba bean, while the dry matter level was found to be very similar in all systems. The organic and conventional crops did not significantly differ in their concentrations in essential plant nutrients when the results were statistically analysed. Nevertheless, the use of chemometrics for the analysis of the multielemental compositions, involving a total of 14 elements allowed the samples to be discriminated. The discrimination power was further increased by analysing up to 25 elements derived from semiquantitative ICP-MS. It was proved that multielemental fingerprinting in combination with semiquantitative ICP-MS and chemometrics can effectively authenticate organic crops.

Elemental analysis can also be used though to determine the contamination of wheat samples due to abrasion of grinding tools, while materials are homogenised. A total of 15 trace elements from samples processed using six different grinding and milling instruments were therefore analysed in the study of Cubadda et al. (2001). The wheat grains tested belonged to one soft and one durum cultivar. The samples underwent different preparative treatments and the analysis was conducted using quadrupole inductively coupled plasma mass spectrometry (Q-ICP-MS). In parallel, an appropriate digestion method was designed and the ArC^+ interference that could affect Cr determination was assessed and corrected. A total of 10 elements (Al, Cd, Co, Cr, Cu, Fe, Mn, Mo, Ni, Pb) presented statistical differences in comparison to the control sample and in the majority of cases it was possible to trace the samples back to the grinding equipment. All devices contaminated the samples to some degree with the elements mentioned above. Abrasion of the grinding tools was higher in the case of durum wheat as expected due to its hardness.

Multi-elemental analysis can also be used to determine the geographical origin of wheat and verify origin claims. In the study of Zhao et al. (2013), high resolution ICP-MS and X-ray fluorescence were used to determine the concentrations of Be, Na, Mg, Ca, Sc, Ti, V, Cr, Mn, Fe, Co, Cu, Zn, Ga, Se, Rb, Sr, Y, Zr, Cd, Cs, and Pb in wheat grains and samples of soil from their cultivation areas in Hebei and Henan provinces in China. Chemometrics were then used to analyse how the levels of these elements in the grain samples were related to the levels in the soil. Discrimination of the origin of the crop was achieved using principal component analysis

(PCA) and linear discriminant analysis (LDA). It was found that the levels of the elements Cr, Mn, Ga, Rb, Sr, Zr, and Cd in wheat grain strongly correlated with the provenance of soil. The models developed using these elements obtained 95% total correct classification, thus indicating that the method is effective for determining the geographical origin of wheat.

The results found in the study of Zörb et al. (2009b) are in agreement with the outcome of the study of Zörb et al. (2006) where metabolites were tested using gas chromatography (GC)-MS. It was mentioned in this study that in some EU countries, organic systems can comprise up to 8% of the total agricultural area. An effort was therefore made to obtain a metabolite profile for organic and conventional wheat (*T. aestivum* L.). Samples from a 2003 Swiss harvest were obtained and analysed. It was demonstrated that GC-MS can be used for the determination of relative levels of a total of 52 metabolites such as amino acids, organic acids, sugars, sugar alcohols, sugar phosphates, and nucleotides. The data were also statistically analysed, and it was proved that the levels of 44 metabolites in samples from organic and mineral farming were not significantly different. This was therefore an indication that the farming system did not or only minimally affected the metabolite composition and qualitative characteristics of wheat grains.

A few studies have also employed near-infrared spectroscopy (NIR) to differentiate bread wheat varieties (Delwiche et al. 1995, Miralbés 2008). In the study of Delwiche et al. (1995) NIR (1,100–2,498 nm) was employed for the identification of hard red winter and hard red spring wheat. For this experiment the researchers used samples from four years of U.S. winter and spring wheats. One hundred and fifty samples per wheat class from the first three years' samples were used for calibration purposes, while the rest (1,325 samples) and all fourth year's samples (778) were used for the verification of the models. A total of four types of classification algorithms were assessed, namely multiple linear regression (MLR), PCA with Mahalanobis distance (MD), partial least squares (PLS) analysis, and artificial neural networks (ANN). In all cases, the models correctly classified 88% or more of samples and usually about 95%. This level of accuracy is considerably higher than the accuracy degrees of discriminant models that solely rely on protein level, NIR-hardness, or protein and hardness combined. It is important to note though that content alone was not essential to obtain successful models. The performance of the ANN model was the highest, being 95–98% accurate even when the number of input nodes was reduced from 223 to 111. Also, in the study of Miralbés (2008), NIR spectroscopy in combination with chemometrics was employed for the discrimination of wheat varieties. A large number of samples representing different wheat varieties and two years of harvest were collected and used for the development of a discriminant equation, using a number of scatters and mathematical treatments within the range of 400–2500 nm. Spanish (Sarina, Bolero, Berdu´n, Sois-son, Chamorro, Artur Nick, Berdun, Marius, Anza, Kalifa), and French (Galibier and Quality) wheat varieties were represented. It was shown that 99.5% and 94% of samples were assigned to the correct varieties for the calibration and the validation sample set, respectively. It was therefore concluded that NIR in combination with chemometrics can rapidly discriminate wheat varieties.

NIR is prevalent in the cereal industry (Cen and He 2007, Porep et al. 2015) and is known to be a good alternative to wet-chemical analyses (Jespersen and Munck 2009). In the case of wheat, is has also been used for detecting samples contaminated with mycotoxins (Gaspardo et al. 2012, Girolamo et al. 2009), to quantify pigments in cereals (Geleta et al. 2014, Atienza et al. 2005), and for calculating factors associated with baking quality (Jirsa et al. 2008, Miralbés 2004, Dowell et al. 2006). It is a relatively inexpensive and rapid technique which is offered for automation and can be effective in authentication testing at all stages of wheat breeding and processing.

3.5 Chromatographic

Chromatography has been employed for decades to enable the detection of wheat components. Miller et al. (1960) for example compared three hard red winter wheat varieties on the basis of components extracted using 80% ethanol. A total of eleven amino-acids, five organic acids and two inorganic acids were detected. The separation of seven sugars was also reported. The most distinct difference identified was in the case of the variety Tenmarq which contained higher levels of allulose than the Ponca variety, while this component was not detected in the rest of the varieties. Sorbitol was also identified in Tenmarq, while it was not present in any other varieties tested. More recently, GC-TMS was used to identify and quantify certain allelochemicals in wheat (*T. aestivum* L.). Levels ranging from 2.9 to 110 mg per kg of dry shoot residues were detected. In comparison to GC–MS, the GC–MS–MS technique was significantly more selective and sensitive, thus allowing a more reliable quantitation to be provided as compounds were determined. The TMS technique allowed unwanted chemical backgrounds to be filtered out. Analysis of phenolic acid and 2,4-dihydroxy-7-methoxy-1,4-benzoxazin-3-one was carried out and a chromatogram of high resolution was given (Wu et al. 1999).

A fuzzy chromatography (FC) MS fingerprinting method in combination with chemometric analysis was described in the study of Geng et al. (2016). The technique was used for the differentiation of breads made using whole wheat flour and refined wheat flour. The profiles of the chemical composition of the samples were given using ultra high-performance LC with high-resolution accurate-mass multistage MS

(UHPLC-HRAM-MS). PCA and soft independent modelling of class analogy (SIMCA) of the FCMS fingerprints indicated that the alk(en) ylresorcinols were the most important marker that could be used to differentiate between whole wheat and refined wheat flour/breads, but diglycerides, and phosphatidylethanolamine could also be used. SIMCA, using whole wheat modelling was found to be very reliable and effective in authenticating breads from whole wheat.

It is also important to mention the work of Russo et al. (2014) in which an ultra-performance LC electrospray ionization (UPLC-ESI)-MS/MS method was employed to detect and quantify durum wheat adulteration. To conduct the targeted analysis, specific transitions at m/z 543.7 > 657.4 and m/z 543.7 > 299.2 of a species-specific marker obtained from a tryptic peptide of puroindoline a (Pin-a) were monitored. This peptide is only found in common wheat. Furthermore, two transitions at m/z 500.4 > 725.4 and m/z 500.4 > 561.9 of a purothionin A-1 peptide found in both species, were also monitored. The limits of detection and quantification for the Pin-a peptide were 0.01% and 0.03%, respectively, while this marker was also very effective when used to quantify common wheat kernels and pasta, which means that it can be used for both raw materials and processed foods.

These research articles therefore indicate that chromatographic methods are nowadays used in combination with spectrometry and chemometrics to authenticate wheat species but also different types of wheat. Chromatographic methods are sensitive and therefore very accurate and can be used with a high degree of confidence. The need for expensive instrumentation though as well as the need for expertise in order to analyse the data and interpret the results, especially when chemometrics are employed, are amongst their biggest disadvantages.

3.6 Crystallization and image processing

When crystallization is the method of choice, sample preparation, crystallization, data evaluation and multivariate statistical analysis are carried out to extract data. Crystallization using additives can be very useful as a complementary fingerprint method to differentiate between organic and conventional wheat based on pattern recognition (Kahl et al. 2014). Szulc et al. (2010) reached this conclusion in their study in which organic and conventional winter wheat samples were analysed using the copper chloride crystallisation method and computerised image analyses. The data were processed using pattern recognition and were classified with the help of multivariate statistics. Suitable discriminant analysis (DA) models were designed, allowing correct predictions to be made of up to 100% of samples tested. The method is inexpensive and could be used for discriminating organic and conventional samples using crystal growth as a marker, including the morphological characteristics of dendritic patterns. The computerised analysis of crystallograms has only recently been examined in a limited number of studies and is therefore worth examining its potential further. It was clearly demonstrated though that crystallograms obtained from conventional samples were characterised by higher complexity than the organic ones that were generally more homogeneous and uniform.

Image processing has been used to verify wheat authenticity for decades. Two of the most important studies detailing the use of this technique were published in the late 80's and are briefly discussed below. Neuman et al. (1987) employed computed whole-grain contours to enable the automatic determination of plan-form spatial shape features and Fourier descriptors of kernel perimeters. Four durum wheat varieties that were tested as part of an admixture with ten common wheat varieties, were accurately classified in accordance with their types. Similarly, samples of five Canada Western Red Spring wheat varieties were also accurately classified according to their class. While in the case of the Canada Western Red Spring, no misclassification was observed, the Canada Western Red Winter, Canada Western Soft White Spring, Canada Utility and Canada Prairie Spring varieties were not classified correctly to the same degree, with only 15 to 96% of samples being classified correctly.

According to the study of Neuman et al. (1989), the development of a colour digital image processing system was reported that aimed at evaluating the colour of cereal grains. Video digitization of red, green and blue signals was used to acquire digital colour images and extract colour features. The method revealed that there were considerable differences between varieties belonging to different classes (e.g., amber durum and hard red) and even certain varieties of the same class.

Similarly, Zayas et al. (1989) used image analysis to enable the discrimination of wheat and non-wheat material and between weed seeds and stones in the nonwheat part of the samples. Multivariate DA was employed to differentiate between wheat and nonwheat and among weed seeds, while a structural prototype was developed to differentiate between wheat and nonwheat. The structural prototype method was very effective in discriminating between wheat and non-wheat, while multivariate DA allowed the differentiation of weed seeds, but stones could not be identified as successfully. As a result, the stones would need to be removed before the analysis to allow the samples to be correctly discriminated.

More recently, Arefi et al. (2011) used integrating machine vision and ANN (Matlab software) to identify 4 wheat varieties based on the morphological characteristics and the colour of the grains. A chamber of imaging was designed, and a program was coded in Matlab to capture images and segment the samples. Average, variance, skewness and kurtosis values of images in RGB and l*a*b* colour spaces were determined and a total of 11 features of the 280 images were taken into account for the training stage of ANN. Also, 40 images were used to validate the results and 80 images to test ANN. In total, 95.86% of the samples were correctly classified, indicating that the method is highly accurate.

Finally, Luo et al. (1999) used a colour machine vision system to identify healthy and damaged kernels. The damaged kernels were further classified into broken, grass-green/green-frosted, black-point/smudged, mil-dewed, heated and bin-/fire-burnt. The development of a software was reported that was used for the extraction of a number of morphological and colour features from the images of the kernels. A number of models were assessed using the SAS procedures, STEPDISC and DISCRIM. It was shown that colour features were effective in identifying healthy and damaged kernels, while using a combination of morphological and colour features, the samples were identified even more accurately. In total, 93% of healthy, 90% of broken, 99% of grass-green/green-frosted, 99% of black-point/smudged, 99% of mildewed, 98% of heated and 100% of bin-/fire-burnt samples were correctly identified.

4. Conclusion

Wheat is among the most important cereals being used worldwide for the production of food. A number of different varieties of wheat are available that can be used, depending on their properties, for the production of products with very distinct physical and end-use characteristics such as bread and pasta. The variety in the viscoelastic characteristics of the gluten that different wheat types can offer but also the different methods of production that are available (organic, conventional, wholemeal, etc.), render the product prone to adulteration. This is particularly important in the case of wheat not only due to the economic factors and loss in consumer trust that are related to this type of fraud, but also due to the fact that the product can cause allergies and intolerances and it is therefore essential that correct information is provided within the supply chains. Various methods have been used to detect adulteration of wheat and wheat-derived products, including DNA-based methods (e.g., real-time PCR) which are considered highly accurate, but also proteomics, lipidomics, spectroscopic, chromatographic methods as well as image processing techniques. Spectroscopic methods have also been used for the analysis of elements present in the crop and the soil, which in combination with chemometrics can allow the verification of claims in relation to the geographical origin of the product. Further advancements in these techniques and a reduction in the relevant costs could allow their more systematic use by the industry and regulatory authorities. A standardization of the techniques used for these purposes could also be the right step towards this direction.

Abbreviations

Artificial neural networks	ANN
Discriminant analyses	DA
Fuzzy chromatography	FC
Gas Chromatography	GC
High-performance liquid chromatography with high-resolution accurate-mass multistage mass spectrometry	UHPLC-HRAM-MS
Inductively coupled plasma-optical emission spectrometry	ICP-OES
Linear discriminant analysis	LDA
Liquid Chromatography	LC
Liquid chromatography-tandem mass spectrometry	LC-TMS
Mahalanobis distance	MD
Mass Spectrometry	MS
Multiple linear regression	MLR
Near-infrared spectroscopy	NIR
Partial least squares	PLS
Polymerase Chain Reaction	PCR
Principal component analysis	PCA
Quadrupole inductively coupled plasma mass spectrometry	Q-ICP-MS
Soft independent modelling of class analogy	SIMCA
Ultra-performance liquid chromatography electrospray ionization	UPLC-ESI

References

Allmann, M., U. Candrian, C. Hofelein and J. Luthy. 1993. Polymerase chain reaction (PCR): a possible alternative to immunochemical methods assuring safety and quality of food. Detection of wheat contamination in non-wheat food products. Zeitschrift für Lebensmittel-Untersuchung und - Forschung 196: 248–251.

Arefi, A., A.M. Motlagh and R.F. Teimourlou. 2011. Wheat class identification using computer vision system and artificial neural networks; Institute of Agrophysics Polish Academy of Sciences Lublin, Poland 25: 319–323.

Atienza, S.G., C.M. Avila, M.C. Ramírez and A. Martín. 2005. Application of near infrared reflectance spectroscopy to the determination of carotenoid content in tritordeum for breeding purposes. Australian Journal of Agricultural Research 56(1): 85–89.

Berke, T.G., S. Baenziger and R. Morris. 1992. Chromosomal location of wheat quantitative trait loci affecting agronomic performance of seven traits, using reciprocal chromosome substitutions. Crop Science 32: 621–627.

Bönick, J., G. Huschek and H.M. Rawel. 2017. Determination of wheat, rye and spelt authenticity in bread by targeted peptide biomarkers. Journal of Food Composition and Analysis 58: 82–91.

Bryan, G.J., A. Dixon, M.D. Gale and G. Wiseman. 1998. A PCR-based method for the detection of hexaploid bread wheat adulteration of durum wheat and pasta. Journal of Cereal Science 28: 135–145.

Cen, H. and Y. He. 2007. Theory and application of near infrared reflectance spectroscopy in determination of food quality. Trends in Food Science and Technology 18(2): 72– 83.

Cubadda, F., M. Baldini, M. Carcea, L.A. Pasqi, A. Raggi and P. Stacchini. 2001. Influence of laboratory homogenization procedures on trace element content of food samples: an ICP = MS study on soft and durum wheat. Food Additives and Contaminants 18(9): 778–787.

Datta, D., S.C. Bhardwaj and M. Prashar. 2008. Development of bread wheat (*Triticum aestivum* L.) lines with specific rust resistance genes and their authentication through molecular markers. Indian Journal of Genetics 68(2): 113–123.

Delwiche, S., Y.R. Chen and W. Hruschka. 1995. Differentiation of hard red wheat by near-infrared analysis of bulk samples. Cereal Chemistry 72(3): 243–247.

Dowell, F., E. Maghirang, F. Xie, G. Lookhart, R. Pierce, B. Seabourn, S. Bean, J. Wilson and O. Chung. 2006. Predicting wheat quality characteristics and functionality using near-infrared spectroscopy. Cereal Chemistry 83(5): 529–536.

Fischer, R.A. and D.B. Byerlee. 1991. Trends of wheat production in the warmer areas: major issues and economic considerations. pp. 3–27. *In*: Fischer, R.A. (ed.). Wheat for the Non-traditional Warm Areas. Proc. of Conf., Iguazu, Brazil, 29 Jul.–3 Aug. 1990. CIMMYT, Mexico, DF.

Gaspardo, B., S. Del Zotto, E. Torelli, S. Cividino, G. Firrao, G. Della Riccia and B. Stefanon. 2012. A rapid method for detection of fumonisins B1 and B2 in corn meal using Fourier transform near infrared (FT-NIR) spectroscopy implemented with integrating sphere. Food Chemistry 135(3): 1608–1612.

Geleta, N., H. Grausgruber and J. Vollmann. 2014. Near-infrared spectroscopy of yellow endosperm pigments in wheat. International Journal of Agriculture and Forestry 4(6): 446–450.

Geng, P., J.M. Harnly and P. Chen. 2016. Differentiation of bread made with whole grain and refined wheat (*T. aestivum*) flour using LC/MS-based chromatographic fingerprinting and chemometric approaches. Journal of Food Composition and Analysis 47: 92–100.

Girolamo, A., V. de Lippolis, E. Nordkvist and A. Visconti. 2009. Rapid and non-invasive analysis of deoxynivalenol in durum and common wheat by Fourier-Transform Near Infrared (FT-NIR) spectroscopy. Food Additives & Contaminants: Part A 26(6): 907–917.

Gupta, R.K. 2004. Quality of Indian wheat and infrastructure for analysis. *In*: Joshi, A.K., R. Chand, B. Arun and G. Singh [eds.]. A Compendium of the Training Program (26–30 December, 2003) on Wheat Improvement in Eastern and Warmer Regions of India: Conventional and non-conventional approaches. NATP project, (ICAR), BHU, Varanasi, India.

Harper, J.L., P.H. Lovell and K.G. Moore. 1970. The shapes and sizes of seeds. Annu. Rev. Ecol. Syst. 1: 327–356.

Hernández, M., T. Esteve and M. Pla. 2005. Real-time polymerase chain reaction based assays for quantitative detection of barley, rice, sunflower, and wheat. Journal of Agricultural and Food Chemistry 53: 7003–7009.

Jespersen, B.M. and L. Munck. 2009. Cereals and cereal products. pp. 275–319. *In*: Sun, D.-W. [ed.]. Infrared Spectroscopy for Food Quality Analysis and Control. Elsevier, Oxford, UK.

Jirsa, O., M. Hrušková and I. Švec. 2008. Near-infrared prediction of milling and baking parameters of wheat varieties. Journal of Food Engineering 87(1): 21–25.

Johnson, R. 1988. Durable resistance to yellow (stripe) rust in wheat and its implications in plant breeding. pp. 63–75. *In*: Simmonds, N.W. and Rajaram, S. [eds.]. Breeding Strategies for Resistance to the Rusts of Wheat. CIMMYT, Mexico City, Mexico.

Joshi, A.K., B. Mishra, R. Chatrath, G. Ortiz Ferrara and R.P. Singh. 2007a. Wheat improvement in India: present status, emerging challenges and future prospects. Euphytica 157: 431–446.

Joshi, A.K., M. Kumari, V.P. Singh, C.M. Reddy, S. Kumar, J. Rane and R. Chand. 2007b. Stay green trait: variation, inheritance and its association with spot blotch resistance in spring wheat (*Triticum aestivum* L.). Euphytica 153: 59–71.

Joshi, A.K., G. Ortiz-Ferrara, J. Crossa, G. Singh, G. Alvarado, M.R. Bhatta, E. Duveiller, R.C. Sharma, D.B. Pandit, A.B. Siddique, S.Y. Das, R.N. Sharma and R. Chand. 2007c. Associations of environments in South Asia based on spot blotch disease of wheat caused by Cochliobolus sativus. Crop Science 47(3): 1071–1081.

Kahl, J., N. Busscher, G. Mergardt, P. Mäder, T. Torp and A. Ploeger. 2014. Differentiation of organic and non-organic winter wheat cultivars from a controlled field trial by crystallization patterns. Journal of the Science of Food and Agriculture 95: 53–58.

Laursen, K.H., J.K. Schjoerring, E.O. Jørgen M. Askegaard, U. Halekoh and H. Søren. 2011. Multielemental fingerprinting as a tool for authentication of organic wheat, barley, faba bean, and potato. Journal of Agricultural and Food Chemistry 59: 4385–4396.

Luo, X., D.S. Jayas and S.J. Symons. 1999. Identification of damaged kernels in wheat using a colour machine vision system; Journal of Cereal Science 30: 49–59.

Michaletti, A., M.R. Naghavi, M. Toorchi, L. Zolla and S. Rinalducci,. 2018. Metabolomics and proteomics reveal drought-stress responses of leaf tissues from spring-wheat. Scientific Reports 8(5710): 1–18.

Miller, B.S. and T. Swain. 1960. Chromatographic analyses of the free amino-acids, organic acids and sugars in what plant extracts. Journal of the Science of Food and Agriculture 11: 344–348.

Miralbés, C. 2004. Quality control in the milling industry using near infrared transmittance spectroscopy. Food Chemistry 88(4): 621–628.

Miralbés, C. 2008. Discrimination of European wheat varieties using near infrared reflectance spectroscopy. Food Chemistry 106: 386–389.

Moon, J., J. Kim and C. Jang. 2016. Development of multiplex PCR for species-specific identification of the Poaceae family based on chloroplast gene, rpoC2. Applied Biological Chemistry 59(2): 201–207.

Nagarajan, S. 2005. Can India produce enough wheat even by 2020. Current Science 89(9): 1467–1471.

Neuman, M.R., H.D. Sapirstein, E. Shwedyk and W. Bushuk. 1987. Discrimination of wheat class and variety by digital image analysis of whole grain samples. Journal of Cereal Science 6: 125–132.

Neuman, M.R., H.D. Sapirstein, E. Shwedyk and W. Bushuk. 1989. Wheat grain colour analysis by digital image processing. I. Methodology. Journal of Cereal Science 10(3): 175–182.

Official Journal of the European Union (OJEU). 2011. Commission Regulation 1169/ 2011/EC of the European Parliament and of the Council of 25 October 2011 on the provision of food information to consumers, amending Regulations (EC) No 1924/2006 and (EC) No 1925/2006 of the European Parliament and of the Council, and repealing Commission Directive 87/250/EEC, Council Directive 90/ 496/EEC, Commission Directive 1999/10/EC, Directive 2000/13/EC of the European Parliament and of the Council, Commission Directives 2002/67/EC and 2008/5/EC and Commission Regulation (EC) No 608/2004. The Official Journal of the European Union, L304, 18e63. URL http:// eur-lex.europa.eu/LexUriServ/LexUriServ.do?uri¼OJ: L:2011:304:0018:0063:EN: PDF (Accessed 23.01.14).

Official Journal of the European Union (OJEU). 2014. Commission Regulation 78/2014/EC of 22 November 2013 amending annexes II and III to Regulation (EU) No 1169/2011 of the European Parliament and of the Council on the provision of food information to consumers, as regards certain cereals causing allergies or intolerances and foods with added phytosterols, phytosterol esters, phytostanols or phytostanol esters. The Official Journal of the European Union, L27, 7e8.

Pandey, S.P., S. Kumar, U. Kumar, R. Chand and A.K. Joshi. 2005. Sources of inoculum and reappearance of spot blotch of wheat in rice-wheat cropping system in eastern India. European Journal of Plant Pathology 111: 47–55.

Pasqualone, A., C. Montemurro, A. Grinn-Gofron, G. Sonnante and A. Blanco. 2007. Detection of soft wheat in semolina and durum wheat bread by analysis of DNA microsatellites. Journal of Agricultural and Food Chemistry 55: 3312–3318.

Pestsova, E., E. Salina, A. Börner, V. Korzun, O.I. Maystrenko and M.S. Röder. 2000. Microsatellites confirm the authenticity of inter-varietal chromosome substitution lines of wheat (*Triticum aestivum* L.). Theoretical and Applied Genetics 101: 95–99.

Porep, J.U., D.R. Kammerer and R. Carle. 2015. On-line application of near infrared (NIR) spectroscopy in food production. Trends in Food Science & Technology 46(2): 211–230.

Primrose, S., M. Woolfe and S. Rollinson. 2010. Food forensics: methods for determining the authenticity of foodstuffs. Trends in Food Science & Technology 21: 582–590.

Prins, T., J. van Dijk, A. Van Hoef, M. Voorhuijzen, S. Broeders, S. Trapmann, R. Seyfarth, A. Pardigol, C. Schoen, H. Aarts and E. Kok. 2010. Towards a multiplex cereal traceability tool using padlock probe ligation on genomic DNA. Food Chemistry 118(4): 966–973.

Raba, D.N., M.V. Ghiulai and G.I. Jianu. 2007. Studies regarding the chemical composition of several wheat species, flour types and pastes assortments. Bulletin of University of Agricultural Sciences and Veterinary Medicine CLUJ-Napoca 63: 1–5.

Ruibal-Mendieta, N.L., D.L. Delacroix and M. Meurens. 2002. A comparative analysis of free, bound and total lipid content on spelt and winter wheat wholemeal. Journal of Cereal Science 35: 337–342.

Ruibal-Mendieta, N.L., A. Dekeyser, D.L. Delacroix, E. Mignolet, Y. Larondelle and M. Meurens. 2004. The operate/palpitate ratio allows the distinction between whole meals of spelt (*Triticum spelta* L.) and winter wheat (*T. aestivum* L.). Journal of Cereal Science 39: 413–415.

Russo, R., E. Cusano, A. Perissi, F. Ferron, V. Severino, A. Parente and A. Chambery. 2014. Ultra-high performance liquid chromatography tandem mass spectrometry for the detection of durum wheat contamination or adulteration. Journal of Mass Spectrometry 49: 1239–1246.

Saini, M., S.N. Jagjit and N.R. Prakash. 2014. Analysis of wheat grain varieties using image processing-a review. International Journal of Science and Research 3(6): 490–495.

Shewry, P.R. 2009. Wheat. Journal of Experimental Botany 60(6): 1537–1553.

Shouche, S.P., R. Rastogi, S.G. Bhagwat and J.K. Sainis. 2001. Shape analysis of grains of Indian wheat varieties. Computers and Electronics in Agriculture 33: 55–76.

Silletti, S., L. Morello, F. Gavazzi, S. Gianì, L. Braglia and D. Breviario. 2018. Untargeted DNA-based methods for the authentication of wheat species and related cereals in food products. Food Chemistry 271: 410–418.

Singh, R.P., J. Huerta-Espino and H.M. William. 2005. Genetics and breeding for durable resistance to leaf an d stripe rusts in wheat. Turkish Journal of Agriculture and Forestry 29: 121–127.

Sonnante, G., C. Montemurro, A. Morgese, W. Sabetta, A. Blanco and A. Pasqualone. 2009. DNA microsatellite region for a reliable quantification of soft wheat adulteration in durum wheat-based foodstuffs by real-time PCR. Journal of Agricultural and Food Chemistry 57: 10199–10204.

Stone, P.J. and M.E. Nicolas. 1995. Effect of timing of heat stress during grain filling on two wheat varieties differing in heat tolerance. I. Grain growth. Australian Journal of Plant Physiology 22(6): 927–934

Szulc, M., J. Kahl, N. Busscher, G. Mergardt, P. Doesburg and A. Ploeger. 2010. Discrimination between organically and conventionally grown winter wheat farm pair samples using the copper chloride crystallisation method in combination with computerised image analysis. Computers and Electronics in Agriculture 74: 218–222.

Tandon, J.P. 1994. Wheat cultivation, research organization and production technology in the hot dry regions of India. pp. 17–23. *In*: Saunders, D.A. and G.P. Hettel [eds.]. Wheat in heat-stressed environments: irrigated, dry areas and rice-wheat farming systems. CIMMYT, Mexico City, Mexico.

Terzi, V., M. Malnati, M. Barbanera, A. Stanca and P. Faccioli. 2003. Development of analytical systems based on real-time PCR for *Triticum* species-specific detection and quantitation of bread wheat contamination in semolina and pasta. Journal of Cereal Science 38: 87–94.

Terzi, V., C. Morcia, A. Gorrini, A. Stanca, P. Shewry and P. Faccioli. 2005. DNA-based methods for identification and quantification of small grain cereal mixtures and fingerprinting of varieties. Journal of Cereal Science 41(3): 213–220.

Tyers, M. and M. Mann. 2003. From genomics to proteomics. Nature 422: 193–197.

Voorhuijzen, M.M., J.P. van Dijk, T.W. Prins, A.M.A. Van Hoef, R. Seyfarth and E.J. Kok. 2012. Development of a multiplex DNA-based traceability tool for crop plant materials. Analytical and Bioanalytical Chemistry 402: 693–701.

Wardlaw, I.F. 1994. The effect of high temperature on kernel development in wheat: Variability related to pre-heading post-anthesis conditions. Australian Journal of Plant Physiology 21(6): 731–739.

Woolfe, M., T. Gurung and M.J. Walker. 2013. Can analytical chemists do molecular biology? A survey of the up-skilling of the UK official food control system in DNA food authenticity techniques. Food Control 33: 385–392.

Wu, H., T. Haig, J. Pratley, D. Lemerle and M. An. 1999. Simultaneous determination of phenolic acids and 2,4-dihydroxy-7-methoxy-1,4-benzoxazin-3-one in wheat (*Triticum aestivum* L.) by gas chromatography–tandem mass spectrometry. Journal of Chromatography A 864: 315–321.

Zayas, I., F.S. Lai and Y. Pomeranz. 1986. Discrimination between wheat classes using image analysis. Cereal Chemistry 63: 52–56.

Zayas, I., Y. Pomeranz and F.S. Lai. 1989. Discrimination of wheat and non-wheat components in grain samples by image analysis; Cereal Chemistry 66: 233–237.

Zhao, H., B. Guo, Y. Wei and B. Zhang. 2013. Multi-element composition of wheat grain and provenance soil and their potentialities as fingerprints of geographical origin. Journal of Cereal Science 57: 391–397.

Zörb, C., G. Langenkämper, T. Betsche, K. Niehaus and A. Barsch. 2006. Metabolite profiling of wheat grains (*Triticum aestivum* L.) from organic and conventional agriculture. Journal of Agricultural and Food Chemistry 54(21): 8301–8306.

Zörb, C., T. Betsche and G. Langenkämper. 2009a. Search for diagnostic proteins to prove authenticity of organic wheat grains (*Triticum aestivum* L.). Journal of Agricultural and Food Chemistry 57(7): 2932–2937.

Zörb, C., K. Niehaus, A. Barsch, T. Betsche and G. Langenkämper. 2009b. Levels of Compounds and metabolites in wheat ears and grains in organic and conventional agriculture. Journal of Agricultural and Food Chemistry 57(20): 9555–9562.

10

Authenticity of Wine

Konstantinos V Kotsanopoulos[1,*] and
Parthena P Vartholomidou[2]

1. Introduction

The importance of wine traceability and authenticity has gained significant attention worldwide. As a result, a number of quality control programs have become mandatory in several countries, including EU and USA, aiming at both protecting consumers from involuntary mistakes taking place during production and limiting the potential of illegal adulteration practices. The rapid development of advanced analytical techniques, used to measure trace compounds or to simultaneously determine a wide array of analytes, has taken place together with the introduction of more comprehensive methodologies for data mining and analysis. Several improvements have therefore been made in the fields of wine traceability, discrimination and classification of grapes and wines (principally in terms of grape varieties, geographical origin and processing conditions). A number of challenges however still exist and numerous problems need to be resolved in order to improve the evaluation of wine authenticity and detect unintentional mistakes or fraud incidents (Versari et al. 2014).

The authenticity of wine has been extensively researched as this product can be easily adulterated as a result of its strong chemical basis (high alcohol content, low pH) and the fact that it is widely available throughout the

[1] School of Agricultural Sciences, Department of Ichthyology and Aquatic Environment, University of Thessaly, Fytoko St, 38446 Nea Ionia Magnesias, Volos, Hellas, Greece.
[2] School of Science & Technology, Hellenic Open University, Parodos Aristotelous 18, 26 335, Patra, Greece.
* Corresponding author: kostaskot@yahoo.gr

world. As a consequence, reliable and sometimes complicated techniques need to be implemented, together with multivariate analysis to verify its authenticity. The majority of established [High performance liquid chromatography (HPLC), Gas Chromatography-Mass Spectrometry (GC-MS)] and modern [Fourier-transform infrared spectroscopy (FT-IR), Nuclear magnetic resonance (NMR), Atomic absorption spectroscopy (AAS), DNA-based, Polymerase chain reaction (PCR)] techniques used in the field of wine authenticity have therefore been extensively examined (Arvanitoyannis 2010).

Several methods for detecting and confirming the adulteration of wine have been reported in literature. Unrecorded alcohol involves illicitly produced alcohol and surrogate alcohol. Rapid-screening methodologies such as IR or NMR have been the focus of the research together with techniques such as gas or liquid chromatography (Lachenmeier 2016). Table 1 presents the details of some of the most important studies that have been conducted for authenticating wine.

This chapter aims at presenting the key studies and techniques that have been published in the past few years introducing developments in the field of wine authenticity and the prevention or detection of adulteration.

2. Methods

2.1 Chromatographic techniques

2.1.1 HPLC

The formation of pyranoanthocyanins can be used as a marker to differentiate different grape varieties and aged red wines, based on the ratio of acetylated and coumaroylated anthocyanins (Rac/coum). Coelution of the pyranoanthocyanin pinotin A and the 3-coumaroyl glucoside of malvidin can influence the data obtained for the Rac/coum values, which in turn could lead to misclassification of wine varieties and rejection of wines by authorities. Research using a number of reversed phase (RF)-HPLC columns highlighted the significance of a good chromatographic resolution of these analytes (von Baer et al. 2008).

The assessment of polyphenols using HPCC is discussed in the study of Goldberg and Soleas (1999). HPLC has been employed since the 1970s and enological laboratories are the facilities that process most of the relevant instrumentation. An HPLC method was therefore described that can measure simultaneously and with high accuracy the concentration of a number of polyphenols in wine. A new method was also designed to measure not only these polyphenols, but also the levels of phenolic acids and certain esters. Very promising results were obtained.

The authenticity of wine as a subject has been a great concern for the world's wine market. In Europe, where varietal wines are principally

Table 1. Examples of techniques used for the authentication of wine.

Technique	Scope/Details	Markers	Effectiveness	References
MIR spectrophotometry	Differentiation between Shiraz, Cabernet Sauvignon, and Merlot	Spectra analysis with similarity index method	98% correct classification	Bevin et al. 2006
MIR spectrophotometry and chemometrics	Differentiation between red and white wines	Spectral data analysis with PCA and LDA	> 95% correct classification	Bevin et al. 2008
ATR-MIR and chemometrics	Classification of sparkling wine per type and quality	MIR spectra	Effective differentiation, appropriate for rapid screening	Culbert et al. 2015
FT-MIR and chemometrics	Discrimination of white, rosé and red wines from different cultivars	phenolic derivatives, carbohydrates, amino acids and organic acid	Rapid, inexpensive and efficient discrimination	Banc et al. 2014
Incremental Support Vector Machine and UV-vis spectroscopy	Discrimination between red wine brands	UV-vis spectra	~ 95% recognition rate, average training time of I-SVM models 0.47 seconds/batch	Liu et al. 2018
UV-vis spectroscopy and pattern recognition	Differentiation/classification of wines as per origin, grape variety and ageing process		75% correct classification	Urbano et al. 2006
UV–vis spectroscopy and chemometrics	Discrimination of Chinese rice wines from different geographical origins	UV–vis spectra	98.96–100% correct classification	Wu et al. 2015

Table 1 contd. ...

...Table 1 contd.

Technique	Scope/Details	Markers	Effectiveness	References
UV-VIS-NIR and chemometrics	Quantitative evaluation of ethanol, glycerol, glucose, tartaric acid, malic acid, lactic acid, and acetic acid in wines	Spectra	Spectra and concentrations were correlated well but interferences were seen in mixtures	Martelo-Vidal and Vázquez 2014
Artificial neural networks and UV–Vis–NIR	Quantitative evaluation of ethanol, glucose, glycerol, tartaric acid, malic acid, acetic acid and lactic acid in aqueous mixtures		Reliable models were generated for ethanol, tartaric acid and malic acid	Martelo-Vidal and Vázquez 2015
HPLC-DAD-ESI-MS and chemometrics	Differentiation between Carménère and Merlot	Quercetin and myricetin	Effective classification with errors of 2.5% and 11.4% for Carménère and Merlot, respectively	Vergara et al. 2011
Reversed phase HPLC	Differentiation of grape varieties of red wines	Acetylated and coumaroylated anthocyanins ratios	Effective with appropriate resolution	Von Baer et al. 2008
HPLC and chemometrics	Differentiation of red wines from different grape varieties and geographical origins	Chromatographic profiles of tartaric, malic, citric, lactic, acetic, and succinic acids	86.7–100% correct classification	Huang et al. 2017
HPLC	Authentication of Sauvignon blanc, Feteasca regala and Chardonnay	Phenolic compounds	Authenticity effectively verified	Lengyel and Sikolya 2014
HPLC-ESI-Q-ToF	Identification of flavonols in grapes and wines	Flavonols	*m/z* values successfully matched with expected formula	Favre et al. 2018
HS-SPME and GC×GC/TOFMS and chemometrics	Differentiation of grapes of Cabernet Sauvignon, Merlot, Chardonnay, Sauvignon Blanc and Pinot Noir	Volatile compounds	100% classification using 12 volatiles, but ^{1}D-GC/MS can lead to misclassification	Welke et al. 2013

SPME-GC–MS and chemometrics	Determination of biogenic amines content	Biogenic amines	Detection limits of 0.009-0.155 μg/L	Płotka-Wasylka et al. 2018
GC with FID, FPD and MS detectors and chemometrics	Differentiation between white grape wine varieties	Volatiles	Perfect classification using the Random Forest algorithm	Gómez-Meire et al. 2014
^{1}H NMR spectroscopy and chemometrics	Differentiation of red and white wine varieties	^{1}H NMR spectra	82–94% correct classification	Fan et al. 2018
Real-time PCR & label free optical biosensor platform and Q-ICP-MS	Identification of varieties of *Vitis vinifera* L. and geographical origin verification	SNPs and $^{87}Sr/^{86}S$ isotope ratio of wines and soil	Highly specific varietal identification, effective origin verification	Fernandes et al. 2015
High resolution melting	Genotype discrimination using leaf, must and wine	DNA strands	Sensitive, rapid and efficient but not all assays suitable for all 3 sample types	Pereira et al. 2017
DNA-based optical biosensor	Differentiation between *V. vinifera* varieties	DNA extracted from leaf, must and wine	Capable of differentiating varieties even when complex matrices used to extract DNA	Barrias et al. 2019

produced, wine authenticity is gaining particular attention. Anthocyanin pigments represent a significant category of phenolic substances that can be employed to assess wine authenticity. Seventeen varietal red wines originating from the village of DolníKounice (wine-growing region Moravia, Czech Republic) were subjected to analysis, covering three varieties and three regions. An estimation of anthocyanin pigments in varietal red wines was given using an HPLC method. The use of chemometrics enabled the determination of grapevine varieties and red wines based on their levels in delphinidin-3-*O*-glucoside (DpGl), malvidin-3-*O*-glucoside (MvGl) and delphinidin-3-*O*-acetylglucoside (DpGlAc) concentrations. Discrimination based on different regions was carried out on the basis of DpGl, MvGl and delphinidin-3-*O*-*p*-coumarylglucoside (DpGlCm) content (Kumšta et al. 2014).

In another study, chemometrics and the chromatographic profiles of red wines' organic acids were used to classify them based on grape varieties and different geographical locations. The detection of tartaric, malic, citric, lactic, acetic and succinic acids in wines was performed using HPLC. Multivariate statistical methods including principal component analysis (PCA) and linear discriminant analysis (LDA) were used and pattern recognition models were developed to classify the samples on the basis of varieties and geographical origins. The PCA was clearly successful in grouping the samples in accordance with their varieties, and the LDA further achieved 100% correct classification based on geographical origin, while the different varieties were differentiated with 86.7–100% accuracy. It was revealed that the use of chromatographic profiles of organic acids offers a great potential when used to classify red wines (Huang et al. 2017).

In recent years, liquid chromatography-based methods have proved to be very promising in assessing wines on the basis of the profiles of their polyphenols. These profiles are very important for grapes and wines as they highly affect parameters such as color, sensory characteristics, and nutritional content. Polyphenols and their levels in wine as it matures can be used as authenticity markers. Furthermore, parameters such as vintage, the region and variety, as well as technological conditions can play a key role in the authentication process of wine through the use of polyphenol profiles (Palade and Popa 2018).

The effective discrimination between different wines is of particular importance for the Chilean wine making industry, and particularly for Carménère variety. The establishment of authentication parameters is therefore essential for this wine as it has a lot of similarities with Merlot. The level of anthocyanins, shikimic acid, and the principal flavonols found in wine can be used to differentiate between the Carménère and Merlot varieties. Myricetin and quercetin are the flavonols found at the highest levels in wines in both free and conjugated forms. These substances are responsible for important wine antioxidant properties. Based on this fact, the concentrations of free and conjugated quercetin and myricetin, were used to differentiate between Carménère and Merlot varieties. Flavonols in Chilean wine were examined using HPLC-DAD-ESI-MS^n. An evaluation of the concentration ranges of flavonols present in 248 Chilean red wines was presented. The average levels of the combination of myricetin and quercetin were lower in Cabernet Sauvignon (53.9 mgL^1) compared to Carménère (81.5 mgL^{-1}) and Merlot (78.9 mgL^1). It was found that these mean levels were higher than most concentrations previously reported. The chemometric analysis indicated that the ratio of quercetin/myricetin and free myricetin allowed the effective differentiation between Carménère and Merlot wines (Vergara et al. 2011).

Another study aimed at determining distinctive elements related to the authenticity of a number of wine types from the same vineyard. The substances responsible for the colour of white wines were reviewed. Therefore, HPLC methods allowed the determination and quantification of phenolics. It was also demonstrated that it was possible to authenticate wines using spectrophotometric methods via determining the color of the products (Lengyel and Sikolya 2014).

2.1.2 GC

Based on the study of Gómez-Meire et al. (2014), the classification of wine can be a very complex exercise involving the assessment of several factors. Volatile substances of different wines were subjected to analysis using GC. It was shown that the samples could be accurately classified using the Random Forest algorithm.

Biogenic amines have been in the center of research in the past few years, yet a lot of questions around these substances still remain to be answered. Although biogenic amines are found in wine and can be toxic for humans, there is no common legislative framework controlling their levels. A comprehensive understanding of their levels in wine is also of high importance for the food trade sector. A study was therefore carried out with the aim of determining the level of certain biogenic amines in wine samples originating from Poland. The correlation between the levels of biogenic amines and certain factors such as pH, alcohol content and fermentation temperature was also determined using chemometric analysis. The biogenic amines were determined using a solid phase micro extraction method followed by GC-MS (SPME-GC-MS), which appeared to have low detection limits ranging from 0.009 µg/L (tyramine) to 0.155 µg/L (histamine). Based on the results of the study, it was indicated that no samples contained biogenic amines at levels exceeding those reported in published studies (with the total concentration of biogenic amines ranging from 7 to 2174 µg/L), and, as a result, these wines do not seem to be posing a risk related to the intake of potentially toxic biogenic amines. Furthermore, a number of correlations existed between occurrence, concentration of biogenic amines, physicochemical characteristics and important aspects of the wine making process. Although data on biogenic amines are at the moment not recorded in wine composition databases, data related to their presence, distribution, levels and correlation between biogenic amines and wine attributes can be very important for the food industry, health professionals and the consumers (Płotka-Wasylka et al. 2018).

In another study, headspace solid-phase microextraction (HS-SPME) in combination with comprehensive two-dimensional GC with time-of-flight MS detection (GC×GC/TOFMS) was employed for the analysis of volatiles in five types of wines produced with Cabernet Sauvignon, Merlot, Chardonnay, Sauvignon Blanc and Pinot Noir grapes. Fisher ratio, PCA and stepwise linear discriminant analysis (SLDA) were employed to design a discriminant model, which allowed the researchers to use 12 volatile substances to differentiate and classify wines in accordance with their grape cultivars. A thorough analysis of GC×GC/TOFMS data indicated that the use of one-dimensional GC with a MS detector (1D-GC/MS) would potentially lead to incorrect classification of some of these 12 compounds, due to them presenting partial coelution with other components in the first chromatographic dimension (Welke et al. 2013).

2.3 Spectroscopic techniques

Certain information about components of grape juice, must, and wine can be very useful for decision support systems that aim at improving, monitoring and adopting grape and wine production to new challenges and opportunities. Numerous sensors that can be used to collect this information have been developed. It is however necessary to adapt these sensors to special requirements, such as robustness, calibration and maintenance, operating costs, duration, sensitivity, and specificity to a certain application. The sensors mainly employed by the wine industry are those that rely on the use of mid-infrared (MIR), near-infrared (NIR), visible (VIS) and ultraviolet (UV) spectroscopy (Cozzolino et al. 2011).

2.3.1 UV-VIS

In the study of Wu et al. (2005), Chinese rice wines originating from four distinct areas ('Zhejiang', 'Jiangsu', 'Shanghai' and 'Fujian') in China were differentiated according to their UV–vis spectra. The UV–vis absorption spectra of 112 samples in the wavelength range of 190–800 nm were used. PCA was employed to analyze the absorption data and indicate differences between samples and potentially the possibility to use multivariate methods to distinguish differences. Classification models were designed using soft independent modeling of class analogy, LDA, discriminant partial least squares (PLS) and support vector machine. Seven mathematical pretreatments were used to allow improvements to be made on the performance of the multivariate classification models. Among the seven pretreatments, standard normal transformation (SNV) was significantly better than the other six. It was demonstrated that in comparison to other models, SNV-treated LDA models performed better, correctly classifying 98.96% of the samples as an average in the training set, and 100% in the testing set. It was therefore proved that UV spectroscopy in combination with chemometric data analysis can be used for the efficient classification of Chinese rice wines according to their geographical origins.

UV-VIS-NIR spectroscopy in combination with chemometrics was used in another study as a non-destructive technique to develop models in order to quantitatively characterize the main compounds of wine. The study of mixtures can give valuable information about how interferences affect the performance of calibrations in wines. Pre-treatments for the spectra and modeling strategies such as PLS regression or Principal Component Regression were subjected to analysis. The pure substances that were considered in this study showed a good relationship between spectra and concentrations. The results of the study therefore indicated that NIR spectroscopy could be used instead of conventional chemical methods to determine ethanol, while UV spectroscopy can be used for determining tartaric acid and malic acid (Martelo-Vidal and Vázquex 2014).

Urbano et al. (2006) reported the use of a qualitative model requiring inexpensive instruments that could allow wine samples to be classified into categories. The parameters the discrimination was made against were origin, grape variety and ageing process. UV–Vis spectroscopy was employed to develop a cost effective and simple methodology. Following the collection of spectra, the data were analyzed aiming at both showing trends hidden in the data matrix from the sample spectra and studying the characteristics of the models. PCA and soft independent modeling of class analogy were employed for the analysis and for developing models. The ultraviolet region used in this study was found to be very promising for differentiating wines in accordance with their origin within the same appellationd' origine and the ageing process followed. In total, false results were given in about 25% of the cases, and the classification of wines in accordance with their origin zone gave the best values (10%). No sample preparation was needed.

As reported by Yu et al. (2018), wine is commercially very important. Effective methodologies are therefore always required for controlling and verifying its quality. UV, Vis, NIR and IR spectroscopy are amongst the most valuable techniques for assessing the qualitative characteristics of the product. Liu et al. (2018) reported the development of a method that aimed at overcoming the increased training time required in cases that a recognition model is updated, based on the condition of new features obtained from new samples. As a common complex chemical mixture, red wine contains a plethora of chemical substances and was therefore well suited for this research. The novel method based on an incremental learning support vector machine in combination with UV-Vis spectroscopy was employed to differentiate between a number of red wine brands. In this method, the introduction of new features included in the new training samples into the recognition model was carried out through iterative learning in each iteration, and rapid update of the recognition model was carried out requiring no significant increase of the training time. It was proved that the recognition model developed in this method could achieve a good balance between training efficiency and recognition accuracy.

2.3.2 NMR

During the last decades, NMR spectroscopy has proven to be a valuable tool for monitoring food authenticity when used to identify specific markers or for entire metabolite profiling. NMR methods have been extensively used to examine how certain factors (genetic background, geographical origin, adulteration, etc.) affect food metabolite composition (Sobolev et al. 2017). A summary of studies that used NMR combined with multivariate statistics was given in the study of Amargianitaki and Spyros (2017). According to these authors, 2D NMR was employed to analyse non-volatile metabolites and phenolics in wine. Metabolomics were subsequently used for examining the metabolome of the product and how it is affected by parameters such as the region of cultivation, vintage, production practices, maturation, etc. The metabolome was used as a marker to allow a holistic analytical approach to be given aiming at authenticating wine in accordance with factors that affect its price.

The ability of ^{1}H NMR in combination with chemometrics to discriminate grape varieties was also assessed. A total of 3 grape varieties used to produce red (Cabernet Sauvignon, Rose Honey, Cabernet Gernischt) and white wines (Ugni Blanc, Long Yan, Chardonnay), were analyzed. A segment-wise peak alignment was used to manage peak misalignments of the recorded ^{1}H NMR spectra. Binning of the aligned ^{1}H NMR spectra was conducted to reduce the data. The resulting bins were used as input variables for PCA and LDA analyses. The combined use of PCA and LDA allowed the grape varieties to be well discriminated. The validity of the PCA/LDA model was verified by both internal leave-one-out cross validation and external repeated double random cross validation. Leave-one-out cross validation and repeated double random cross validation correctly classified 82% and 83%, as an average, of red wine varieties, respectively, and 94% and 90%, respectively in the case of white wines. It was proved that ^{1}H NMR spectroscopy in combination with multivariate analysis can effectively verify the authenticity of Chinese wines (Fan et al. 2018).

Fully-automated high resolution ^{1}H-NMR spectroscopy has unique screening potential in the fields of food quality and safety by taking advantage of short analysis time (about 20 min). This technique allows for reproducible and transferable data to be produced amongst laboratories. Its reproducibility allows statistical studies to be conducted, e.g., for detecting the variety, geographical origin and adulterants, in cases where records of minimal changes of various ingredients need to be taken simultaneously. The reproducibility of the solutions depends on the user, instrument and the laboratory used. Sample must be prepared, measured and processed following well defined processes. The non-targeted approach enabled the detection of unknown deviations, assuming they are visible in the ^{1}H-NMR spectra (Spraul et al. 2015).

2.3.3 MIR

The use of MIR was reported in the study of Culbert et al. (2015). This study evaluated the effectiveness of attenuated total reflection (ATR)-MIR spectroscopy, in combination with PCA and PLS regression, in rapidly classifying sparkling wine in accordance with its type and qualitative characteristics. Sparkling wines of Australian origin were subjected to analysis using ATR-MIR in combination with chemmetrics (multivariate data). Evaluation of spectra of a number of sparkling white wines was also conducted on the basis of quality ratings and sensory characteristics as determined and rated by an expert panel. pH, titratable acidity, sugar residues, alcoholic content and phenols were measured. Classification of the samples as per the wine type on the PCA score plot was performed according to their MIR spectral data, while the wines could be classified in accordance with their production method, which highly affected the wine type and properties.

In another study, FT-MIR in combination with chemometrics was employed to discriminate between a number of Romanian wines as per their type and region of origin. PCA and hierarchical cluster analysis were used. The wines were separated into four distinct regions that corresponded to phenolic derivatives, carbohydrates, amino acids and organic acids, which are all substances that define the quality and authenticity of wines. Each component category was qualitatively and quantitatively evaluated allowing each wine category to be discriminated (from red, to rosé and white colors, to dry, half-dry and half-sweet flavors). The multivariate data analysis based on absorption peaks from FT-MIR spectra indicated that the samples could be very well clustered in accordance with phenolics, carbohydrates, amino acids and organic acids. Thus, the ATR-FT-MIR analysis proved to be a rapid, inexpensive and efficient tool when used for evaluating wines and differentiating them based on colour and level of sweetness, which varied depending on the cultivar (Banc et al. 2014).

Another study examined the use of MIR spectroscopy in combination with PCA and LDA for the classification of wines according to their varieties. Seven different varieties obtained from different regions in Australia were used for this study. All samples belonged to the 2006 vintage and no blending with other varieties took place. Spectral data were limited to a low number of components and LDA was then employed for the separation of the samples into varieties. The accuracy of the LDA models constructed in the case of red wines was tested using a set of red wines scanned in 2005. Over 95% of samples were accurately assigned (Bevin et al. 2008).

Finally, in the study of Bevin et al. (2006), a very quick and simple method for authenticating red wines was developed. The method was intended to be used during transportation and production. Specifically, when wine is transported between two sites, a sample is collected and a MIR spectrum is obtained. One hundred sixty-one samples from three main red wine varieties cultivated in Australia, Shiraz, Cabernet Sauvignon, and Merlot, were obtained from six wineries across Australia and scanned in transmission on two MIR spectrophotometers (Foss WineScan FT 120) (926–5012 cm^{-1}). A similarity index (SI) method was employed for the classification of the wine samples based on their spectral data. The SI was proved to allow an acceptable measurement to be taken for authentication of red wine integrity during transportation. In five out of six winery data sets, the SI led to correct classification of 98% of the wines. It was shown that misclassification between the different wineries was limited to < 1% of wines. Further research is required though to assess the applicability of the SI at a commercial scale and the evaluation of its potential as a rapid quality control tool that can routinely authenticate wine samples during transport.

2.3.4 FTIR

The use of Fourier transforms infrared (FTIR) spectroscopy for wine authentication has also been reported in literature. In the study of Pennington et al. (2007), two methods were evaluated for their effectiveness in authenticating wines of different varieties using two complementary analytical approaches. A mass spectrometer was employed for the detection of the headspace volatiles and FTIR spectroscopy was employed to take measurements of phenolic compounds of four red and four white wine varietals sourced from 18 different wineries across the United States. Chemometrics were employed to analyze the spectra generated by each technique differentiating the samples according to other varieties. The development of K-Nearest Neighbor and Soft Independent Modeling of Class Analogy models was conducted to allow varietal authentication. PLS models were designed to model the binary mixtures between varietal wines and determine the contributions of the varietals in wine mixtures. Chemometric models based on MS volatiles spectra and FTIR phenolics spectra offered a simple and fast approach to analyze and classify the varietal nature of wines.

Furthermore, FTIR single bounce micro-attenuated total reflectance (mATR) spectroscopy combined with multivariate and artificial neural network (ANN) data analysis, were used to identify adulteration with glycerol in red wines. Following intentional adulteration of wine samples with industrial grade glycerol in concentrations ranging from 0.1 to 15%, calibration models were developed and validated. Records of spectra of wine samples contaminated with glycerol were taken in the region of 900–1500 cm^{-1}. PLS and PLS first derivatives were employed to quantitatively analyze the data), while LDA and canonical variate analysis (CVA) were employed to classify and discriminate between the samples. The standard error of prediction in the validation set was found to be from 1.44 to 2.25%. Classification of glycerol adulterants in the different brands of red wine using CVA allowed 94–98% of samples to be accurately classified. Analysis using a quick back propagation network (BPN) and the radial basis function network (RBFN) algorithms enabled 93% to 100% correct classification (Dixit et al. 2005).

2.4 Electronic tongue

An electronic nose is a device that analyzes the chemical structure of certain odor components. It makes use of a chemical detector and a tool for pattern recognition. Electronic noses have been used for a number of years but their use has been restricted due to their large size and their cost. Current research has focused on reducing the size of the devices, lowering their cost and increasing their sensitivity (https://whatis.techtarget.com/definition/electronic-nose-e-nose). An electronic tongue using chemically modified voltammetric electrodes allowed the successful detection of model fraudulent red wines obtained via the addition of a number of adulterants (often used in the industry to improve the organoleptic characteristics of the product) to wine samples. Two different sensors were employed. The electrochemical response of both families of electrodes was linked to the ions and to the electroactive molecules that existed in the solution and particular sensitivity was observed towards pH and antioxidants. The use of sensitive materials with distinct chemical properties allowed complementary information to be obtained, enhancing the capability of discrimination of the system. Prediction models were effectively employed for estimating the chemical parameters of wines, such as alcoholic degree and acidity by means of PLS. The sensor array coupled with pattern recognition techniques could be used to differentiate between adulterations according to their chemical nature (Parra et al. 2006).

2.5 DNA techniques

It is clear that methods of high speed and reliability are required to identify incidents of fraud in the wine market. The development of a number of methodologies relying on the chemical characteristics of wines has been reported, but they are limited due to the effect of environmental factors. Within the emerging area of DNA-based detectors, various food products have been analyzed and the methods are still reliable when it comes to identifying varieties. However, these methods cannot be used for geographical determination. For the latter, soil-related fingerprints provide the most reliable results (Fernandes et al. 2015). Wine production has a global economic impact. Its market reliability though is at great risk due to frequent fraud and adulteration incidents and attempts. Therefore, the establishment of robust traceability systems is of paramount importance, especially in cases of wines with certified labels, in which only a small number of grapevine varieties are permitted in a restricted well-defined geographical area. Molecular markers have been widely used in the food sector, and particularly, High-Resolution DNA Melting (HRM) technology has been used with a lot of success in the area of wine authentication, as part of traceability systems (Pereiraa et al. 2018).

Wine authenticity methods are increasingly required, principally in Denomination of Origin designations. The DNA-based methodologies can be used to reliably track food/wine varietal composition. A HRM application was evaluated as a screening method for must and wine authenticity. Three sample types (leaf, must and wine) were used for validating three HRM assays (Vv1–705 bp; Vv2–375 bp; and Vv3–119 bp). The Vv1 HRM assay allowed successful results to be given only when applied to leaf and must samples. The Vv2 HRM assay led to successful application on all sample types, allowing the genotypes to be discriminated based on melting temperature values. The smallest amplicon, Vv3, yielded a coincident melting curve shape in all sample types (leaf and wine) with corresponding genotypes (Pereira et al. 2017).

The diversity arising among the *V. vinifera* L. species allows wines of very different characteristics to be produced. Platforms for food composition analysis are currently being developed. This emerging area includes the development of DNA biosensors that can be used in various applications that range from food safety to authenticity. The detection capacity of a DNA-based optical biosensor was studied using different *V. vinifera* matrices (leaf, must and wine). In this process, genomic DNA was extracted from leaf, must and wine of three *V. vinifera* varieties and was analyzed on the long-period grating (LPG) DNA-based biosensor. The biosensor allowed the varieties to be distinguished even using DNA extracted from complex matrices, thus proving its potential in wine authenticity (Barrias et al. 2019).

Fernandes et al. (2015) reported the development of the WineBioCode project that aims at defining the optimum methodology for authenticating wine. Specifically, two DNA-based strategies were followed making use of real-time PCR and a label free optical biosensor platform. The strategies allowed specific DNA-targets to be successfully identified when applied to *V. vinifera* L. The geographical evaluation was assessed via determining the strontium $^{87}Sr/^{86}Sr$ isotope ratio.

2.6 Combined methods

Favre et al. (2018) reported that flavonols are polyphenol substances that are believed to be very important for the quality of wine. The properties and characteristics of these compounds are now better understood. The identification of derivatives of the flavonol 3-O-glucosides of isorhamnetin, laricitrin and syringetin was achieved for the first time in *V. vinifera* skin and wines. In the case of the p-coumaroylated derivatives a signal was shown that corresponded to a phenolic acid loss. Confirmation of the structural characteristics of the aglycones was given in individual MS3 experiments. Additionally, the fragmentation signals that corresponded to the aglycone radical ions could also be used as markers, and could be used in future studies employing ESI-MS. When HPLC-ESI-Q-ToF was used, the *m/z* values of the substances examined led to discrimination of samples.

Finally, as Martelo-Vidal and Vázquez (2015) reported, ultraviolet–visible (UV–VIS) and near-infrared (NIR) spectroscopy in combination with artificial neural networks (ANNs) can be employed to non-destructively quantify ethanol, sugars and acids in solutions. This study indicated that it is very important to determine which factors are the best for training the ANN to construct models of high reliability.

3. Conclusions

It is therefore clear that wine is amongst the most vulnerable categories of foods for fraud and its market value and characteristics make it often the target of fraudsters. Various techniques have traditionally been used to verify the authenticity of wine while a number of newly developed or improved techniques are currently examined as potential tools. Chromatographic and spectroscopic techniques are among the most common techniques used nowadays, while electronic tongues and DNA techniques offer new opportunities for increasing the accuracy and effectiveness of authentication. Developments in the use of these methods are expected to focus on improving the accuracy and speed of the methods and reducing their costs.

Abbreviations

Atomic absorption spectroscopy	AAS
Attenuated total reflection	ATR
Back propagation network	BPN
Canonical variate analysis	CVA
Fourier-transform infrared spectroscopy	FT-IR
Gas Chromatography - Mass Spectrometry	GC-MS
High performance liquid chromatography	HPLC
Headspace solid-phase microextraction	HS-SPME
Linear discriminant analysis	LDA
Mid-infrared	MIR
Near-infrared	NIR
Nuclear magnetic resonance	NMR
Principal component analysis	PCA
Polymerase chain reaction	PCR
Partial Least Squares	PLS
Radial basis function network	RBFN
Standard normal transformation	SNV
Solid phase micro extraction method followed by GC-MS	SPME-GC-MS
Ultraviolet	UV
Visible	VIS
High Resolution Melting	HRM

References

Amargianitaki, M. and A. Spyros. 2017. NMR-based metabolomics in wine quality control and authentication. Chemical and Biological Technologies in Agriculture 4(9): 1–12.

Arvanitoyannis, I. 2010. Wine authenticity, traceability and safety monitoring. pp. 218–270. *In*: Reynolds, A.G. [ed.]. Managing Wine Quality. Woodhead Publishing, Cambridge, UK.

Banc, R., F. Loghin, D. Miere, F. Fetea and C. Socaciu. 2014. Romanian wines quality and authenticity using FT-MIR spectroscopy coupled with multivariate data analysis. Notulae Botanicae Horti Agrobotanici Cluj-Napoca 42(2): 556–564.

Barrias, S., J.R. Fernandes, J.E. Eiras-Dias, J. Brazão and P. Martins-Lopes. 2019. Label free DNA-based optical biosensor as a potential system for wine authenticity. Food Chemistry 270: 299–304.

Bevin, C.J., A.J. Fergusson, W.B. Perry, L.J. Janik and D. Cozzolino. 2006. Development of a Rapid "Fingerprinting" system for wine authenticity by mid-infrared spectroscopy. Journal of Agricultural and Food Chemistry 54(26): 9713–9718.

Bevin, C.J., R.G. Dambergs, A.J. Fergusson and D. Cozzolino. 2008. Varietal discrimination of Australian wines by means of mid-infrared spectroscopy and multivariate analysis. AnalyticaChimicaActa 621(1): 19–23.

Cozzolino, D., W. Cynkar, N. Shah and P. Smith. 2011. Technical solutions for analysis of grape juice, must, and wine: the role of infrared spectroscopy and chemometrics. Analytical and Bioanalytical Chemistry 401: 1475–1484.

Culbert, J., D. Cozzolino, R. Ristic and K. Wilkinson. 2015. Classification of sparkling wine style and quality by MIR spectroscopy. Molecules 20(5): 8341–8356.

Dixit, V., J.C. Tewari, B.-K. Cho and J.M.K. Irudayaraj. 2005. Identification and quantification of industrial grade glycerol adulteration in red wine with fourier transform infrared spectroscopy using chemometrics and artificial neural networks. Applied Spectroscopy 59(12): 1553–1561.

Fan, S., Q. Zhong, C. Fauhl-Hassek, M.K.-H. Pfister, B. Horn and Z. Huang. 2018. Classification of Chinese wine varieties using ^{1}H NMR spectroscopy combined with multivariate statistical analysis. Food Control 88: 113–122.

Favre, G., G. González-Neves, D. Piccardo, S. Gómez-Alonso, J. Pérez-Navarro and I. Hermosín-Gutiérrez. 2018. New acylatedflavonols identified in Vitisvinifera grapes and wines. Food Research International 112: 98–107.

Fernandes, J.R., L. Pereira, P. Jorge, L. Moreira, H. Gonçalves, L. Coelho, D. Alexandre, J. Eiras-Dias, J. Brazão, P. Clímaco, M. Baleiras-Couto, S. Catarino, A. Graça and P. Martins-Lopes. 2015. Wine fingerprinting using a bio-geochemical approach. 38th World Congress of Vine and Wine, pp. 1–4.

Goldberg, D.M. and G.J. Soleas. 1999. Analysis of antioxidant wine polyphenols by high-performance liquid chromatography. Methods in Enzymology 299: 122–137.

Gómez-Meire, S., C. Campos, E. Falqué, F. Díaz and F. Fdez-Riverola. 2014. Assuring the authenticity of northwest Spain white wine varieties using machine learning techniques. Food Research International 60: 230–240.

https://whatis.techtarget.com/definition/electronic-nose-e-nose Accessed 07 Jul 2019.

Huang, X.-Y., Z.-T. Jiang, J. Tan and R. Li. 2017. Geographical Origin Traceability of Red Wines Based on Chemometric Classification via Organic Acid Profiles 2017: 1–7.

Kumšta, M., P. Pavloušek and P. Kárník. 2014. Use of anthocyanin profiles when differentiating individual varietal wines and terroirs. Food Technol. Biotechnol. 52(4): 383–390.

Lachenmeier, D.W. 2016. Advances in the detection of the adulteration of alcoholic beverages including unrecorded alcohol. pp. 565–578 *In*: Downey, G. [ed.]. Advances in Food Authenticity Testing. Woodhead Publishing, Cambridge, UK.

Lengyel, E. and L. Sikolya. 2014. Authenticity tests of white wines from the apold depression. Management of Sustainable Development Sibiu 6(2): 55–59.

Liu, J., T.-J. Pan and Z.-Y. Zhang. 2018. Incremental support vector machine combined with ultraviolet-visible spectroscopy for rapid discriminant analysis of red wine. Journal of Spectroscopy 2018: 1–5.

Martelo-Vidal, M.J. and M. Vázquez. 2014. Evaluation of ultraviolet, visible, and near infrared evaluation of ultraviolet, visible, and near infrared. Czech. J. Food Sci. 32: 37–47.

Martelo-Vidal, M.J. and M. Vázquez. 2015. Application of artificial neural networks coupled to UV–VIS–NIR spectroscopy for the rapid quantification of wine compounds in aqueous mixtures. CyTA-Journal of Food 13(1): 32–39.

Palade, L.M. and M.E. Popa. 2018. Polyphenol fingerprinting approaches in wine traceability and authenticity: assessment and implications of red wines. Beverages 4(75): 1–18.

Parra, V., A.A. Arrieta, J.A. Fernández-Escudero, M.L. Rodríguez-Méndez and J.A. De Saja. 2006. Electronic tongue based on chemically modified electrodes and voltammetry for the detection of adulterations in wines. Sensors and Actuators B: Chemical 118: 448–453.

Pennington, N., F. Ni, Md.A. Mabud and S. Dugar. 2007. A simplified approach to wine varietal authentication using complementary methods: Headspace mass spectrometry and FTIR spectroscopy. ACS Symposium Series 12: 180–199.

Pereira, L., S. Gomes, C. Castro, J.E. Eiras-Dias, J. Brazão, A. Graça, J.R. Fernandes and P. Martins-Lopes. 2017. High Resolution Melting (HRM) applied to wine authenticity. Food Chemistry 216: 80–86.

Pereira, L., S. Gomes, S. Barrias, J.R. Fernandes and P. Martins-Lopes. 2018. Applying high-resolution melting (HRM) technology to olive oil and wine authenticity. Food Research International 103: 170–181.

Płotka-Wasylka, J., V. Simeonov, C. Morrison and J. Namieśnik. 2018. Impact of selected parameters of the fermentation process of wine and wine itself on the biogenic amines content: Evaluation by application of chemometric tools. Microchemical Journal 142: 187–194.

Sobolev, A.P., S. Circi, D. Capitani and L. Mannina. 2017. Molecular fingerprinting of food authenticity. Current Opinion in Food Science 16: 59–66.

Spraul, M., M. Link, H. Schaefer, F. Fang and B. Schuetz. 2015. Wine analysis to check quality and authenticity by fully-automated ^{1}H-NMR. BIO Web of Conferences 5: 1–5.

Urbano, M., M.D. Luque de Castro, P.M. Pérez, J. García-Olmo and M.A. Gómez-Nieto. 2006. Ultraviolet–visible spectroscopy and pattern recognition methods for differentiation and classification of wines. Food Chemistry 97: 166–175.

Vergara, C., D. Von Baer, C. Mardones, L. Gutiérrez, I. Hermosín-Gutiérrez and N. Castillo-Muñoz. 2011. Flavonol profiles for varietal differentation between Carménère and Merlot Wines produced in Chile: HPLC and Chemometric analysis. Journal of the Chilean Chemical Society 56: 827–832.

Versari, A., V.F. Laurie, A. Ricci, L. Laghi and G.P. Parpinello. 2014. Progress in authentication, typification and traceability of grapes and wines by chemometric approaches. Food Research International 60: 2–18.

von Baer, D., M. Rentzsch, M.A. Hitschfeld, C. Mardones, C. Vergara and P. Winterhalter. 2008. Relevance of chromatographic efficiency in varietal authenticity verification of red wines based on their anthocyanin profiles: Interference of pyranoanthocyanins formed during wine ageing. AnalyticaChimicaActa 621(1): 52–56.

Welke, J.E., V. Manfroi, M. Zanus, M. Lazzarotto and C.A. Zini. 2013. Differentiation of wines according to grape variety using multivariate analysis of comprehensive two-dimensional gas chromatography with time-of-flight mass spectrometric detection data. Food Chemistry 141(4): 3897–3905.

Wu, Z., H. Li, J. Long, E. Xu, X. Xu, Z. Jin and A. Jiao. 2015. Discrimination of Chinese rice wines of different geographical origins by UV–vis spectroscopy and chemometrics. Journal of The Institute of Brewing 121: 167–174.

Yu, J., H. Wang, J. Zhan and W. Huang. 2018. Review of recent UV–Vis and infrared spectroscopy researches on wine detection and discrimination. Applied Spectroscopy Reviews 53 (1): 65–86.

11

Authenticity of Olive Oil

Konstantinos V Kotsanopoulos[1,*] and
Parthena P Vartholomidou[2]

1. Introduction

The traceability of olive oil products is essential in order to both assess their origin but also to limit the risk of fraud incidents (Consolandi et al. 2008). Olive (*Olea europaea* L.) is amongst the oldest tree crops worldwide and a valuable source of oil, rich in substances beneficial to human health. Olive oil can only be produced from the fruit of the olive tree (*O. europaea* L.) and is distinctly different to most other vegetable oils when the extraction method is considered, allowing it to be consumed raw, thus conserving its vitamins and other beneficial compounds. Olive oil is commonly more expensive than most other vegetable oils and is therefore more prone to adulteration with cheaper oils. A number of different grades of olive oil are also available on the market, which also leads to different prices. There is therefore also the possibility of adding cheaper grades of olive oil to better graded ones in order to increase profit (Rabiei and Enferadi 2011).

The high price of olive oil and the fact that it is considered to offer several health benefits make it a particularly good choice for fraudsters. Thus, adulteration may take place not only accidentally, but also deliberately in the form of mislabeling of cheaper olive oil categories or by mixing cheaper oils with virgin olive oil thus lowering the cost of the

[1] School of Agricultural Sciences, Department of Ichthyology and Aquatic Environment, University of Thessaly, Fytoko St, 38446 Nea Ionia Magnesias, Volos, Hellas, Greece.
[2] School of Science & Technology, Hellenic Open University, Parodos Aristotelous 18, 26 335, Patra, Greece.
* Corresponding author: kostaskot@yahoo.gr

product (Aparicio et al. 2013). Extra-virgin olive oil (EVOO) is a natural product that is very popular among consumers worldwide due to its nutritional value and beneficial effects on health. In the last few years though, the market of EVOO has suffered a significant loss of consumer trust, mainly due to increasing cases of fraud incidents. This fact explains the increasing demand for analytical methods that are able to accurately verify claims of origin (Paolinia et al. 2017).

This chapter presents the main techniques and methodologies used for detecting adulteration of olive oil and ensuring the traceability of the product in an effort to prevent adulteration and prevent fraud. Figure 1 gives an overview of the main techniques currently used to verify the authenticity of detecting adulteration in olive oil.

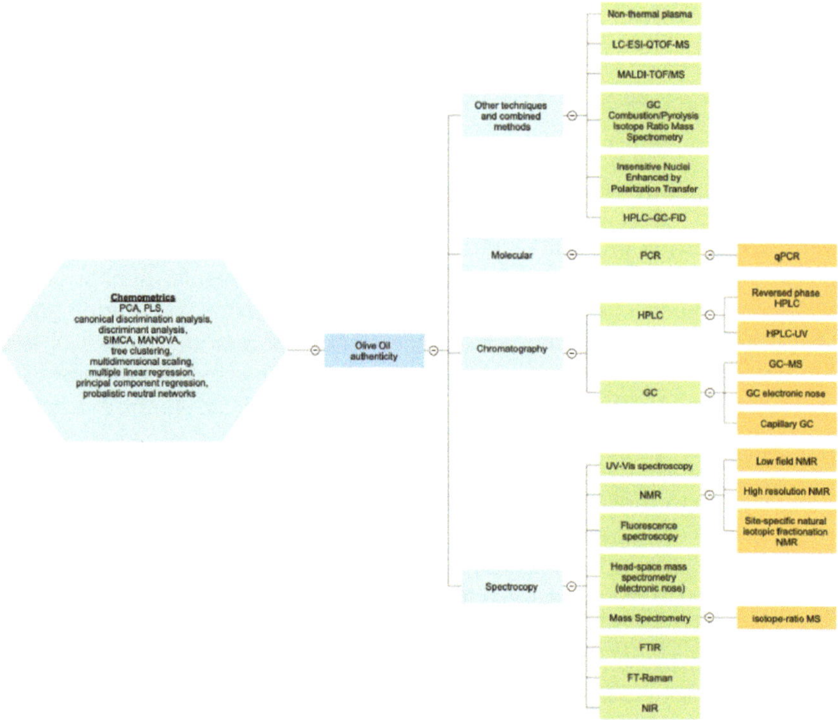

Figure 1. Overview of the main techniques used to verify the authenticity or detect adulteration in olive oil.

2. Methods

Trade standards are updated constantly to allow solutions to be given to adulteration phenomena driven by fraudsters who continuously use more and more sophisticated procedures. Clustered inside targeted and profiling approaches, methods using spectroscopy or isotopic and chromatographic techniques have been extensively applied. Chromatographic methods, which include a lot of official methods, as well as different spectroscopic, isotopic and trace element techniques are all used to ensure that the progress of research in detecting adulteration is quick enough. The speed that analyses can be carried out with, the requirements of the use of statistics, the qualitative characteristics of the methods, limits of detection and the applicability range are, among others, essential parameters that are taken into account. The in-depth and comprehensive understanding of the methods used to quantify the chemical substances responsible for authenticating olive oil will also allow the complex nature of olive oil to be better understood rendering analytical results more informative (Aparicio et al. 2013).

2.1 Chromatographic techniques

2.1.1 HPLC

High performance liquid chromatography (HPLC) combined with ultraviolet detection (HPLC-UV) was used in the study of Carranco et al. (2018) to analyze and characterize olive oils. More than 130 edible oils, among which monovarietal EVOOs, were subjected to analysis. Principal component analysis (PCA) allowed olive oils to be clearly discriminated from vegetable oils. Nevertheless, specific HPLC-UV chromatographic segments had to be used to discriminate EVOOs of the same variety. Partial least square (PLS) regression was used to authenticate olive oil contaminated with other types of EVOO, refined olive oil, and vegetable oil. The results taken after PLS were of very good quality, with quantification errors of less than 2.9%.

The detection of adulteration of olive oils with as little as 1% of linoleic-rich vegetable oils could be quantitatively performed by reversed-phase HPLC on an octyl-bonded silica stationary phase (Supelcosil-LC 8). The mobile phase used was acetone-acetonitrile (70:30, v/v). A comparison of the chromatogram of pure olive oil with chromatograms of soybean, sunflower and corn oils with olive oil was performed. It was indicated that the detection of adulteration with less than 1% of linoleic-rich vegetable oils in olive oil is possible qualitatively and quantitatively in less than 15 minutes. The establishment of an olive oil authenticity factor was achieved that could be used to rapidly indicate adulteration and a simple equation was used to determine the extent of that adulteration (El-Hamdy et al. 1995).

A method involving reversed-phase HPLC with amperometric detection was used by Dionisi et al. (1995) to analyze tocopherols and tocotrienols in vegetable oils. The recovery of α-tocotrienol and γ-tocotrienol in extra virgin olive oil was 97.0 and 102.0%, respectively.

No tocotrienols were found in olive, hazelnut, sunflower, and soybean oils. However, it was shown that relatively high levels of tocotrienols were present in palm and grape seed oils. This method can be used for the detection of small quantities (1–2%) of palm and grape seed oils in olive oil or in any tocotrienol free vegetable oil and could therefore allow an easier assessment of vegetable oil authenticity.

In another study, 120 authentic olive oil samples of two Cretan olive cultivars, Koroneiki and Mastoidis, collected at different maturity stages from different oil producing fields were analyzed using isocratic HPLC according to the official EC method. Multivariate analysis, including PCA and canonical discrimination analysis were employed to differentiate cultivars, regions and sampling dates. The use of triglyceride compositional data was proven to be sufficient to differentiate between the two varieties by means of discriminant analysis (DA). The samples of each variety were clearly discriminated according to their geographical origin. Samples originating from neighboring locations presenting no significant differences in geographic morphology, or samples originating from locations with certain extremes in the climatic conditions, led to some discrepancies in classification (Stefanoudaki et al. 1997).

A comparison of 3 HPLC systems for determining tocopherols and tocotrienols in olive oil was also reported. Fluorescence and diode array connected in series, ultraviolet, and evaporative light scattering were the systems chosen, with the fluorescence detection allowing the optimum results to be extracted. The system was tested for quantifying tocopherols and tocotrienols in olive oil. The method was verified by evaluating linearity, limits of detection, repeatability potential, and recovery. Similar qualitative profiles were obtained from all samples with six identified substances: α-T, β-T, γ-T, δ-T, α-T3, and γ-T3. α-Tocopherol (α-T) was the principal vitamin E isomer found in all samples and its quantity ranged from 93 to 260 mg/kg. The levels of tocopherols and tocotrienols were in the range of 100 to 270 mg/kg. The geographical origin did not affect the tocopherol and tocotrienol content of the samples (Cunha et al. 2006).

2.1.2 GC

GC was used in the study of Yan et al. (2018) for the analysis of the contaminants 2- and 3-monochloropropanediol (2- and 3-MCPD) esters and glycidyl esters (GEs). The analyses were carried out in 84 oil samples using GC–MS/MS for discriminating different grades of olive oil. Concentrations of 2- and 3-MCPD esters and GEs ranged between 0–6 mg/kg, 0–1.5 mg/kg, and 0–1 mg/kg oil, respectively. Significantly higher ($P < 0.001$) concentrations of these compounds were found in lower-grade olive oils than EVOO. Similar differences were found in the cases of refined and cold-pressed oils. The method could detect adulterants in quantities as low as 2%–14% depending on the adulterants. It was concluded that the MCPD esters appeared to be highly specific and therefore very effective in detecting lower processing grade oils in EVOO.

The effectiveness of flash GC electronic nose and multivariate data analysis employed to rapidly screen commercial EVOOs sold with a different geographical origin declaration was evaluated in another study. An analysis with solid phase micro extraction coupled to GC mass spectrometry (GC-MS) was also performed. The method was suitable for verifying the geographic origin of EVOOs based on PCA and DA applied to the volatile profile of the headspace as a fingerprint. The variables chosen effectively led to discrimination between "100% Italian" and "non-100% Italian" oils. PLS DA also allowed for accurate classification to be achieved (Melucci et al. 2016).

In the study of Andrikopoulos et al. (2001) it was reported that individual triglyceride species of olive oil and a number of seed oils can be differentiated on a WCOT TAP CB fused-silica capillary column by capillary GC. The presence of < 5% seed oils (except peanut oil) in olive oil can be detected by detecting the increasing amounts of trilinolein or tripalmitin. Adulteration of olive oil with > 20% peanut oil can be found by evaluating the levels of palmitodilinolein. Separation of triglyceride species in the reversed-phase HPLC (RP-HPLC) mode can be achieved using the capillary GC technique, while their structural characteristics can be determined using selective ion monitoring mass spectrometry. It was shown though that mixtures of virgin olive oil with refined or residue olive oil cannot be differentiated from the pure virgin oil using this method.

In another study, 21 admixtures of olive oil with virgin oil and refined hazelnut oils obtained via a number of processing techniques from different varieties and geographical locations were analyzed for filbertone using solid phase microextraction and multidimensional GC. The results indicated that the method was sensitive enough to effectively detect filbertone and can therefore be used for detecting adulteration of olive oil with virgin hazelnut oils at levels of up to 7%. The fact that filbertone is present at very low levels can make its detection very difficult in certain admixtures. Regardless of that though, the lowest adulteration level that can be detected strongly depends on the oil varieties present in the adulterated samples. In this study, occasional detection of R- and S-enantiomers of filbertone in olive oils adulterated with 10–20% of refined hazelnut oil was achieved (Flores et al. 2006).

Ruiz-Samblás et al. (2012) reported the development of a method that can be used to reliably identify and quantify adulteration of olive oils with vegetable oils. The process relies on determining triacylglycerols in oils via the use of high temperature GC-MS. The pre-treatment of the chromatographic profiles was necessary before constructing the models.

In the beginning, a class-modeling approach, Soft Independent Modeling of Class Analogy (SIMCA) was employed to identify vegetable oil. After that, a separate calibration model for each blending type was developed using PLS. The correlation coefficients of actual versus predicted levels given by multivariate calibration models ranged from 0.95 to 0.99. Additionally, genetic algorithms were employed, as a variable selection method, improving the models and allowing R^2 values higher than 0.90 for calibration set to be yielded. This model had better predictive capabilities than the PLS. It was therefore shown that the method was effective in quantifying blends of olive oil with the vegetable oils tested when the latter contained at least 10% olive oil.

The study of Mannina et al. (2002) focused on investigating whether there is direct evidence relating the fatty acid composition of olive oils to certain cultivars grown within a well-specified geographical area. To group olive oils in accordance with their cultivars, ^{13}C high-field nuclear magnetic resonance (NMR) and GC were employed to analyze 60 extra virgin olive oils originating from the same region in Italy (southwestern Sicily) and obtained from four monovarietal cultivars. The ^{13}C NMR spectrum can be used to extract information about glycerol triesters of olive oils. GC can be used to give the complete fatty acid profiles of olive oil samples. NMR and GC peaks were selected based on the sensitivity they presented for different cultivars, using multivariate analysis of variance (MANOVA). PCA, tree clustering analysis, multidimensional scaling, and linear DA were then performed on the MANOVA-selected peaks. Results obtained from ^{13}C NMR and GC techniques in combination with the multivariate statistical procedure indicated that the method can be useful for the analysis of fatty acids in order to group the monovarietal olive oils that belong to the same cultivars. Grouping of olive oils in accordance with their cultivars was achieved for particular ^{13}C resonances all belonging to fatty chains in the *sn* 1,3 position of the glycerol moiety.

2.2 Spectroscopic techniques

2.2.1 UV-VIS

Spectroscopic techniques have been used in several studies to assist in the quantification of adulteration of olive oil and in designing models to achieve this purpose.

A method aiming at quantifying adulteration of EVOO with soybean edible oil using fluorescence and UV Vis spectroscopy was presented by Milanez and Nobrega (2017). Different levels of soyabean oil (10, 50, 100, 150, 200, 250 and 300 g/kg) were used to adulterate samples for testing purposes. A number of multivariate regression strategies were subjected to analysis using PLS with full spectrum; PLS with significant regression coefficients selected by the Jack-Knife algorithm and multiple linear regression (MLR) with pre-selected variables by stepwise algorithms, successive projections algorithm and genetic algorithm. An assessment of the predictive ability of the models was carried out for each spectroscopic technique. In the case of fluorescence spectroscopy, the prediction results obtained were satisfactory for all regression models with Root Mean Square Error of Prediction values ranging from 14.0 to 17.5 g/kg. The evaluation of the regression methods for UV-Vis spectra gave higher Root Mean Square Error of Prediction values that ranged from 13.3 to 30.4 g/kg. It was therefore indicated that the two spectroscopic techniques perform similarly when it comes to their predictive ability of the regression models.

Casale et al. (2010) reported that the combined use of data given by different analytical instruments can be very complex. They therefore studied the potential of coupling three analytical techniques for designing a class model for EVOO originating from Liguria. A sampling plan for the Ligurian EVOO was designed by selecting a representative subset from all EVOO samples of this region. Two algorithms were used, namely Kennard–Stone and Potential Function, aiming at achieving uniform distribution on the production area and allowing the subset to be representative of the production density. Head-space mass spectrometry (electronic nose), UV–visible and Near-Infrared (NIR) spectroscopy were used for the analysis of the samples. It was shown that there is a high potential in using chemometrics to effectively extract and combine (fusion) multivariate and non-specific data that could be used to design a class model for Liguria EVOO.

Fluorescence spectra of different vegetable oils, such as olive oil, olive residue oil, refined olive oil, corn oil, soybean oil, sunflower oil, and cotton oil, were studied by Kyriakidis and Skarkalis (2000) in their original state, with a wavelength of 360 nm used as excitation radiation. All oils taken into account, with the exception of EVOO, exhibited a strong fluorescence band at 430–450 nm. EVOO gave an interesting fluorescence spectrum with 3 bands (a low intensity doublet at 440 and 455 nm, a strong one at 525 nm, and a medium intensity at doublet 681 nm). The band at 681 nm was reported as the chlorophyll band. The band at 525 nm to some extent derived from vitamin E, while the low intensity doublet at 440 and 455 nm was in correlation with the absorption intensity at 232 and 270 nm of olive oil. These fluorescence spectra were quickly and easily measured (in about 5 min) and could potentially be employed to authenticate samples of EVOO.

2.2.2 NIR

NIR was the method of choice in the study of Jiménez-Carvelo et al. (2018). EVOO is used extensively as a food ingredient or in food preparation as it offers a number of health benefits. This fact, together with its higher prince in comparison to other vegetable oils renders it a good candidate for adulteration. NIR and fluorescence excitation–emission matrix spectroscopies, coupled to chemometric tools were therefore employed to assist in the authentication and validation of the geographical origin of Argentinean EVOO. For each spectral data set, PCA was employed to both first–order NIR and second–order fluorescence data, partial least squares–DA (PLS–DA) to NIR data, and the multidimensional version of the latter (NPLS–DA) to fluorescence data. Sixty EVOO samples of both known and unknown registered designation of origin (RDO) were analyzed, together with samples adulterated with other edible oils. It was shown that fluorescence spectroscopy did not have the ability to determine the RDO of all EVOO samples. NIR, however,led to correct classification in 100% of the cases. It was proved that fluorescence data provides only slightly better results than NIR spectroscopy when it comes to detecting EVOO adulteration with other vegetable oils.

The accuracy of NIR spectra in detecting adulteration of olive oil was also mentioned in the study of Galtier et al. (2007). In this study, chemometric treatment of NIR spectra was employed for the quantification of fatty acids and triacylglycerols in virgin olive oil samples (n = 125) and for their classification (PLS1-DA) into five very geographically closed registered designations of origin (RDOs) of French virgin olive oils ("Aix-en-Provence", "Haute-Provence", "Nice", "Nyons" and "Vallée des Baux"). It was shown that there was a correlation between each RDO and one or two specific components of virgin olive oils in accordance with their cultivar compositions. Satisfactory results were given despite the similar compositions of the cultivars of two denominations of origin ("Aix-en-Provence" and "Vallée des Baux"). The use of chemometrics on the NIR spectra allowed the extraction of similar results, as those obtained using techniques such as GC and HPLC and can therefore be used to rapidly and reliably authenticate French virgin olive oils.

In the study of Woodcock et al. (2008), the effectiveness of NIR transflectance spectroscopy (1100–2498 nm) in combination with chemometrics in confirming the geographical origin of European olive oil samples was assessed using 913 extra virgin olive oil samples (210 Ligurian and 703 non-Ligurian). The collection of the samples was carried out over three consecutive harvests (2005, 2006, and 2007). A multivariate spectral fingerprint for Ligurian olive oil was designed and used to confirm or reject a claim that any given sample was Ligurian. Samples were pseudo-randomly split into calibration (n = 280) and validation sets (n = 633) and the only selection constraint put in place was to have only equal numbers of Ligurian and non-Ligurian samples in the calibration set. The authors of the study noted that following the use of chemometrics, the most successful models were as accurate as 92.8 to 81.5% in predicting the origins of samples of Ligurian and non-Ligurian olive oils, respectively.

2.2.3 NMR

NMR Spectroscopy has been extensively used to analyze olive oil and it is considered to be a valuable tool for assessing the quality and authenticity of the product (Dais and Hatzakis 2013). The use of a rapid time-domain NMR (TD-NMR) process for detecting olive oil adulteration with polyunsaturated vegetable oils in filled bottles was suggested by Santos et al. (2017). Measurements of the ^{1}H transverse relaxation times (T_2) of 37 commercial samples were taken using a low-field NMR (LF-NMR) spectrometer and a unilateral NMR (UNMR) sensor. The use of LF-NMR revealed better feasibility in comparison to the UNMR sensor, with higher signal-to-noise ratio and larger difference in the T_2 decays. PCA gave tight and clearly separated clusters of pure olive oil, pure soybean oil, and blends of the two. A soft independent modeling of class analogies (SIMCA) analysis classification model showed that in the case of five brands of olive oil marketed in Brazil, adulteration with polyunsaturated fatty acids was detected, which was also further confirmed by high-resolution NMR. Overall, LF-NMR allowed rapid screening for olive oil adulteration directly in the sealed bottles.

A review published by Ogrinc et al. (2003) examined the use of high-resolution NMR spectroscopy and mass spectrometry (MS) for the characterization of wine, fruit juices, and olive oil and detection of potential adulteration. Particular importance was given on the complementary use of SNIF-NMR (site-specific natural isotopic fractionation nuclear magnetic resonance) and IRMS (isotope-ratio mass spectrometry) in association with chemometric methods. The authors stated that improvements in methods involving NMR have led to new applications for the detection of wine adulteration. One- and multidimensional ^{1}H and ^{13}C NMR methods combined with chemometrics can effectively detect adulteration in terms of varieties, geographical origin and vintage, as well as the addition of low-quality product to higher-grade wine. Although a lot of progress has been made, there is a number of significant challenges that still remain, which in the case of the use of isotopes these can be (1) the use of more isotopes (^{15}N, ^{18}O, ^{2}H) for the analysis of the minor trace substances (proteins, phenolic compounds, trace oligosaccharides; (2) routine application for detecting intermolecular isotope patterns; (3) and routine CSIR of hydrogen isotopes after GC separation. Also, in the study of Sacchi et al. (1997) the authentication and quality assessment of virgin olive oil were conducted using high-resolution ^{1}H and ^{13}C NMR spectroscopy. The study presented the numerous determinations that were evaluated, paying particular attention on detecting adulteration with foreign (seed) oils and esterified or refined olive and olive pomace oils.

A combination of ^{1}H NMR and ^{31}P NMR spectroscopy and multivariate statistical analysis was employed in another study, to allow the classification of 192 samples from 13 types of vegetable oils, namely, hazelnut, sunflower, corn, soybean, sesame, walnut, rapeseed, almond, palm, groundnut, safflower, coconut, and virgin olive oils collected from different geographical locations in Greece. 1,2-Diglycerides, 1,3-diglycerides, the ratio of 1,2-diglycerides to total diglycerides, acidity, iodine value, and fatty acid composition recorded by analyzing the respective ^{1}H NMR and ^{31}P NMR spectra, were chosen as variables to develop a classification/prediction model using DA. This model, obtained from the training set of 128 samples, significantly discriminated between the different classes of oils, while 100% correct classification was achieved for 64 of the samples. Preparation and analysis of different artificial mixtures of olive–hazelnut, olive–corn, olive–sunflower, and olive–soybean oils were carried out using ^{1}H NMR and ^{31}P NMR spectroscopy. The DA that followed allowed detection of adulteration levels as low as 5% w/w, in all cases that fresh virgin olive oil samples were analyzed, as reflected by their high 1,2-diglycerides to total diglycerides ratio ($D \geq 0.90$) (Vigli et al. 2003).

In another study, ^{1}H high-field NMR was employed for the analysis of 216 extra virgin olive oils collected in a period of three years (1996, 1997, and 1998) from different regions in Italy. The aim of the study was the evaluation of the effectiveness of this technique in the geographical characterization of olive oils. A statistical procedure performed on the intensity of selected NMR peaks was also proposed. Tree clustering analysis of NMR data conducted without a priori hypothesis indicated that reliable parameters were present that could be used to classify the olive oils in accordance with the location of their production. Linear DA was used taking into account certain NMR parameters of olive oils of the same production year, allowing the classification of samples in accordance with their geographical origin with a very low number of errors. Furthermore, the samples were satisfactorily grouped via combining the NMR data of olive oils from the years 1996 and 1997. Using appropriate sampling, a detailed analysis of data led to the conclusion that the place of production could be singled out as a discriminating factor regardless of the cultivars the olive oils were derived from (Mannina et al. 2001).

^{1}H NMR spectroscopy can be used instead of conventional chromatography to determine the composition of oils. A variety of oils from olive, hazelnut and sunflower were therefore subjected to analyses using ^{1}H NMR spectroscopy. The conditions of the experiments were carefully chosen in order to reduce the experimental time. It was shown that multivariate statistical methods, and particularly DA, applied to certain predominant peaks in the ^{1}H NMR spectra of oils can lead to a good separation between these three oils and allow their mixtures to be detected (Guillou 2000). Authentic EVOOs of different geographical origins were analyzed using ^{1}H NMR fingerprinting in conjunction with multivariate statistical analysis. Monte-Carlo embedded cross-validation was employed to indicate that the combined use of PCA, canonical analysis, and classification via the nearest class mean allows the prediction of the origin of olive oil samples from ^{1}H NMR data. Taking into account the relatively low number of samples analyzed, correct prediction probabilities of 78% were obtained with region specific correct predictions between 53% and 100% (Longobardi et al. 2012).

The combined use of ^{1}H NMR fingerprinting with multivariate analysis can be used to differentiate olive oils according to their geographical origin and processing as well as production year. Multivariate (chemometric) techniques can be used to isolate the most significant spectral data. Linear DA of the first 50 PC's was used to classify the olive oil samples (97 or 91) in accordance with their geographical origin and production year. The data analysis was conducted both with and without outliers. Significant improvements of the success rate can be reached using probalistic neural networks. Correct classification by probalistic neural networks ranged from 58 to 100% on the external validation sets. The use of other chemometric techniques did not lead to better results (Rezzi et al. 2005).

2.2.4 Fourier-transform infrared spectroscopy (FTIR)

In the study Maggio et al. (2010), a procedure that could be used to qualitatively and quantitatively determine vegetable oils (canola, hazelnut, pomace and high linoleic/oleic sunflower) as adulterants in EVOO samples was developed. PLS was used to analyze data obtained using FTIR. Calibration models were developed for EVOO purity, with wavelength selection in the infrared region, in accordance with their predictive ability, with first derivative and mean centering applied as data pretreatment. Internal validation of the PLS models was performed by the leave-one-out procedure. The method was proved to be very effective for determining modeled adulterants, but it could also reveal adulteration even if other adulterants were present.

FTIR spectroscopy was also used in the study of Tay et al. (2002) for identifying the adulteration of olive oils. Single-bounce attenuated total reflectance measurements were made on pure olive oil and olive oil samples that contained various levels of sunflower oil (20–100 mL vegetable oil/L of olive oil). DA using 12 principal components allowed the pure and adulterated samples to be discriminated based on their spectra. A partial least squares model was applied for verifying the concentrations of the adulterant. Moreover, the DA could be used for the classification of olive oil samples, distinct from other oils based on their infrared spectra.

An FTIR spectroscopy method was also used to analyze EVOO adulterated with palm oil. Measurements were made on pure EVOO and that mixed with various levels of palm oil (1.0–50.0% wt./wt. in EVOO). Optimization of two multivariate calibrations, namely PLS and principal component regression was carried out to construct the calibration models, either for normal spectra or its first and second derivatives. The DA was employed for classification analysis between EVOO and EVOO mixed with palm oil and other vegetable oils (palm oil, corn oil, canola oil, and sunflower oil). Frequencies at fingerprint region, and particularly at 1500–1000 cm^{-1}, were studied for both quantification and classification. Either PLS or principal component regression at first derivative spectra revealed the best calibration models for the prediction of the levels of adulterated EVOO samples, with coefficient of determination (R^2) of 0.999 and root mean standard error of cross validation (RMSECV) of 0.285 and 0.373, respectively. DA correctly classified pure and adulterated samples based on their FTIR spectra and no misclassification was observed. Furthermore, DA was also effective enough to discriminate between EVOO samples and samples of other vegetable oils (Rohman and Man 2010).

2.3 DNA techniques

2.3.1 Polymerase chain reaction (PCR)

The olive oil market has undoubtedly a worldwide economic impact. The reliability of the market is of paramount importance and, as a result, every effort should be made to control the increasing number of fraud and adulteration attempts. The implementation of traceability systems is crucial especially for olive oil with certified labels. In these cases, only a small number of olives varieties can be used in a restricted well-defined geographical region. Molecular markers have been widely used in the food sector, while high-resolution DNA Melting technology, in particular, has been used with success for olive oil authentication (Pereiraa et al. 2018).

With regards to the use of PCR, it is important to mention the studies of Uncu et al. (2017) and Alonso-Rebollo et al. (2017). In the study of Uncu et al. (2017), a comparison was made between a DNA-barcode assay and lipidomics. PCR capillary electrophoresis (PCR-CE) was applied on blends of olive and seed oils. The samples were also tested using GC to analyze their fatty acid composition. While the PCR-CE assay was as efficient as the GC in the detection of most vegetable oils, it was clearly more powerful than the commonly used analytical chemistry techniques in detecting foreign species and especially corn and safflower. Furthermore, the DNA-based test allowed correct identification of all tested olive oil: hazelnut oil blends, while it did not allow the detection of hazelnut oil adulteration via fatty acid profile analysis. Therefore, this report demonstrated that a PCR-CE barcode assay could be used for the effective detection of adulteration in olive oil.

Alonso-Rebollo et al. (2017) reported that the effectiveness of qPCR when used for detecting adulteration of olive-oil is heavily determined by the DNA extracted and the amplification primers. As a result, four trnL gene amplification assays were developed (A-, B-, C- and D-trnL systems). Optimization of the qPCR conditions, primer concentration and annealing temperature, were conducted. Tests for the efficiency and sensitivity of the systems were carried out to choose the optimum method. The system chosen (D-trnL) was highly specific. This qPCR system allowed olive DNA processed using different processing methods, to be detected with very high accuracy, thus indicating that the method is suitable for authenticating olive oil of all types.

2.4 Other techniques

A new approach for authenticating olive oil samples of different quality grades was proposed by Vaclavik et al. (2009). A new type of ion source, direct analysis in real time coupled to a high-resolution time of flight mass spectrometer (TOFMS) was used to build a triacylglycerols and/or polar compounds profile extracted with a methanol–water mixture. The principal factors affecting the ionization efficiency of triacylglycerols were the type of solvent used, the degree of sample dilution, ion beam temperature, and presence of ammonia vapors. The ionization yield of polar substances heavily depended on the water content in the extract and ion beam temperature. The use of direct analysis in real time-TOFMS, allowed the EVOO, olive pomace oil and olive oil samples to be easily differentiated, but also allowed EVOO adulteration with hazelnut oil to be detected. Based on the linear DA, the proposed method allowed HO addition of 6 and 15% (v/v) to be detected when evaluating direct analysis in real time–TOFMS mass profiles of polar compounds and triacylglycerols, respectively.

In another study, non-thermal plasma (NTP) was evaluated as a preparative analytical technique that could allow adulterated and pure olive oils to be classified. Non-thermal plasma discharges produce highly oxidative species such as singlet oxygen, and atomic oxygen. It can therefore be expected that NTP-induced oxidation triggers certain lipid oxidation processes that depend on the specific composition of the oil matrix and minor constituents. After adding 1–3% sunflower oil in EVOO samples, the researchers submitted them to NTP treatment. It was demonstrated that while authentication of untreated samples from the pure olive oil could not be achieved, NTP treatments of 60 min (Ar/O_2 0.1%) on the oil batches led to the formation of a unique set of secondary volatile lipid oxidation compounds, which then allowed the adulterated samples to be classified. Specifically, the researchers took advantage of the fact that sunflower oil commonly contains higher levels of poly-unsaturated fatty acids than virgin olive oil, and therefore adulteration of the latter with sunflower oil could lead to the production of oxidation volatiles at a higher rate. This assumption was proved to be correct for some oxidation markers such as nonanal (ratio 0.642), 6-methyl-5-hepten-2-one (0.708), (E)-4-undecenal (0.805) and decanal (0.816). However, a number of significant lipid oxidation compounds such as 5-nonanol (ratio 4.450), 2-butanone (3.097), 2-pentyl furan (ratio 2.174) and (E)-2-octenal (ratio 1.588) were not found in pure oxidized olive oil (Durme and Vandamme 2016).

2.5 Combined methods

An optimized and validated liquid chromatography-electrospray ionization-quadrupole-time of flight-mass spectrometry (LC-ESI-QTOF-MS) method with an integrated non-target screening workflow was employed with the aim of investigating the metabolomic profile of 51 Greek monovarietal EVOOs from the varieties: Manaki, Ladoelia, Koroneiki, Amfissis, Chalkidikis and Kolovi. R language and an XCMS package were employed for data processing. A local database that consists of 1608 substances naturally found in *O. Europa* L. was developed aiming at accelerating the identification workflow. The preliminary study of the distribution of EVOOs toward their cultivars was carried out with PCA. Ant Colony Optimization-Random Forest (ACO-RF) was employed to allow the prioritization of over 250 characteristics and develop a classification tree. Apigenin, vanillic acid, luteolin 7-methyl ether and oleocanthal were proposed as the most suitable markers for classifying Greek EVOOs' cultivars. It was also suggested that the method can be used for food authenticity purposes and can successfully carry out classification of Greek EVOOs using a non-target screening ultra-high performance liquid chromatography coupled with electrospray ionization-quadrupole-time of flight-mass spectrometry (RP-UHPLC-ESI-QTOFMS) method combined with ACO-RF (Kalogiouri et al. 2018).

A combination of chromatographic, mass spectrometric and spectroscopic methods supplemented with chemometric tools was applied in a pilot study in Croatia, aiming at authenticating EVOOs. Determination of fatty acids was carried out by GC-MS, triacylglycerols by matrix-assisted laser desorption ionization time-of-flight/mass spectrometry (MALDI-TOF/MS), and non-selective analysis was conducted using near-infrared spectroscopy. Chemometric methods were employed to analyze the information where TAG and NIR spectra appeared to be the most promising for discriminating between the samples, according to their geographical origin. Triglycerides profiling data were the most effective classification markers for EVOOs in the PCA. Moreover, near-infrared spectroscopy data were very useful for predicting quality parameters. It was therefore concluded that NIR is a rapid, inexpensive and reliable technique that can be used to authenticate EVOOs, but the use of MALDI-TOF/MS is essential if the samples need to be fully characterized (Peršurić et al. 2018).

Sixty-seven olive oil samples of different qualities and seventy-nine samples of other vegetable oils were collected in the study of Jimenez-Carvelo et al. (2017). Transesterification was carried out before employing spectroscopy. k-nearest neighbours (kNN), PLS-DA, one-class partial least squares (OCPLS), support vector machine classification (SVM-C), and soft independent modelling of class analogies (SIMCA) were used. Partial least square-regression (PLS-R) was employed as a multivariate quantification method. FTIR fingerprints very effectively differentiate between pure olive and non-olive oil. Correct classification was achieved for all olive oil samples and more than 90% of other oils when PLS-DA or SVM-C were used. Generally, FTIR fingerprints appeared to have higher discrimination potential than Raman's in both classification and regression scenarios. As in the case of similar studies though, a low number of oils was used to prepare the blends.

The detection of low concentrations of refined hazelnut oil in refined olive oil can be quite challenging. FT-Raman and FT-MIR spectroscopies were therefore employed in the study of Baeten et al. (2005) to determine the level of detection of the presence of hazelnut oil in olive oil. Spectroscopy was used to analyze not only the entire oil but also its unsaponifiable matter. Univariate and multivariate statistical models were developed, and it was shown that it was possible to successfully discriminate between olive and hazelnut oils, while the detection was possible when the presence of hazelnut oil's concentration in olive oil was > 8%. The limit of detection was higher when blends of edible oils from diverse geographical origins were used.

The main biomolecules found in olive oil are fatty acids. An analytical method was therefore described that can measure the $\delta^{13}C$ and the $\delta^{2}H$ values of fatty acid methyl esters obtained from olive oil triglycerides, following base-catalyzed transesterification using methanol. Analysis of the ester was conducted using GC Combustion/Pyrolysis Isotope Ratio Mass Spectrometry (GCC/Py-IRMS). The method allowed a baseline separation between the peaks of the main esters (C16, C18, C18:1, C18:2) to be obtained. The accuracy, precision and uncertainty of the method were determined. Moreover, the effects of methylation on $\delta^{13}C$ and $\delta^{2}H$ values were examined. The method gave satisfactory results, allowing reliable data to be extracted. As it is known already that $\delta^{13}C$ and $\delta^{2}H$ determined in bulk olive oils can be used, to some extent,to distinguish samples in accordance with their geographical origin, it was expected that the method could improve this differentiation potential. Further studies would need to be conducted though, based on the analysis of real samples obtained from different geographical areas, to confirm these findings (Paolinia et al. 2017).

A number of analytical methods can currently be employed for evaluating virgin olive oil minor compounds. It is true, however, that multiclass methodologies have not yet been widely used. Therefore, LC-MS and GC-MS platforms were employed to develop two methods that targeted the simultaneous determination of more than 40 substances that belonged to different virgin olive oil secondary compounds. A very effective liquid-liquid extraction methodology was used, following optimization, for the isolation of the substances. The chromatographic analysis times were 31 and 50 min, respectively. The two methods were compared in terms of performance, complexity, cost and effectiveness in analyzing the compounds. The development and optimization of this type of multi-class analytical methodologies highly increases throughput and renders the processes less expensive and simpler, while reduces the need for single-chemical class determinations (Olmo-García et al. 2018).

In the study of Merchaka et al. (2013), a rapid adiabatic ^{13}C-INEPT (Insensitive Nuclei Enhanced by Polarization Transfer) experiment for the isotopomic analysis of olive oil samples, which had previously been optimized in order to quantify individual fatty acids within triacylglycerols using multivariate linear regression models, was evaluated. The study aimed at evaluating the effectiveness of ^{13}C-INEPT in authenticating olive oils when compared to GC and ^{1}H NMR. A new set of olive oil samples was therefore subjected to analysis using the above-mentioned techniques. A comparison was made between analytical variables and repeatability. The accuracy of the fatty acid quantification models was verified, while ^{13}C-INEPT achieved the best classification of olive oil samples in accordance with the altitude the plants grew at and the color of the olives.

Finally, with the use of 4,40-desmethylsterols derivatives chromatographic fingerprint obtained from an off-line HPLC–GC-FID system together with a preprocessing step that eliminated the influence of the sample size, two models were obtained based on the Un-FpM or in the R-FpM that could be used for the quantification of EVOO in mixtures of edible oils. It was suggested that the model that was based on the R-FpM, with a reduced number of initial variables (431), 4LVs and a prediction error of about 10% (suitable for this type of samples), could be suitable for the analysis of samples in all cases that the EVOO concentration must be known (Pérez-Castaño et al. 2018).

3. Conclusions

Olive oil is a highly nutritious food rich in vitamins and unsaturated fatty acids. Different grades can be produced, with EVOO being the purest and most expensive grade. Adulteration of olive oil can occur by either mixing it with other vegetable oils or by adding cheaper olive oil to more expensive grades (most commonly EVOO). A variety of methods are currently used to verify the authenticity of the product and detect adulterants or contaminants, including chromatographic (e.g., HPLC, GC), spectroscopic (e.g., UV-Vis, NIR, NMR, FTIR), or DNA-based methods (e.g., PCR). Finally a series of methodologies combining the above-mentioned techniques have recently been trialed, leading to potential improvements in the accuracy and effectiveness of the authentication processes.

Abbreviations

Ant Colony Optimization-Random Forest	ACO-RF
Discriminant analysis	DA
Extra-virgin olive oil	EVOO
Flame-ionization detector	FID
Extra-virgin olive oil	EVOO
Flame-ionization detector	FID
Fourier-transform infrared spectroscopy	FTIR
Gas chromatography	GC
Gas chromatography-mass spectrometry	GC-MS
Gas chromatography Combustion/Pyrolysis Isotope Ratio Mass Spectrometry	GCC/Py-IRMS
High performance liquid chromatography	HPLC
Isotope-ratio mass spectrometry	IRMS
k-nearest neighbours	kNN
Liquid chromatography-electrospray ionization-quadrupole-time of flight-mass spectrometry	LC-ESI-QTOF-MS
Low-field nuclear magnetic resonance	LF-NMR
Matrix-assisted laser desorption ionization time-of-flight/mass spectrometry	MALDI-TOF/MS
Multivariate analysis of variance	MANOVA
Multiple linear regression	MLR
Mass spectrometry	MS
Nuclear magnetic resonance	NMR

Non-thermal plasma	NTP
One-class partial least squares	OCPLS
Principal component analysis	PCA
Polymerase chain reaction	PCR
Polymerase chain reaction capillary electrophoresis	PCR-CE
Partial least square	PLS
Partial least square-regression	PLS-R
Insensitive Nuclei Enhanced by Polarization Transfer	INEPT
Registered designation of origin	RDO
Reversed-phase ultra-high performance liquid chromatography coupled with electrospray ionization-quadrupole-time of flight-mass spectrometry	RP-UHPLC-ESI-QTOFMS
Soft independent modeling of class analogies analysis	SIMCA
Soft independent modelling of class analogies	SIMCA
Site-specific natural isotopic fractionation nuclear magnetic resonance	SNIF-NMR
Support vector machine classification	SVM-C
Time-domain nuclear magnetic resonance	TD-NMR
Time of flight mass spectrometer	TOFMS
Unilateral nuclear magnetic resonance	UNMR
Ultra-violet	UV

References

Alonso-Rebollo, A., S. Ramos-Gómez, M.D. Busto and N. Ortega. 2017. Development and optimization of an efficient qPCR system for olive authentication in edible oils. Food Chemistry 232: 827–835.

Andrikopoulos, N.K., I.G. Giannakis and V. Tzamtzis. 2001. Analysis of olive oil and seed oil triglycerides by capillary gas chromatography as a tool for the detection of the adulteration of olive oil. Journal of Chromatographic Science 39(4): 137–145.

Aparicio, R., M.T. Morales, R. Aparicio-Ruiz, N. Tena and D.L. García-González. 2013. Authenticity of olive oil: Mapping and comparing official methods and promising alternatives. Food Research International 54(2): 2025–2038.

Baeten, V., J.A.F. Pierna, P. Dardenne, M. Meurens, D.L. García-González and R. Aparicio-Ruiz. 2005. Detection of the presence of hazelnut oil in olive oil by FT-Raman and FT-MIR spectroscopy. Journal of Agricultural and Food Chemistry 53(16): 6201–6206.

Carranco, N., M. Farrés-Cebrián, J. Saurina and O. Núñez. 2018. Authentication and quantitation of fraud in extra virgin olive oils based on HPLC-UV Fingerprinting and Multivariate Calibration. Foods, MDPI 7(44): 1–15.

Casale, M., C. Casolino, P. Oliveri and M. Forina. 2010. The potential of coupling information using three analytical techniques for identifying the geographical origin of Liguria extra virgin olive oil. Food Chemistry 118(1): 163–170.

Consolandi, C., L. Palmieri, M. Severgnini, E. Maestri, N. Marmiroli, C. Agrimonti, L. Baldoni, P. Donini, De G. Bellis and B. Castiglioni. 2008. A procedure for olive oil traceability and authenticity: DNA extraction, multiplex PCR and LDR-universal array analysis. European Food Research and Technology 227: 1429–1438.

Cunha, S.C., J.S. Amaral, J.O. Fernandes and M.B.P.P. Oliveira. 2006. Quantification of tocopherols and tocotrienols in portuguese olive oils using HPLC with three different detection systems. Journal of Agricultural and Food Chemistry 54(9): 3351–3356.
Dais, P. and E. Hatzakis,. 2013. Quality assessment and authentication of virgin olive oil by NMR. Analytica Chimica Acta 765: 1–27.
Dionisi, F., J. Prodolliet and E. Tagliaferri. 1995. Assessment of olive oil adulteration by reversed-phase high-performance liquid chromatography/amperometric detection of tocopherols and tocotrienols. Journal of the American Oil and Chemists' Society 72: 1505–1511.
Durme, J.V. and J. Vandamme. 2016. Non-thermal plasma as preparative technique to evaluate olive oil. Food Chemistry 208: 185–191.
El-Hamdy, A.H. and N.K. El-Fizga. 1995. Detection of olive oil adulteration by measuring its authenticity factor using reversed-phase high-performance liquid chromatography. Journal of Chromatography A 708(2): 351–355.
Flores, G., M. Luisa, del R. Castillo, G.P. Blanch and M. Herraiz. 2006. Detection of the adulteration of olive oils by solid phase microextraction and multidimensional gas chromatography. Food Chemistry 97(2): 336–342.
Galtier, O., N. Dupuy, Y. Le Dréau, D. Ollivier, C. Pinatel, J. Kister and J. Artaud. 2007. Geographic origins and compositions of virgin olive oils determinated by chemometric analysis of NIR spectra. Analytica Chimica Acta 595: 136–144.
Jiménez-Carvelo, A.M., V.A. Lozano and A.C. Olivieri. 2018. Comparative chemometric analysis of fluorescence and near infrared spectroscopies for authenticity confirmation and geographical origin of Argentinean extra virgin olive oils. Food Control 96: 22–28.
Kalogiouri, N.P., R. Aalizadeh and N.S. Thomaidis. 2018. Application of an advanced and wide scope non-target screening workflow with LC-ESI-QTOF-MS and chemometrics for the classification of the Greek olive oil varieties. Food Chemistry 256: 53–61.
Kyriakidis, N.B. and P. Skarkalis. 2000. Fluorescence spectra measurement of olive oil and other vegetable oils. Journal of AOAC International 83(6): 1435–1439(5).
Longobardi, F., A. Ventrella, C. Napoli, E. Humpfer, B. Schütz, H. Schäfer, M.G. Kontominas, and A. Sacco. 2012. Classification of olive oils according to geographical origin by using ^{1}H NMR fingerprinting combined with multivariate analysis. Food Chemistry 130(1): 177–183.
Maggio, R.M., L. Cerretani, E. Chiavaro, T.S. Kaufman and A. Bendini. 2010. A novel chemometric strategy for the estimation of extra virgin olive oil adulteration with edible oils. Food Control 21(6): 890–895.
Mannina, L., M. Patumi, N. Proietti, D. Bassi and A.L. Segre. 2001. Geographical characterization of italian extra virgin olive oils using high-field ^{1}H NMR spectroscopy. Journal of Agricultural and Food Chemistry 49(6): 2687–2696.
Mannina, L., G. Dugo, F. Salvo, L. Cicero, G. Ansanelli, C. Calcagni and A. Segre. 2002. Study of the cultivar–composition relationship in sicilian olive oils by GC, NMR, and statistical methods. Journal of Agricultural and Food Chemistry 51(1): 120–127.
Melucci, D., A. Bendini, F. Tesini, S. Barbieri, A. Zappi, S. Vichi, L. Conte and T.G. Toschi. 2016. Rapid direct analysis to discriminate geographic origin of extra virgin olive oils by flash gas chromatography electronic nose and chemometrics. Food Chemistry 204: 263–273.
Merchaka, N., T. Rizk, V. Silvestre, G.S. Remaud, J. Bejjani and S. Akoka. 2018. Olive oil characterization and classification by 13C NMR with a polarization transfer technique: A comparison with gas chromatography and 1H NMR. Food Chemistry 245: 717–723.
Milanez, K.D.T.M., T.C.A. Nobrega and D.S. Nascimento. 2017. Multivariate modeling for detecting adulteration of extra virgin olive oil with soybean oil using fluorescence and UVeVis spectroscopies: A preliminary approach. LWT - Food Science and Technology 85: 9–15.

Ogrinc, N., I.J. Košir, J.E. Spangenberg and J. Kidrič. 2003. The application of NMR and MS methods for detection of adulteration of wine, fruit juices, and olive oil. A review. Analytical and Bioanalytical Chemistry 376: 424–430.

Olmo-García, L., J.J. Polari, X. Li, A. Bajoub, A. Fernández-Gutiérrez, S.C. Wang and A. Carrasco-Pancorbo. 2018. Deep insight into the minor fraction of virgin olive oil by using LC-MS and GC-MS multi-class methodologies. Food Chemistry 261: 184–193.

Paolini, M., L. Bontempo and F. Camin. 2017. Compound-specific $\delta^{13}C$ and $\delta^{2}H$ analysis of olive oil fatty acids. Talanta 174: 38–43.

Pereiraa, L., S. Gomesa, S. Barrias, J.R. Fernandes and P. Martins-Lopes. 2018. Applying high-resolution melting (HRM) technology to olive oil and wine authenticity. Food Research International 103: 170–181.

Pérez-Castaño, E., M. Sánchez-Viñas, D. Gázquez-Evangelista and M.G. Bagur-González. 2018. Quantification of extra virgin olive oil in dressing and edible oil blends using the representative TMS-4,40-desmethylsterols gas-chromatographic-normalized fingerprint. Food Chemistry 239: 1192–1199.

Peršurić, Ž., L. Saftić, T. Mašek and S.K. Pavelić. 2018. Comparison of triacylglycerol analysis by MALDI-TOF/MS, fatty acid analysis by GC-MS and non-selective analysis by NIRS in combination with chemometrics for determination of extra virgin olive oil geographical origin. A case study. LWT—Food Science and Technology 95: 326–332.

Rabiei, Z. and S.T. Enferadi. 2011. Traceability of origin and authenticity of olive oil. pp. 163–184. *In*: Boskou, D. [ed.]. Olive Oil Constituents, Quality, Health Properties and Bioconversions. InTech-Oppen Access Company, Rijeka, Croitia.

Rezzi, S., D.E. Axelson, K. Hébergera, F. Reniero, C. Mariani and C. Guillou. 2005. Classification of olive oils using high throughput flow ^{1}H NMR fingerprinting with principal component analysis, linear discriminant analysis and probabilistic neural networks. Analytica Chimica Acta 552: 13–24.

Rohman, A. and Y.B.C. Man. 2010. Fourier transform infrared (FTIR) spectroscopy for analysis of extra virgin olive oil adulterated with palm oil. Food Research International 43(3): 886–892.

Ruiz-Samblás, C., F. Marini, L. Cuadros-Rodríguez and A. González-Casado. 2012. Quantification of blending of olive oils and edible vegetable oils by triacylglycerol fingerprint gas chromatography and chemometric tools. Journal of Chromatography B 910: 71–77.

Sacchi, R., F. Addeo and L. Paolillo. 1997. ^{1}H and ^{13}C NMR of virgin olive oil. An overview. MRC 35(13): 133–145.

Santos, P.M., F.V.C. Kock, M.S. Santos, C.M.S. Lobo, A.S. Carvalho and L.A. Colnago. 2017. Non-Invasive detection of adulterated olive oil in full bottles using time-domain NMR Relaxometry. Journal of the Brazilian Chemical Society 28(2): 385–390.

Stefanoudaki, E., F. Kotsifaki and A. Koutsaftakis. 1997. The potential of HPLC triglyceride profiles for the classification of cretan olive oils. Food Chemistry 60(3): 425–432

Tay, A., R.K. Singh, S.S. Krishnan and J.P. Gore. 2002. Authentication of olive oil adulterated with vegetable oils using fourier transform infrared spectroscopy. LWT - Food Science and Technology 35(1): 99–103.

Uncu, A.T., A.O. Uncu, A. Frary and S. Doganlar. 2017. Barcode DNA length polymorphisms vs. fatty acid profiling for adulteration detection in olive oil. Food Chemistry 221: 1026–1033.

Vaclavik, L., T. Cajka, V. Hrbek and J. Hajslova. 2009. Ambient mass spectrometry employing direct analysis in real time (DART) ion source for olive oil quality and authenticity assessment. Analytica Chimica Acta 645: 56–63.

Vigli, G., A. Philippidis, A. Spyros and P. Dais. 2003. Classification of edible oils by employing ^{31}P and ^{1}H NMR spectroscopy in combination with multivariate statistical analysis. A proposal for the detection of seed oil adulteration in virgin olive oils. Journal of Agricultural and Food Chemistry 51(19): 5715–5722.

Woodcock, T., G. Downey and C.P. O'Donnell. 2008. Confirmation of declared provenance of European extra virgin olive oil samples by NIR spectroscopy. Journal of Agricultural and Food Chemistry 56(23): 11520–11525.

Yan, J., S.B. Oey, S.P.J. van Leeuwen and S.M. van Ruth. 2018. Discrimination of processing grades of olive oil and other vegetable oils by monochloropropanediol esters and glycidyl esters. Food Chemistry 248: 93–100.

12

Overview of the Legislation Regarding Traceability and Authenticity of Foods of Plant Origin in EU, USA, Canada, Australia-New Zealand

Montet Didier,[1,2,*] *Dey Gargi*,[3] *Guibert Benjamin*,[1,2] and *Temple Henri*[4]

1. Introduction

Food traceability existed before the first publication of international regulations on food safety. Old records from the Middle Ages reported incidents of human poisoning due to the consumption of rye bread made from grain infected with ergot fungi (Montet and Dey 2018), the consumption of spoiled meat (Ferrières 2002) and mineral intoxication in 1857 (Snow 2003). A long list of food issues retracing the history of traceability were reported by Montet and Dey (2018). The triggering factor

[1] Cirad, UMR Qualisud, TA B-95/16 73, rue JF Breton 34398 Montpellier Cedex 5, France, Email: didier.montet@cirad.fr
[2] UMR 95 QualiSud, Université de Montpellier, CIRAD, Montpellier SupAgro, Université d'Avignon, Université de la Réunion, Montpellier, France.
[3] Kalinga Institute of Industrial Technology, Deemed to be University, Bhubaneswar, India, Email: drgargi.dey@gmail.com
[4] ICABE-AEMAC, 18, rue des Ormeaux, 12230 La Cavalerie, France. Email: icabe.temple@gmail.com
* Corresponding author: didier.montet@cirad.fr

that attracted the attention of lawyers involved in food safety towards traceability, was the European episode of mad cow disease in 1996 which lead to the development of the European regulation 178/2002 that contains a full section on traceability of foodstuffs (article 17).

Food traceability became a worldwide concern and many countries adopted different levels of regulation to cover it (Viruega 2013). US customs produced its US Bioterrorism Act (2001) and its Food Modernization and Safety Act (USA-FDA FSMA 2011). Canada published its National Agriculture and Food Traceability system (Center for food in Canada, 2012). In addition to its Food law,the EU created its Rapid Alert system for Food and Feed (RASFF) that tracks and traces-related issues for food imported into the European Union. Temple (2018) examined in detail French and Europeans regulations concerning traceability. In Asia, the Full Beef traceability system of Korea (GAIN Report N° KS1033. 2010), the Grapenet for export of table grapes from India to EU (https://www.unescap.org/sites/default/files/Brief15.pdf) and a future cloud computing center in Shanghai, China have been established, while the Recall and withdrawal system for food products was published by the Australian Food & Grocery Council (2021).

In the developing world, a few countries have a food act involving an article on traceability. As an example, we can mention the Philippino Act N° 10611 (Article 16 and section 27), which gives a definition for traceability stating that "Traceability refers to the ability to follow the movement of a food through specified stages of production processing and distribution" (Article II, sec 4, Definition of terms). Article 8, sec 27, gives the following complete definition of traceability:

Traceability shall be established at relevant stages of production, post-harvest handling, processing and distribution, when needed to ensure compliance with food safety requirements. The rule of traceability shall also cover production inputs such as feeds, food additives, ingredients, packaging materials and other substances expected to be incorporated into a food or food product. To ensure traceability, food business operators are required to:

a) *Be able to identify any person or company from whom they have been supplied with food, food-producing animals, production chemicals as pesticides and drugs.*
b) *Establish and implement systems and procedures, which allow the above information to be available to the regulatory authorities on demand.*
c) *Establish systems and procedures to identify the other business to which their products have been supplied. This information shall be made available to the regulatory authorities upon demand.*

The *Codex Alimentarius* Commission defines Traceability as "the ability to follow the movement of a food through specified stage(s) of production, processing and distribution" (FAO 2017). The Commission has also adopted standards on traceability and imported food inspections (FAO 2006). The *Codex Alimentarius* Commission is the United Nations body charged with setting international standards related to food safety. At its meeting in Geneva, Switzerland, on 3–7 July (2006), it approved principles for tracing food through production and distribution processes, as well as guidelines for ensuring that imported food is safe for human health and in compliance with importing countries' food safety requirements. The Codex Commission also approved two proposals for the elaboration of new standards on animal and plant biotechnology. FAO also published a Food traceability guidance (2017) and also defined some principles for traceability based on product tracing as a tool within a food inspection and certification system (CAC/GL60-2006).

The International Organization of Standardization (ISO) gave one of the first robust definitions of traceability. It defined traceability as the "ability to trace the history, application or location of an entity by means of recorded identifications" (ISO 8402, 1994 revised by ISO 9000:2000). This definition was not only related to food but to all goods. If we apply this ISO definition to food, traceability may relate to:

- The origin of food could be the place of production, but it could also mean the place of transformation or packaging.
- *The product processing history* that could be the transformation of raw agricultural products into food or the mixture of ingredients to produce a ready to eat food.
- *The distribution and location of the product after delivery.*

Furthermore, in ISO 22005:2007, traceability was defined as the "*ability to follow the movement of a feed or food through specified stage(s) of production, processing and distribution*".

Charlebois et al. (2014) made an attempt to rank countries in order of the quality of their regulations in food traceability. They found that European countries as well as pan-European countries such as Norway and Sweden received an overall world ranking of Superior. Canada, Australia, New Zealand, United States, Japan, Brazil received an overall rank of average. China received an overall world ranking of Poor, while insufficient data were available for Russia.

2. Traceability in European Union Regulation

On the 28th of January 2002, Europe published for the first time a common regulation for all European countries that defined food and its level of safety required for the European consumers (Temple 2018). This regulation is EU N° 178/2002 and can be found on the European Union server in all the languages of the community (http://eur-lex.europa.eu). It was published in 2002 and has been in effect since 1 January 2005. It includes, among other things, the definition of food, the safety level expected in the market, the notion of traceability and the creation of the European Food Safety Authority (EFSA). It establishes the principles governing food and feed, and the safety expectations for food at community and national level. It dictates common principles and responsibilities and establishes the principle of managing food crises and the principle of precaution. Article 18 of the EU Regulation 178/2002 requires that companies should be able to trace food consumed or exported to the EU. It states that:

1. *The traceability of food, feed, food-producing animals, and any other substance intended to be, or expected to be, incorporated into a food or feed shall be established at all stages of production, processing and distribution.*
2. *Food and feed business operators shall be able to identify any person from whom they have been supplied with a food, a feed, a food-producing animal, or any substance intended to be, or expected to be, incorporated into a food or feed. To this end, such operators shall have in place systems and procedures, which allow for this information to be made available to the competent authorities on demand.*
3. *Food and feed business operators shall have in place systems and procedures to identify the other businesses to which their products have been supplied. This information shall be made available to the competent authorities on demand.*
4. *Food or feed which is placed on the market or is likely to be placed on the market in the Community shall be adequately labelled or identified to facilitate its traceability, through relevant documentation or information in accordance with the relevant requirements of more specific provisions.*
5. *Provisions for the purpose of applying the requirements of this Article in respect of specific sectors may be adopted in accordance with the procedure laid down in Article 58(2).*

The application of this regulation in all national legal systems of European Member States and associated countries such as Norway or Switzerland is automatically transposed (Temple 2018).

2.1 EU legislation specific to traceability of fruits and vegetables

It is common knowledge that the increasing speed of agro-food flows following the increasing speed of transportation and inventory management, as well as the perishable aspect of food, most products are consumed before they can be analyzed. Therefore, the different stakeholders rely on the expertise of the industry and the best practices commonly followed. Laboratories can only inform about and confirm the quality status of products. They are also strongly involved in serious crises.

The EU has adopted a regulation [N° 1308/2013 (2013, applied on 1 January 2014)] that established a common organization of markets of agricultural products. This regulation makes the following reference to traceability: "In order to ensure product quality and traceability, the power to adopt certain acts should be delegated to the Commission in respect of the conditions under which product specifications may include additional requirements" while it also states that "*Products of the fruit and vegetables sector intended to be sold fresh to the consumer should be marketed only if they are sound, fair and of marketable quality and if the country of origin is indicated*". The same regulation also includes text covering dates of harvesting, stating that: "*Marketing years should be fixed for cereals, rice, sugar, dried fodder, seeds, wine, olive oil and table olives, flax and hemp, fruit and vegetables, processed fruit and vegetables, bananas, milk and milk products, and silkworms, and adapted as far as possible to the biological production cycles of each of those products*". Inline 36, this Regulation also makes "*a distinction between fruit and vegetables on the one hand, comprising fruit and vegetables for direct consumption and fruit and vegetables intended for processing, and processed fruit and vegetables on the other hand*". Finally, it specifies in Article 23 that provisions will be made to ensure the supply of fruit and vegetables, processed fruit and vegetables and banana products to children.

Traceability is particularly useful in the protection of geographical indication (GI). GI is a sign used on fruits and vegetables that have a specific geographical origin and possess qualities or a reputation that are due to that origin. A GI sign allows a product to be identified by its origin/place of production. EU Regulation N° 510/2006 (2006) lays down the rules for the protection of Designations of Origin and Geographical indications (PDOs and PGIs). It establishes a European Community registration system for geographical names that gives food legal protection in the European Union.

2.2 EU regulations on GM foods

Organic production of fruits and vegetables is also subject to European Community regulation. EU Regulations N° 834/2007 and N° 889/2008 define the rules for production, preparation, labeling and control in organic farming. EU Regulation N° 1830/2003 concerns the traceability and labelling of genetically modified organisms including fruits and vegetables and the traceability of food and feed produced from genetically modified organisms. This regulation states that "*Traceability requirements for GMOs should facilitate both the withdrawal of products where unforeseen adverse effects on human health, animal health or the environment, including ecosystems, are established, and the targeting of monitoring to examine potential effects on, in particular, the environment. Traceability should also facilitate the implementation of risk management measures in accordance with the precautionary principle*".

3. Traceability in US Regulations—USA Food and Drug Administration (FDA)

FDA food safety modernization act (FSMA) is a public law (ref. 111–353) that was published on 4 January 2011. The FSMA recognized that FDA must have traceability tools to respond effectively when food health problems emerge despite preventive controls. It indicates that "*FDA is directed to establish a system that will enhance its ability to track and trace both domestic and imported foods*". In addition, FDA could create pilot projects to explore and evaluate methods to rapidly prevent or control food borne illness outbreaks. This program has to include at least 3 types of fruits or vegetables that have been the subject of outbreaks during the past five years (Senate Bill 510, the FSMA).

This system of traceability includes provisions for mandatory recalls when a company fails to voluntarily recall unsafe food after being asked to by FDA and gives powers to FDA to suspend registration of a company if they determine that there is a reasonable probability that consumption of the food can lead to serious adverse health consequences or death. In chapter (e) Evaluation and recommendation, the product tracing requirements include guidelines that provide adequate assurance of traceability in the event of intentional adulteration or acts of terrorism. In chapter (f) concerning the Traceback and surveillance report, The Secretary included recommendations to enhance surveillance, outbreak response, and traceability in order to relate outbreak identification and traceback. The needs for communication with the public, industry, and State and local governments are also covered.

In the chapter Eligibility, it is mentioned that the Secretary shall consider the risk of the food to be imported based on following traceability factors:

1. *The known safety risks of the food to be imported.*
2. *The compliance history of foreign suppliers used by the importer.*
3. *The capability of the regulatory system of the country of export to ensure compliance with United States food safety standards for a designated food.*
4. *The record keeping, testing, inspections and audits of facilities, traceability of articles of food, temperature controls, and sourcing practices of the importer.*
5. *The potential risk for intentional adulteration of the food.*

Federal regulations such as the Public Health Security and Bioterrorism Preparedness Response Act of 12 June 2002 dictate that the manufacturers must provide information including a description of the food, dates of production, packaging, quantity, and lot numbers that render the food traceable. They have also established procedures for preparation for bioterrorism and public health emergencies.

3.1 US regulations specific to the traceability of fruits and vegetables

Javier Martinez, food safety manager in Fresh Concepts reported in the Food safety magazine (2016) that: "*When the Food Safety Modernization Act (FSMA) was signed into law in 2011, the goal was to strengthen the U.S. food safety system. FSMA put the power into the hands of each grower to ensure their food supply was safe and traceable if an outbreak should occur. It is important to note that there is a difference in traceability between fresh and processed fruits and vegetables, which makes the tracking of fresh produce even more important*". He cited that there was a Produce Traceability Initiative (PTI) where supply chain partners agreed to work towards a standardized approach to traceability measures. All aspects of the produce supply chain were covered including growing, storing, cooling, packing, processing and distributing operations.

The implementation of PTI followed 7 Milestones in order to establish an electronic traceability system of products and as part of that a checklist was launched for all growers/packers/shippers as well as for receivers and buyers (https://www.producetraceability.org/).

In 2010, Global standards 1 of US published the "Traceability for Fresh Fruits and Vegetables Implementation Guide" (GS1 2010). This document on traceability is a business process that enables trading partners to tracefresh fruit and vegetables for human consumption from farm to fork. Based on this guide, each traceability actor must be able to identify the direct supplier and direct customer of their product in the case of a food-related incident. The document was created to protect consumers as rapidly as possible against dangerous food by withdrawing it from the supply chain. It serves as a best practice guide to implement traceability in the fresh fruit and vegetable industry and is applied to all levels of production from shipping containers, including pallets, to consumers. Also, a specific report from the USDA Advisory Committee on Biotechnology and 21st Century Agriculture was published in 2005 on *Global Traceability and Labeling Requirements for Agricultural Biotechnology-Derived Products*. This report focuses on the impacts of the various labeling and/or traceability frameworks currently in place. These frameworks reflected the scientific, political, and cultural perspectives worldwide.

3.2 US regulations on GM foods

Up to now, only the European Union has issued regulations (EU Regulation N° 1830/2003) on labeling of GMO and specific rules for their traceability. The expert committees chose to study plant cases one by one. The US does not require that labels of food derived from genetically modified plants indicate that this is the case. In fact, in many countries, there is no regulation concerning GMO. Traceability regulations and systems are currently evolving to meet different market needs.

4. Traceability in Canadian Legislation

Canada recently published its most recent regulations that were applied on January 15, 2019, the Safe Food for Canadians Act (SFCA 2012) and the Safe Food for Canadians Regulations (SFCR 2018). The SFCR requires that *"certain food businesses track the movement of their food in the supply chain—forward to the immediate customer and back to the immediate supplier. These requirements are applied to a broad scope of food businesses"*. In other words, traceability for SFCR is *"the ability to track the movement of a food or a food commodity, one step back and one step forward"*.

These new traceability requirements will be applied to food businesses that:

- *import or export food*
- *distribute or send food products across provincial or territorial borders*
- *manufacture, process, treat, preserve, grade, store, package or label food to be exported or sent across provincial or territorial borders*
- *grow and harvest fresh fruits or vegetables to be exported or sent across provincial or territorial borders*
- *slaughter food animals from which meat products are derived, where the meat product is exported or sent across provincial or territorial borders*
- *store and handle edible meat products in their imported condition for inspection by the Canadian Food Inspection Agency (CFIA)*
- *sell food to consumers at retail, which would need to be traced one step back but not forward to the consumer*

An overview of the regulatory requirements on traceability, found in Part 5 of the Safe Food for Canadians Regulations (2018) is given below.

As in the case of many other countries, the traceability system of Canada as dictated by SFCR is based on the international standard established by *Codex Alimentarius*. The aim of this regulation is to:

- reduce costs due to recalls,
- protect consumers against risk of injury,
- increase consumer trust in the safety of the food they buy.

Also, as in the cases of other countries, Canada uses traceability documents that allow the identification of the source of food and as well as to whom it was provided. All people involved in food production have to maintain traceability documents that allow officials to identify the scope of a recall and ensure that food will not harm consumers.

Traceability documentation is essential when it comes to food businesses. Food producers need to be able to identify their food products by indicating their *common name, lot code and name and principal place of business of the person by whom or for whom the food was manufactured, prepared, produced, packaged or labelled and stored.*

They need to follow the recommendations of "trace the food one step back", by supplying the "*date on which it was provided to you, name and address of the person who provided it to you*" and the same for tracing the food one step forward.

4.1 Canadian regulations specific to the traceability of fruits and vegetables

SFCR includes specific labelling exceptions for fresh fruits or vegetables at the time of export (Paragraph 92(3)(a,b,c)). Certain foods at the time of sale to consumers at retail stores do not need to bear a lot of code or unique identifiers; and name and principal place of business. These foods are:

- *Fresh fruits or vegetables packaged in a wrapper or confining band less than 13 millimeters in width; and*
- *Fresh fruits or vegetables packaged in a protective wrapper, or a protective bag, that is clear and transparent and on which no information is shown other than a price, bar code, code number, product treatment symbol or environmental statement.*
- *Prepackaged fresh fruits or vegetables that are visible and identifiable in their container* (Subsection 92(5)).

The Canadian Food Inspection Agency adopted Organic Products Regulations in June 2009 to comply particularly with the stricter European requirements on exported produce (Organic Products Regulations, SOR/2009-176, 2009). Concerning pesticides or other contaminants, Canada published a permitted substances list (2015). The imported organic agricultural products may be certified to Canadian Organic Standards. Equivalency arrangements between Canada and exporting countries also exist. The Canadian government has cited that: *Where an equivalency arrangement is in place, organic products may be certified by a certification body accredited by that country and recognized by Canada.* Imported certified organic products containing 95% or more organic ingredients may display the Canada organic logo on the labels.

4.2 Canadian regulations on GM foods

The Standards Council of Canada officially adopted the National Standard for *Voluntary Labelling and Advertising of Foods that are and are Not products of Genetic Engineering* (2004, Reaffirmed 2016). The Canadian General

Standards Board (CGSB) is responsible for the administration of the standard-development process. Health Canada is responsible for setting food labelling policies including foods that have been derived through genetic engineering.

It is interesting to see that for Canada "*Voluntary labelling is permitted in order to provide consumers with information that is not related to the safety of the product*". The Canadian government supported the development of a national standard to provide guidance on the voluntary labelling of products of genetic engineering.

5. Traceability in Australian Regulations

In Australia, as in the rest of the world, the best practices for implementing traceability in the fresh fruit and vegetable (produce) industry are given by Global Standards GS1 Australia (GS1 Implementation Guide 2010). The GS1 guidelines apply to:

1. Fresh fruits and vegetables for human consumption
2. Traceability practices throughout the supply chain from grower to retail stores
3. All products and shipping containers

The GS1 specifies the critical information to be recorded, collected and shared to ensure 'one step up and one step down' traceability. A 'traceable' item identified by the company could be a product (e.g., case/carton, consumer item), a logistic unit (e.g., bin, container) or a shipment. The traceable item should be assigned a GS1 global trade item number (GTIN) and a batch/lot. The identification number is then to be communicated throughout the physical flow of the traceable item in the supply chain. It is also mandatory that all the supply chain partners collect/record and share the information. As mentioned earlier, the GS1 traceability system operates on the system of "one step up and one step down", that is to ensure that the company is able to identify the supplier and the customer or recipient of the traceable item. The GS1 guideline further states that both internal and external traceability must be maintained throughout the supply chain. It is necessary for the traceable identification number to stay intact on the label until the item is consumed or destroyed.

Further on, the GS1 dissects and stipulates the traceability guidelines for each partner in the supply chain. For instance, if the partner is only a grower, not packing the product, then for them the traceable item is the logistic unit (e.g., bins, totes, containers, trailers). It is required that they maintain logistic unit identification, commodity name and variety (where applicable), receiver identification, ship from location, ship to location, ship date, grower record details (such as fields, seeds, production details),

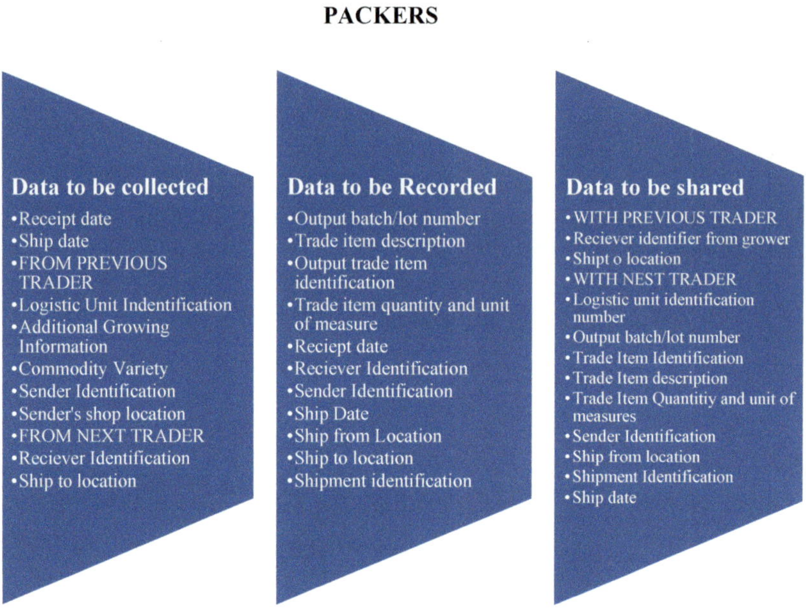

Figure 1. GS1 guidelines for packers/repackers.

additional grower information (e.g., harvest crew, date of harvest) as well as sender identification. Similarly, GS1 also stipulates guidelines for packers/re-packers (Fig. 1); distributors (Fig. 2) and food service operators/retail stores (Fig. 3).

5.1 Australian regulation specific to the traceability of fruits and vegetables

As per an Australian organic market report (https://austorganic.com/industry/publications/market-report-2021/), the organic products industry is worth AU$ 2.4 million. The most popular organic items among the consumers are fruits and vegetables followed by eggs, dairy and meat products. Due to the commercial importance of these products, it is expected that the nation will have stringent regulations and standardized certification procedures.

In Australia, the regulation and certification process for organic food products is implemented as per the Australian Certified Organic Standards (ACOS) guidelines (Australian Certified Organic Standard 2017). The ACOS guidelines are based on Australian standards for Organic and Biodynamic Products (AS6000-2009) and the National Standard for Organic and Bio-dynamic Produce. The guidelines are in conformance with standards from North America, EU, Asia, the International Federation

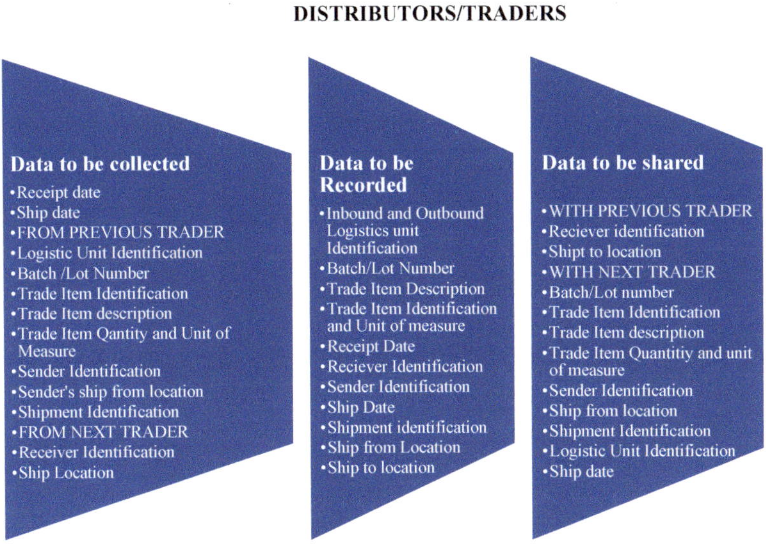

Figure 2. GS1 guidelines for distributor/trader.

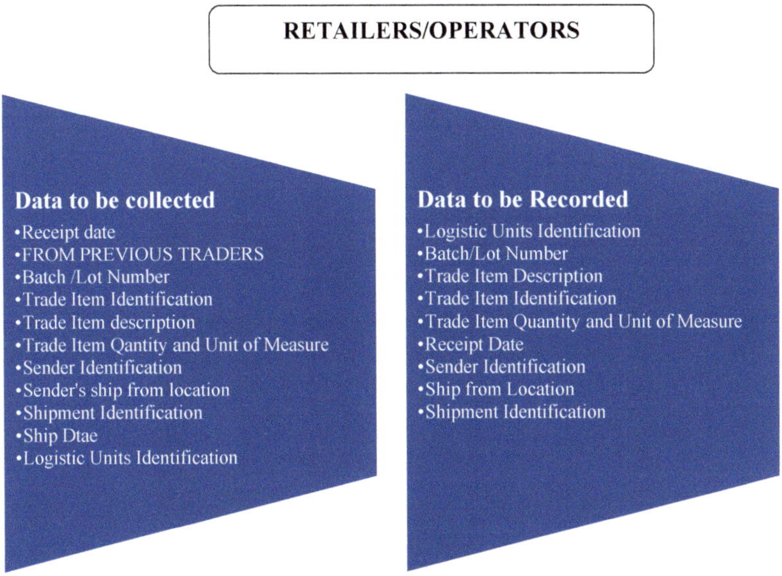

Figure 3. GS1 guidelines for retailers/operators.

of Organic Agriculture Movements (IFOAM) and the *Codex Alimentarius*. The ACOS guidelines may be used for reference especially for products targeted for the Australian Domestic market. Producers that have applied for certification under the ACO IFOAM program need to conform to additional requirements. In the case that organic foods are intended to be exported to the EU, for instance, then one must apply for certification under the ACOS EU program.

For the domestic Australian market, the basic certification requirements involve audit, certification and licensing via an independently accredited certifier and conformance to ACO standards.

5.1.1 Labelling of organic products

Certified organic produce including both raw and processed produce should be labelled appropriately with the following information on the packages:

- the name of product
- appropriate reference to certification (e.g., Organic, Organic in conversion, etc.)
- certification name and relevant ACO logo and
- name and address and or/registered mark of the certified operator or owner of the product and or label as required by law.

It must be noted here that for processed products like juice where:

- 100% of all ingredients come from certified organic sources, the may be labelled as '100% organic'
- minimum of 95% wt/wt of ingredients are organic, they may be labelled us 'certified organic' and where
- less than 95% but more than 70% wt/wt of ingredients are organic, they may be labelled as 'Made with organic ingredients'. However, the ACO logo cannot be used for these products.

It is understood that certified organic products should be GMO free. However, a GMO-free label cannot be used to make claims unless the product conforms to relevant regulatory requirements.

5.2 Regulations for GM foods in Australia

Manipulation of the genetic material of a plant results in the production of a GM plant. Over the past decade, the topic of GM food has been highly debated globally with strong proponents both in favor and against their consumption. In Australia by 2005 twenty-five GM foods were approved. The genetic engineering on these crops was mainly carried out to achieve herbicide tolerance, pest resistance and improvement of growth characteristics. In Australia, the food safety standards in general are set in the Australia New Zealand Food Standards Code and this applies to all the Australian states. The regulatory body responsible is called Food Safety Australia and New Zealand (FSANZ). The Australia New Zealand Food Standard Code:

1. Prohibits the use of additives, processing aids and nutritive substances in food unless there is a specific permission for these substances following a safety assessment.
2. Prohibits the use of novel foods, irradiated foods and foods produced using gene technology unless there is a specific permission for these foods following a safety assessment.
3. Specifies maximum limits for contaminants and natural toxicants in food.
4. Establishes microbial and processing requirements for food.
5. Specifies the composition and labelling requirement for many commodities.

Similarly, for GM foods, specific sets of guidelines are given in standard 1.5.2 of the Australia New Zealand Food Standard code (Food Standards Australia New Zealand 2005), under Food produced using gene technology. The major focus of the regulations with respect to GM foods is on their safety assessments. The safety assessments are carried out by the FSANZ on behalf of the Australian government. It should also be mentioned that there are several other governmental agencies that are also involved, such as the Office of the Gene Technology Regulator (Public health matters and environment issues including live release of GMOs); State and territory department (Enforcement of Std 1.5.2); Australia Govt. Dept, of the Environment and heritage, Australian Quarantine and Inspection Service (Imports and exports) Therapeutic Goods Administration, Dept of Health and Ageing and Biotech Australia.

The office of the Gene Technology Regulator (OGTR) sits within the Australian Govt. Dept of Health and Ageing and it operates under the Gene Technology Act 2000, which was established on 21st June 2001. The Act prohibits any use of GMO unless it is:

- Licensed by the Regulator for a particular use or intentional release into the environment
- A Notifiable Low Risk Dealing
- An exempt dealing
- On the register of GMOs

The office of Gene Technology Regulator (OGTR) is further supported by the Gene technology Ethics Committee and The Gene technology Community Consultative Committee.

Cross-Governmental coordination between FSANZ and OTGR operates for food approvals and commercial releases of GM foods. From a practical point of view, all technical requirements of GM foods in Australia have to be assessed by FSANZ and require its approval. Some commercially available GM fruits and vegetables are soybean (herbicide tolerant and high oleic acid content); canola (herbicide tolerant); Corn (Insect protection, herbicide tolerant); potato (insect protectant and virus protectant); sugar beet (herbicide tolerant). Globally, there are other GM fruits and vegetables like papaya, radicchio, rice, squash, tomato, chicory, flax, rock melon that have been approved for sale in EU, USA, Canada, Japan. But these are currently not available in Australia.

A company wishing to sell GM foods in Australia needs to provide the following list of information to allow a safety assessment to be carried out:

1. The identity of host and donor organisms. Any known pathogenicity in the host or donor organisms.
2. The previous use of host and donor organisms in food production.
3. The new genetic material that has been introduced through genetic modification:
   - origin, nature, purpose, function.
   - method of introduction into the host organism.
4. The new genetic material in the GM organism:
   - the DNA sequence of the novel genetic material and border regions with the plant's genome.
   - number of complete or incomplete copies present.
   - stability.
5. The new protein in the GM organism:
   - purpose, physical and biological characteristics.
   - expression profile (which tissues the protein is found in and when the protein is present).
6. Potential adverse effects of the new protein, such as allergenicity and toxicity:
   - similarity of the new protein to known allergens or toxins.
   - physical features that are characteristic of allergens.
   - acute toxicity (animal studies).
7. Composition compared to conventional counterparts—levels of nutrients, anti-nutrients, natural allergens and toxins.
8. Impact on human health from potential transfer of new genetic material to cells in the human digestive tract.

9. End uses of the food, including any requirement for processing prior to consumption.
10. If required, ability of the food to promote typical growth and wellbeing (animal feeding studies).
11. Any other relevant information.

After the necessary due diligence to verify that the GM food under assessment is safe, the product can be released into the market for sale or may be used as an ingredient. For instance, GM soybean can further be processed into oil, flour, lecithin and these may be used in breads, pastries and snack foods. The final processed foods, however, do not need not be assessed individually for safety.

An important part of regulatory requirements for GM foods is their labelling. The standard 1,5.2 stipulates that the statement "genetically modified" must be used in conjunction with the name of the food or in association with specific ingredients in the ingredient list. For unpackaged foods, the "genetically modified" indication must be displayed on or in connection with the display of the food. In cases that the GM food has different properties, the labelling should include the phrase "Altered characteristics" and information related to composition or nutritional values, antinutritional factors or related toxicants, factors known to cause allergic responses in particular sections of the population and its intended use. In cases that a GM food is found to be unintentionally present in non-GM foods, there may be an exclusion from labelling, however, the quantity of GM should be no more than 10 g/kg (1%).

6. Traceability in New Zealand Regulations

The Parliament of New Zealand published its current Food Act in 2014. The Act came into force by 1 March 2016. The Ministry for Primary Industries is in charge of its implementation. It progressively replaces the Food Act 1981 and a new version was reprinted in March 2018. This regulation (Part 5 s 383) specified that *"the criteria that the composition of food, and the sampling and testing of food to determine its composition, must meet to ensure that food is safe and suitable"*.

In Part 2s 74, a general description of a National program is given. This was defined as a *"program designed to identify, control, manage,and eliminate or minimize hazards or other relevant factors for the purpose of achieving safe and suitable food"* and imposes *"controls on a food business in relation to matters such as (without limitation) good operating practice, documentation requirements, verification requirements, registration requirements, and traceability and recall requirements"*.

In Part 5s 397, it is specified that the Ministry may *by notice* adopt joint food standards *"for food manufactured or prepared for sale or sold in New Zealand, or imported into, or exported from, New Zealand"*.

The Ministry must take into account the following matters:

(a) the need to protect public health,

(b) the desirability of avoiding unnecessary restrictions on trade,

(c) the desirability of maintaining consistency between the common food standards adopted and the standards applied at international level,

(d) the need to give effect to New Zealand's obligations under any relevant international treaty, agreement, convention, or protocol,

(e) any other matters considered as relevant by the Minister.

6.1 New Zealand regulations specific to the traceability of fruits and vegetables

The Ministry of Primary Industry (MPI) of New Zealand specified that *"All food and food-related products imported for personal use or sale must meet New Zealand standards and requirements"*. Its role is to make sure that imported fresh fruits, vegetables, herbs as well as fungi are correctly labelled, safe to eat, and free from pests and harmful organisms.

MPI on its website (https://www.mpi.govt.nz/food-safety/food-safety-for-consumers) gives a lot of information concerning the export of fresh fruits successfully to other countries. In fact, some fresh products have to meet extra requirements when they are exported by New Zealand. These additional requirements are:

- *phytosanitary, food safety, quality grade, and labelling requirements for the destination country.*
- *complying with extra requirements for specific commodities, where necessary.*
- *using MPI-authorised Independent Verification Agencies (IVAs) or MPI-approved service providers or facilities for phytosanitary activities.*
- *the export requirements set by the industry sector.*

6.2 Regulations for GM foods in New Zealand

All food sold in New Zealand must comply with the Australia New Zealand Food Standards Code for labelling (https://www.mpi.govt.nz/food-safety-home).

Certain foods do not require any label. In the case of fruits and vegetables, the products mentioned are:

- *unpackaged food*
- *food made and packaged where it is sold, or food packaged in front of you*

- *whole or cut fresh fruit and vegetables in transparent packages*
- *food in an inner package not designed for sale without the outer package.*

GMOs are regulated via the regulation called "The Hazardous Substances and New Organisms Act 1996 (HSNO Act)" that regulates research on GMO and release of all living existing GMO or not animals or plants that do not already exist in New Zealand.

This regulation explains that new organisms (including a GMO) have an approval from the Environmental Risk Management Authority (ERMA) before being imported, developed, field tested or released into the environment. ERMA is an independent body specifically set up for this purpose. It considers each application "*on a case-by-case basis, reflecting the fact that each organism is different and will therefore pose different potential risks and/or benefits*".

7. Conclusions

By closely examining the specific guidelines set by all countries, it may be concluded that regulations on traceability are very similar. Interestingly, the US traceability regulation is included in the Bioterrorism act. Countries that do not have or do not use a regulation on traceability can use the guidelines of *Codex alimentarius* or the FAO recommendations and/or the regulations of importing countries for international trade. In Australia and New Zealand, fruits are covered by a particular regulation that simplifies the food regulations. GMOs are commonly covered by specific regulations. Some countries have specific regulations on organic fruits and vegetables. The most comprehensive text covering traceability seems to be that of the European Union (EU regulation 178/2002 of 2005), which is in accordance with what Charlebois et al. (2014) found in their classification of countries. This regulation integrates food and feed and includes local, imported and exported food. The clarity of the guidelines facilitates International trade. It must also be pointed out that, presently, the EU is the only region that has implemented a regulation on the geographical origin-traceability of foodstuffs.

Abbreviations

Food Safety Modernization Act	FSMA
Rapid Alert System for Food and Feed	RASFF
Global Agriculture Information network	GAIN
Food and Agriculture Organization	FAO
International Organization of Standardization	ISO
Protected Designation of Origin	PDO
Protected Geographical Indication	PGI
Genetically modified foods	GM foods

Food and Drug Administration	FDA
Produce Traceability Initiative	PTI
Global standards	GS
Safe Food for Canadians Act	SFCA
Canadian Food Inspection Agency	CFIA
Safe Food for Canadians Regulations	SFCR
Australian Certified Organic Standards	ACOS
International Federation of Organic Agriculture Movements	IFOAM
Food Safety Australia and New Zealand	FSANZ
Office of the Gene Technology Regulator	OGTR
Ministry of Primary Industry of New Zealand	MPI

References

[ACOS] Australian Certified Organic standard 2017, v1. (https://aco.net.au/Pages/Operators/ACOStandards.aspx).

Australian Food and Grocery Council. 2021. https://www.afgc.org.au/.

Australia New Zealand Food Standards, Safety assessment of genetically modified foods 2005. ISBN 0642 345996. (http://www.foodstandards.gov.au/publications/documents/AR%2006%20Text.pdf).

Canadians Regulations "Organic Products Regulations" 2009 (SOR/2009-176). Laws Website Home.

Canadians Regulations, Safe Food for Canadians Act, S.C. 2012, c. 24, Justice Laws Website.

Canadians Regulations "Labelling of Genetically Engineered Foods in Canada" National standard of Canada: CAN/CGSB-32.315-2004 (Reaffirmed 2016).

Canadian Permitted substances list in organic agriculture. National standard of Canada: CAN/CGSB-32.311-2015.

Charlebois, S., B. Sterling, S. Haratifar and S.N. Kyaw. 2014. Comparison of global food traceability regulations and requirements. Comprehensive Reviews in Food Science and Food Safety 13: 1104–1123.

EU Regulation No 178/2002 of the European Parliament and of the council of 28 January 2002 laying down the general principles and requirements of food law, establishing the European Food Safety Authority and laying down procedures in matters of food safety.

EU Regulation No 1830/2003 of the European Parliament and of the council of 22 September 2003 concerning the traceability and labelling of genetically modified organisms and the traceability of food and feed products produced from genetically modified organisms and amending Directive 2001/18/EC.

EU Regulation No 510/2006 of 20 March 2006 lays down the rules for the protection of Designations of origin and geographical indications (PDOs and PGIs).

EU Regulation No 834/2007 of 28 June 2007 on organic production and labelling of organic products and repealing Regulation (EEC) No 2092/91.

EU Regulation No 889/2008 of 5 September 2008 laying down detailed rules for the implementation of Council Regulation (EC) No 834/2007 on organic production and labelling of organic products with regard to organic production, labelling and control.

EU Regulation No 1308/2013 of the European Parliament and of the council of 17 December 2013 establishing a common organization of the markets in agricultural products and repealing Council Regulations (EEC) No 922/72, (EEC) No 234/79, (EC) No 1037/2001 and (EC) No 1234/2007.

FAO, *Codex Alimentarius* Commission. 2006. Principles for traceability/Product tracing as a tool within a food inspection and certification system. International center for trade and sustainable development, CAC/GL 60.

FAO, Food traceability guidance 2017. ISBN 978-92-5-109876-9.

Ferrières, M. 2002. Histoire des peurs alimentaires. Du Moyen-Âge à l'aube du XXe siècle. L'univers historique, Éd. du Seuil, Paris, France.

GS1. 2010. Traceability for Fresh Fruits and Vegetables Implementation. Issue 2, May 2010. Guide. Availablefrom: (http://www.producetraceability.org/documents/Global_Traceability_Implementation_Fresh_Fruit_Veg.pdf).

http://archive.dpi.nsw.gov.au/__data/assets/pdf_file/0011/353297/organic-standards-and-certification-in-Australia.pdf.

https://austorganic.com/industry/certification-and-standards/standards/acos/.

ISO 8402:1994 revised by ISO 9000:2000. Management de la qualité et assurance de la qualité.

ISO 22005:2007. Traceability in the feed and food chain - General principles and basic requirements for system design and implementation.

Martinez, J. 2016. Traceability and Food Safety in Produce. Food Safety Magazine Digest 12, 6. https://www.food-safety.com/articles/5103-traceability-and-food-safety-in-produce.

Montet, D. and G. Dey. 2018a. History of traceability. Chapter 1, pp. 1–30. *In:* Montet, D. and R.C. Ray [eds.]. Food Traceability and Emerging Technologies. Food Biology series, Science Publishers/CRC Press, USA.

Montet, D., C. Kouakou, Y. Hamdouche, C. Teyssier, T. Rychlik and E. FroederArcuri. 2018b. Geographical origin traceability of foodstuffs using a molecular technique PCR-DGGE. Chapter 7, pp. 137–154. *In:* Montet, D. and R.C. Ray [eds.]. Food Traceability and Emerging Technologies. Food Biology series, Science Publishers/CRC Press, USA.

Montet, D. and R.C. Ray. 2018c. Food traceability and emerging technologies. Food Biology series, Science Publishers/CRC Press, USA.

New Zealand Hazardous Substances and New Organisms Act 1996. Public Act, No 30, Date of assent 10 June 1996.

New Zealand Food Act 2014. Public Act 2014, No 32, Date of assent 6 June 2014.

Philippino Republic Act N°10611, 2012. Fifteen congress of the Philippines, Third regular session, Metro Manila. An act to strengthen the food safety regulatory system in the country to protect consumer health and facilitate market access of local foods and food products, and for other purposes.

Snow, J. 2003. On the adulteration of bread as a cause of rickets. Lancet 1857: 4–5. (Reprinted Int J Epidemiol 32: 336–37.)

Temple, H. 2018. Traceability in French and European laws. Chapter 3, pp. 48–65. *In:* Montet, D. and R.C. Ray [eds.]. Food Traceability and Emerging technologies. Food Biology series, Science Publishers/CRC Press, USA.

[UNESCO] United Nations Educational, Scientific and Cultural Organization 1985. Mabahiss-John Murray 50th Anniversary: Marine Sciences in the North West Indian Ocean and Adjacent Waters. UNESCO Technical Papers in Marine Science, Place de Fontenoy, Paris: 153pp.

U.S customs and boarder protection. The Bioterrorism Act, 2001. https://www.cbp.gov/trade/priority-issues/import-safety/bioterrorism.

USA-FDA. The FDA Food Safety Modernization Act (FSMA), enacted on January 4, 2011. https://www.fda.gov/.

USDA Advisory Committee on Biotechnology and 21st Century Agriculture report on Global Traceability and Labeling Requirements for Agricultural Biotechnology-Derived Products: Impacts and Implications for the United States, 8 February 2005.

Viruega, J.L. 2013. La traçabilité. pp. 377–428. *In:* Multon, J.L., H. Temple and J.L. Viruega. [eds.]. Traité pratique dedroit alimentaire, Lavoisier édition, Tec et Doc, France.

13

Trends and Suggestions in Improving Authenticity and Traceability of Foods of Plant Origin

Montet Didier,[1,2,*] *Davrieux Fabrice,*[1,2] *Guyon François*[3] and *Vaillant Fabrice*[2,4]

1. Introduction

ISO 8402:1994 defines traceability as the ability to trace the history of a product, its uses or location of an entity (covering activity, processes, product, etc.) by means of recorded identification. Article 18 of EU Regulation 178/2002 details the requirements of traceability in the food supply chains. According to this regulation:

1. Food and feed traceability is established at all stages of production, processing and distribution.

[1] Cirad, UMR Qualisud, TA B-95/16 73, rue J.F. Breton 34398 Montpellier Cedex 5, France.
[2] UMR 95 QualiSud, Université de Montpellier, CIRAD, Montpellier SupAgro, Université d'Avignon, Université de la Réunion, Montpellier, France.
[3] Service Commun des Laboratoires, Laboratoire de Marseille, 146 traverse Charles Susini, 13388 Marseille cedex 13, France.
[4] Agrosavia Corporación Colombiana de Investigación Agropecuaria, Research Center "La Selva", Rionegro, Colombia.
* Corresponding author: didier.montet@cirad.fr

2. Food and feed business operators must be able to identify any person who has supplied them with food, or any substance intended to be incorporated or likely to be incorporated into food or into animal feed.
3. Food business operators have systems and procedures in place to identify the companies to which their products have been supplied. This information is made available to the competent authorities at their request.
4. Food and feed placed on the market in the EU or likely to be so shall be labeled or identified in an appropriate manner to facilitate traceability, using the relevant documents or information.

In fact, few analytical techniques can currently be used to meet the requirement of these regulations and manufacturers, regardless of their size, use mainly administrative monitoring techniques, with the best known as bar code. The bar code is a very convenient and inexpensive tool used throughout Europe and other industrialized countries. However, it is not widely used in developing countries. From a consumers' perspective, traceability assists companies in meeting their expectations for safety and allows consumers to gain access to various information. In fact, consumers' purchasing motives are believed to be linked to their financial status, use-by-dates, the composition of a product, its origin but also food safety or nutritional characteristics such as food's nutritional contribution, risks of allergens, ecological, ethical, religious reasons and others.

In practice, from an analytical chemistry point of view, only plant or animal-derived ingredients used to prepare the food can be precisely determined. The analytical techniques used for these purposes are usually based on PCR techniques and/or genomic analyses. To date however we have very few analytical techniques that enable traceability based on the geographical origin of the food (Peres 2007, Montet and Ray 2018). This chapter aims at providing an insight into research using and development of analytical techniques not yet widely employed by the industry or for inspection purposes.

2. Analytical Tools for Species Identification

When fruits and vegetables are concerned, innovative techniques that could be used for the determination of species are expected to be based on different genomic techniques. Genomics rely on the analysis of differences in genomic material. Analyses based on DNA are very efficient when used for raw plants but may be less efficient when processed foods are analyzed due to the degradation of DNA that can happen during heating or freezing processes, mechanical (mixing, grinding, chopping, etc.) and biochemical processing (fermentation, acidification, addition of sugar, etc.). Nevertheless, techniques based on DNA analyses have evolved significantly in recent years and small DNA fractions are sufficient for accurate species identification (Mohammadian et al. 2018) even when mixtures are analyzed. There is only a limited number of studies examining mixtures and identification currently is most frequently made based on metabolic functions or the appearance of traits (Wei et al. 2016).

New advances in molecular biology have led though to the development of new technologies that are likely to revolutionize the analyses of foods of vegetable origin, with one of the most important ones being high throughput sequencing (or next generation sequencing). Techniques based on DNA analyses allow the identification of the genetic origin (clone, variety, cultivars, hybrids, clones, etc.) of a plant, the determination of composition of foods (% of purity, etc.), any pathogenicity potential (toxins, bacteria, etc.) as well as the presence of GMOs (presence/ absence, nature) (Zhang and Hao 2020). It is expected that the use of high-throughput automatic sequencing as well as the use of even more sophisticated technologies will start being frequently employed within the next few years for species identification in food products of vegetable origin. To overcome the challenges that currently prohibit the wide use of these methods, researchers would need to work on the development of convenient techniques and improved biosensors, capable of detecting highly stable and shorter-sized DNA markers or allow nucleic acids to be directly detected in complex food samples (Madesis et al. 2014).

3. Tools for Determining the Geographical Origin of Foods of Plant Origin

The *Codex Alimentarius* committee defines food traceability as the ability to follow the movements of a food through specific stages of production, processing and distribution (FAO 2020). Traceability in logistics has above all been used by companies to eliminate issues such as virtual loss of input/ data and transmission errors, reduced speed of data transmission and low productivity, and is therefore a tool that aims at better controlling costs and time. Companies have therefore the capacity to react faster, taking quick decisions when faced with new information. Its use in better controlling stock levels is also of paramount importance. The most common logistics techniques can be divided into five main categories: linear barcodes, 2D codes, matrix codes, radio-frequency identification techniques and other technologies, some of which may be qualified as innovative (contactless cards, chips) and are briefly described below.

The use of barcodes is a fast and efficient way to register products. In Europe, these codes are placed on all types of foods or on their packs or crates when their use on the actual food is not possible for practical reasons, such as in the case of raw fruits and vegetables. The GTIN 13 code is the barcode most commonly used in distribution in Europe (Fig. 1).

Figure 1. Global Trade Item Number 13 barcode.

There are 13 digits associated with the code. The first three digits indicate the country of origin of the product (e.g., 300 to 379 for France), the digits 3–7 correspond to a manufacturer's number, the digits 8–12 correspond to the product number while the 13th digit represents a key used to validate the barcode (Avisé 2019). The use of a bar code allows the elimination of manual entry errors, reduces the time needed to verify information and provides precise and instant information about the status of a product in a production line, sorting, storage, etc. It also facilitates the precise and rapid monitoring of stocks and therefore optimizes storage management, while a simple printer is sufficient to issue it.

3.1 Future trends in geographical origin traceability

3.1.1 Smart labels and electronic chips

The biggest advantage of smart labels and electronic chips is that they allow instantaneous exchange of digital data by radio frequency, read and write. Numerous data could be recorded, depending on their memory capacity, such as identification number, origin, size, weight, color, encrypted anti-counterfeiting code and expiration date, while in the future they could be used to also hold analytical results, food safety data, etc. The technology allows for selected data to be fixed, written or deleted at each stage of the supply chain, or be retrieved and used in real time using a computer system. Active intelligent labels can be operational in any natural or industrial environment. They can be easily incorporated or affixed to any type of product because of their small, often microscopic size. They allow remote and contactless inventory management in storage locations. Access to content is secure and they have a large memory capacity with the possibility of updating data. Applying these types of chips to crates, for example, can prevent them from being lost or stolen.

The development of an electronic chip, for example, was reported in the study of Elsaadani et al. (2020). Its purpose was the continuous monitoring of the concentrations of mycotoxins in fruits and vegetables. The active part of the chip consisted of an aptamer specific to Ochratoxin A linked to an electronic probe called apta-sensor. This type of aptamer could be potentially used for most mycotoxins or other undesirable molecules (Elsaadani et al. 2020). The surface of such an apta-sensor can be seen in Fig. 2.

Figure 2. Surface of an apta-sensor with its two electrodes (total size 5 mm).

Aptamers are short single strands of artificial DNA or RNA oligonucleotides. They can take a variety of shapes due to their tendency to form helices and single strand loops. They are able to bind or react with a target molecule with high affinity and high selectivity (Famulok et al. 2014). They are identified *in vitro* by an iterative selection process called SELEX (Systematic Evolution of Ligands by Exponential Enrichment) (Stoltenburg et al. 2007). In another recently published study, a fast, highly sensitive and label-free colorimetric aptasensor employing high affinity truncated aptamers was designed to detect bisphenol A. Truncated 38-mer and 12-mer aptamers were obtained through rational truncation from a 63-mer bisphenol A aptamer, rendering them bisphenol A-specific. The use of truncated aptamers in label-free colorimetric detection assays was based on gold nanoparticles (AuNPs). The limit of detections of aptasensors using 38-mer and 12-mer aptamers were 7.60 pM and 14.41 pM. The recovery rates in different foods tested were between 93.88% and 107.30% (Jia et al. 2020).

As the use of this technology has very recently started being examined, it is expected that studies will follow in future to develop aptasensors capable of detecting other contaminants in various foods.

3.1.2 Determination of geographical origin using stable isotope analysis

Authentic claims on the geographical origin of foods are of paramount importance for both trade and consumer confidence. It is therefore important to be able to verify such claims. Indeed, consumers often associate the geographical origin of food with the quality of production, while it can also have a real impact on the price of the products (Carter et al. 2006, Verbeke and Roosen 2009). One way to verify an origin claim is using stable isotope analysis (Dennis 1998, Kelly 2003, Guyon et al. 2020). Stable isotopes are atoms, organic or mineral, incorporated into the plants as they grow (Barbour 2007, Bart 2011). While some elements come from the atmosphere (CO_2 gas, for example), most of them are absorbed by the roots.

The isotope ratios in plants are linked to ratios found in the soil the plants grow on, and this can therefore be used to identify the geographical origin of foods of plant origin. The distribution of isotopic elements is not homogeneous across the surface of the globe (Craig 1961) and the minerals' distribution is linked to the soil composition (Lynn et al. 2002). As a result, the isotopic ratios found in plants (mineral elements and synthesized molecules) reflect all these characteristics (Doner 1985, Bontempo et al. 2014, Camin et al. 2017). Measurements of the isotope ratio of oxygen value ($\delta^{18}O$, expressed in ‰ vs. V-SMOW, Vienna Standard Mean Ocean Water) can provide information about the geographical origin of a plant but also allow distinguishment to be made between, e.g., a pure juice and a juice made from concentrate or rehydrated fruits (Guyon et al. 2015a). These measurements can be supplemented by the analysis of $\delta^{18}O$ of ethanol isolated by distillation after fermentation of fruit sugars (Jamin et al. 2003). This ethanol can also be studied to quantify the content of deuterium (D/H in ppm) and carbon 13 ($\delta^{13}C$, expressed in ‰ vs. PDB, Pee Dee Belemnite). These data can provide information about the geographical origin of a plant. The isotopic ratio of carbon in particular, is also influenced by hydric stress (Gilbert 2011, Guyon et al. 2015b). In recent years, studies employing liquid (LC) or ionic chromatography linked to mass spectrometry (MS) via a liquid interface allowing oxidation of organic matter (HPLC-co-IRMS), demonstrated that these techniques can be used for the determination of ^{13}C isotope ratios organic acids and sugars (Guyon et al. 2013, 2014).

The quantities of mineral-incorporated elements depend on their concentrations in the soil and plants. As a result, the mineral concentrations are representative of the geographic origin of the plant. Numerous studies demonstrated that such measurements in association with stable isotopes of light elements provide accurate classification based on geographical origin (Magdas et al. 2012, 2018). Finally, the quantification of the isotopic ratios of certain mineral elements can also be used to determine the geographical origin of products. The number of studies is limited due to the specific devices and the technical expertise required to carry out the analyses and interpret the data. Nevertheless, it was shown that the isotopic analysis of strontium, $^{87}Sr/^{86}Sr$, enabled the precise determination of the geographical origin of plants (Fortunato et al. 2004, Rummel et al. 2010, Epova et al. 2019). Likewise, the isotopic ratios of lead of anthropogenic origin (derived from tetraethyl lead used in internal combustion engines) can be used even if their value decreases due to the use of unleaded gasoline in vehicles (Augagneur et al. 1996, Médina et al. 2000). All of these studies take advantage of databases built using results obtained from analyzing authentic samples in combination with the development of mathematical models obtained from multidimensional processing of measured data. It is expected therefore that further development of such databases will increase the use of multi-elemental analysis in verifying geographical authentication claims.

3.1.3 Determination of geographical origin using metabolomics

A plant's metabolome is characterized by thousands of metabolites and its specific composition is influenced not only by genetics, but also by environmental conditions and agricultural practices. For this reason, metabolomics can also be used as a tool to trace the origin of foods of plant origin and frequently also determine the crop management systems used (Guyon et al. 2020). Discrimination of geographical origins and cultivation conditions is based on the combined use of statistical analysis and high throughput analytical methods able to partially analyze the metabolome. Generally, Nuclear Magnetic Resonance (NMR) or MS are both methods of choice for analyzing metabolomes as they are able to generate a specific signal for each metabolite (Cientifico et al. 2017, Dickson et al. 2020). NMR is more specific than MS but less sensitive and conversely, MS is less specific but can generate a signal for metabolites in concentrations as low as 10–15 M. The signal profile can be used as a fingerprint of the metabolome or treated and converted into a database containing basic information about a specific metabolite and its concentration. Even though the interpretation of signals differs significantly between NMR and MS, the variation of intensity of a given signal can be correlated with the concentration of a given metabolite in a sample. At this stage, the information acquired is not sufficient to identify or elucidate the structure of metabolites, but this is irrelevant as the objective is solely to identify the signals that allow discrimination between samples. The analysis of discriminant signals is performed using statistical tools such as univariate and multivariate analysis. The construction of extensive databases of signals from different carefully selected samples allows the development of a specific pattern for a given metabolome. For example, fruits grown in a specific region can share a certain signal that is not present in fruits grown in another region. The use of these specific patterns implies that a complete sampling design is needed in order to construct a database representative of the metabolome of a product. In practice, the signals deemed discriminant must be robust and repeatable independently of the variations of uncontrolled variables. If the sampling design is correct, allowing for sufficient repetitions to enable the recovery of all possible variations, a statistical model can be constructed and validated to obtain discriminant signals. Among the most common statistical methods used for these purposes are the partial least square discriminant analysis (PLS-DA) and models such as PLS and Orthogonal Projections to Latent Structures (OPLS). These statistical tools are based on the projection of the data to latent structures which allows the extraction of signals that contribute to the differentiation between groups of samples. If the discrimination is found to be statistically robust, a model can categorize samples based only on analysis of primary signals.

It is possible, for example, to use ^{1}H-NMR signals in order to discriminate potatoes' varieties and crop management (e.g., organic versus conventional) with 100% accuracy (Pacifico 2013). Recently, metabolomic approaches based on LC or GC coupled with MS were used to discriminate between berry fruit juices (Zhang et al. 2018), apple and orange (Xu et al. 2020, Uhao et al. 2020, Shujing et al. 2020) and olive oil (Fayek et al. 2021). It was proved that in all cases MS signals can effectively discriminate between the varieties or authentic samples and adulterants. These studies demonstrated that such fingerprint methods using either NMR or MS are very effective in tracing and authenticating plants and derived products enabling discrimination between varieties, geographical origins, crop management or even production seasons and post-harvest processes.

Nonetheless, it may be necessary in some cases to identify the origin of the discriminating signal and then go further in order to elucidate the identity of the responsible metabolite(s). This can be more easily achieved when the signal comes from NMR than MS. Additional analyses can often reveal the identity of those metabolites, which play the role of biomarkers as they are characteristic of a given group. As an example, Diaz et al. (2014) used an MS-based untargeted metabolomic to discriminate between oranges from different geographical origins. They discovered a discriminant signal in orange peels and through further analysis the signal was tentatively assigned to citrusin D. The sequence of the methods that were followed in this study is known as untargeted metabolomic. This approach could be quite exhaustive and has allowed the discovery of various biomarkers. The NMR-based approach works well when discriminant metabolites are quite concentrated in the sample as, for instance, in the case of primary metabolites. An MS-based approach is more suited when discriminant metabolites are present at low concentrations. In all cases, the metabolomic approach appears to be one of the methods with high potential for allowing traceability of foods of plant origin and authentication of derived products.

3.1.4 Determination of geographical origin using Near Infra-Red spectrometry (NIR)

NIR relies on the properties of the absorption of light radiation by molecules of organic compounds contained in plants. This analytical technology is based on the vibrational properties of molecules. The constraints and limits of NIR can be summarized in the need to establish robust spectral bases and specific calibrations per product and component. It is often unsuitable for the determination of traces and residues. Its main advantages are its speed (from a few msec to 1 min), its precision and its ease of implementation. This technique is non-destructive, non-polluting, quantitative and qualitative. In addition, it can be used in the field, on a large number of samples of different forms. Vermeulen et al. (2018) recommended the use of NIR to track agricultural products and food. The technique has also been used for the detection of unwanted substances in food (FernándezPierna et al. 2012) or for qualitative analyses, such as to determine the amounts of phenolic compounds in fresh apples (Pissard et al. 2018). Portable NIR detectors have been developed by several researchers for applications on foods of plant origin, e.g., pineapple authentication (Amuah et al. 2019), phenotypic characterization of sweet potato (yam) and cassava (Alamu et al. 2020), assessment of shelf life and ripening of fruits (Nordey et al. 2019) or biochemical phenotyping of green coffee during ripening (Dos Santos Scholz et al. 2014). The use of the technique now in combination with modelling techniques (Zareef et al. 2020) and chemometrics (Zalacain et al. 2005) has started finding more and more applications in the industry and more research is expected to follow to cover more plant derived products and models/algorithms.

3.1.5 Determination of geographical origin via analysis of microbial ecology using molecular methods

Research on the microbial ecology of fresh fruits has proved that the composition and abundance of these microbial communities are influenced by multiple factors such as cultivation type (conventional or organic for example), storage and post-harvest treatments (fertilizers, pesticides). In addition, as microorganisms compete for resources, non-pathogenic microorganisms that naturally develop on the surface of fruits can inhibit certain fruit or tree pathogens. In addition, fruits possess many natural defense mechanisms such as their skin which acts as a barrier against plant pathogens (Bigot et al. 2015).

It was mentioned by Montet et al. (2011) that DNA profiles of bacteria from mandarins and DNA profiles of yeasts and molds from physalis were distinct when the plants grew in different locations and could therefore be used as a biological barcode to verify the origin of produce. Molecular techniques in conjunction with image analysis can be used to link the microbial community structure of a produce with its geographical origin (El Sheikha and Montet 2012). Variations in the structures of microbial communities (bacteria, yeasts, fungi) can be detected using molecular techniques based on 16S, 26S, and 28S rDNA profiles. Such profiles can be generated using PCR-DGGE and an example of a gel created using this method is shown in Fig. 3.

Yeasts and molds extracted via washing physalis fruits originated from Egypt were analyzed to establish a correlation between their microbial profiles and their regional origins in Egypt (El Sheikha et al. 2009, 2011). Physalis yeast and fungal band profiles were specific to each sampling

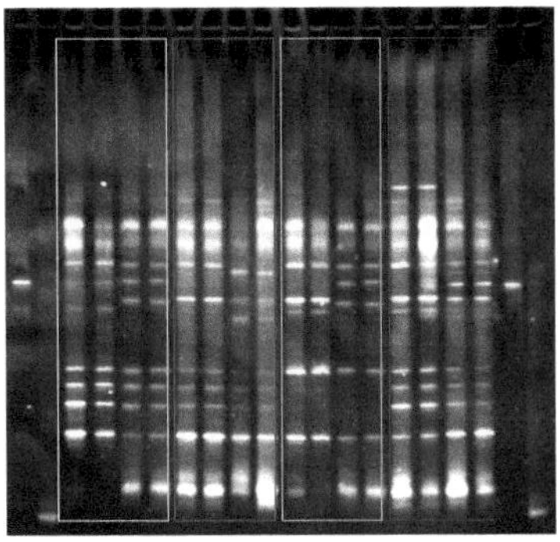

Figure 3. DGGE DNA gel of yeasts extracted from Physalis of different geographical origins (El Sheikha 2010).

site. These bands could be used as a barcode to certify the origin of the fruits (El Shobaky et al. 2015). This method could therefore be an emerging traceability tool that provides a unique organic barcode allowing foods of plant origin to be traced.

Treatment associated with different types of agriculture has a measurable impact on the microflora of foods of plant origin. PCR-DGGE can also therefore be used to discriminate conventional and organic produce using bacterial and fungal rDNAs. Analysis of the microbial genetic profiles of nectarines, peaches, bananas and apples has revealed that it is possible to differentiate fruits according to their production type. Study of "intra-plot" variability has demonstrated that organic fruits could be differentiated from conventionally produced regardless of their position in the plot (center vs. edge) or the tree. The differences observed in the microbial ecology can therefore be large enough to conclude which type of cultivation method was used (Bigot et al. 2015).

3.1.6 Determination of geographical origin via analysis using electronic nose

Devices based on electronic-nose technology have been developed in the past twenty years. This was largely due to the discovery of numerous developments around sensor technology. Recent applications of electronic nose technologies have risen thanks to advances in sensor design, material improvements, software innovations and progress in microcircuitry design and systems integration. The invention of many new e-nose sensor types and arrays, based on different detection principles and mechanisms, is closely correlated with the expansion of new applications (Wilson and Baietto 2009). Statistical analysis software allows researchers to process and interpret the measurements of electronic noses and build qualitative models. They are already used for non-invasive diagnosis, or for the rapid search for harmful compounds or contaminants, in particular within the context of quality control and monitoring of fruit composition (Vidic and Pajot-Augy 2014). Lebrun (2002) for instance studied the different maturation degrees of mangoes of the Lirfa variety, using an electronic nose. This technique was shown to be both sensitive and reliable for the analysis of ripening. Tan and Xu (2020) concluded also that e-nose or e-tongue combining pattern recognition algorithms are very effective, accurate, fast and relatively inexpensive tools that can be used to take both in-line and off-line measurements and could therefore find practical applications in the food industry. As Ali et al. (2020) mentioned however, the use of this technology has certain disadvantages such as the fact that they are not suitable for recognizing individual chemical substances within gas odors or the effects of humidity and temperature to the sensory stimuli. As a result, it is not expected that electronic noses will completely substitute analytical techniques, but further research on the field can improve the methodology around them and enable their wider use in the verification of authenticity of foods of plant origin.

4. Conclusions

Consumer-driven demands and legislative requirements for healthy, "natural" products free from food safety hazards as well as market trends for use of more environmentally friendly agricultural methods, have driven scientific research towards the development of robust analytical tools that are able to ensure traceability and can verify the authenticity of food. However, the effectiveness of these methodologies in a wide range of products and applications remains to be verified even if long and precise studies have already been carried out in certain cases. In the case of foods of plant origin, the identification of species can be performed relatively easily in most cases but the determination of the geographical origin of products is a bigger challenge. Generally speaking, scientific research has led to the development of a wide range of analytical methods for food authentication (Montet and Ray 2018). The use of several types of markers (isotopic, metabolomic and proteomic mainly) seems promising, particularly when combined with multivariate statistical analyses. A lot of research is still needed to fully adopt these methods in official inspection systems and in the industry.

Abbreviations

Gas Chromatography	GC
Liquid Chromatography	LC
Mass Spectrometry	MS
Near Infrared Spectroscopy	NIR
Nuclear Magnetic Resonance	NMR

Orthogonal Projections to Latent Structures	OPLS
Partial Least Square Discriminant Analysis	PLS-DA
Polymerase chain reaction-denaturing gradient gel electrophoresis	PCR-DGGE

References

Alamu, E.O., E. Nuwamanya, D. Cornet, K. Meghar, M. Adesokan, T. Tran, J. Belalcazar, L. Desfontaines and F. Davrieux. 2020. Near-Infrared spectroscopy (NIRS) applications for high throughput phenotyping (HTP) for cassava and yam: A review. International Journal of Food Science and Technology 2020: 1–11.

Ali, M.M., N. Hashim, S.A. Aziz and O. Lasekan. 2020. Principles and recent advances in electronic nose for quality inspection of agricultural and food products. Trends in Food Science & Technology 99: 1–10.

Amuah, C.L.Y., E. Teye, F.P. Lamptey, K. Nyandey, J. Opoku-Ansah and P. Osei-WusuAdueming. 2019. Feasibility study of the use of handheld NIR spectrometer for simultaneous authentication and quantification of quality parameters in intact pineapple fruits. Journal of Spectroscopy 5975461: 1–9.

Augagneur, S., B. Médina and F. Grousset. 1996. Measurement of lead isotope ratios in wine by ICP-MS and its applications to the determination of lead concentration by isotope dilution. Fresenius Journal of Analytical Chemistry 357: 1149–1152.

Avisé. 2019. https://www.avise-info.fr/informations-entreprise/comment-obtenir-un-code-barres-pour-mes-produits. Accessed 22.12.2020.

Barbour, M.M. 2007. Stable oxygen isotope composition of plant tissue: a review. Functional Plant Biology 34: 83–94.

Bart, H.J., B.V. Tangahu, S.R. Sheikh Abdullah, H. Basri, M. Idris, N. Anuar and M. Mukhlisin. 2011. A review on heavy metals (As, Pb, and Hg) uptake by plants through phytoremediation. International Journal of Chemical Engineering 939161: 1–31.

Bigot, C., J.C. Meile, A. Kapitan and D. Montet. 2015. Discriminating organic and conventional foods by analysis of their microbial ecology: An application on fruits. Food Control 18: 123–129.

Bontempo, L., R. Caruso, M. Fiorillo, G.L. Gambino, M. Perini, M. Simoni, P. Traulo, R. Wehrens, G. Gagliano and F. Camin. 2014. Stable isotope ratio of H, C, N, and O in Italian citrus juices. Journal of Mass Spectrometry 49: 785–791.

Camin, F., M. Boner, L. Bontempo, C. Fauhl-Hassek, S.D. Kelly, J. Riedl and A. Rossmann. 2017. Stable isotope techniques for verifying the declared geographical origin of food in legal cases. Trends in Food Science & Technology 61: 176–187.

Carter, C., B. Krissoff and A.P. Zwane. 2006. Can country-of-origin labeling succeed as a marketing tool for produce? Lessons from three case studies. Canadian Journal of Agricultural Economics 54(4): 513–530.

Científico, A., G. Torrivilla, C. Hernández, V. Barka, M. Mora, C. Gonzalez, L. Hernández, and F. Vaillant. 2017. Analysis of juice blackberry (*Rubusadenotrichos*) sweetened with stevia (*Stevia rebaudina Bertoni*), A metabolomic approach. Revista U D C A Actualidad & Divulgación Científica 20(1): 121–129.

Craig, H. 1961. Standard for reporting concentrations of deuterium and oxygen-18 in natural waters. Science 133: 1833–1834.

Dennis, J. 1998. Recent developments in food authentication. Analyst 123: 151R–156R.

Díaz R., O.J. Pozo, J.V. Sancho and F. Hernández. 2014. Metabolomic approaches for orange origin discrimination by ultra-high performance liquid chromatography coupled to quadrupole time-of-flight mass spectrometry. Food Chemistry 157: 84–93.

Dickson, L., M. Tenon, L. Svilar, P. Fança-Berthon, J.C. Martin, H. Rogez and F. Vaillant. 2020. Genipap (*Genipa americana* L.) juice intake biomarkers after medium-term consumption. Food Research International 137: 1–8.

Doner, L.W. 1985. Carbon isotope ratios in natural and synthetic citric acid as indicator of lemon juice adulteration. Journal of Agricultural and Food Chemistry 33: 770–772.
Dos Santos Scholz, M.B., C.S. Good Kitzberger, L.F. Protasio Pereirâ, F. Davrieux, D. Pot, P. Charmetant and T. Leroy. 2014. Application of near infrared spectroscopy for green coffee biochemical phenotyping. Journal of Near Infrared Spectroscopy 22(6): 411–421.
El Sheikha, A.F. 2010. Détermination de l'origine géographique des fruits du Karité et du Physalis par l'utilisation d'empreintes génétiques. Étude de la communauté microbienne par PCR/DGGE. Analyse des activités biologiques d'extraits de fruits. PhD Université de Montpellier II, 13 December 2010.
El Sheikha, A.F., A. Condur, I. Métayer, D.D. Le Nguyen, G. Loiseau and D. Montet. 2009. Determination of fruit origin by using 26S rDNA fingerprinting of yeast communities by PCR-DGGE: preliminary application to Physalis fruits from Egypt. Yeast 26(10): 567–573.
El Sheikha, A.F., I. Métayer and D. Montet. 2011. A Biological barcode for determining the geographical origin of fruit by using 28S rDNA fingerprinting of fungi communities by PCR-DGGE: an application to Physalis fruits from Egypt. Food Biotechnology 25(2): 115–129.
El Sheikha, A.F. and D. Montet. 2012. The determination of geographical origin of foodstuffs by using innovative biological bar-code. Journal of Life Sciences 6: 1334–1342.
El Shobaky, A., J.C. Meile and D. Montet. 2015. New traceability strategies based on a biological bar code by PCR-DGGE using bacterial & yeast communities for determining farming type of peach. Egyptian Journal of Basic and Applied Sciences 2(4): 327–333.
Elsaadani, M., N. Durand, B. Sorli, B. Guibert, P. Alter and D. Montet. 2020. Aptamer assisted ultrafiltration cleanup with high performance liquid chromatography-fluorescence detector for the determination of OTA in green coffee. Food Chemistry 310(125851): 1–6.
Epova, E.N., S. Bérail, F. Séby, V. Vacchina, G. Bareille, B. Médina, L. Sarthou and O.F.X. Donard. 2019. Strontium elemental and isotopic signatures of Bordeaux wines for authenticity and geographical origin assessment. Food Chemistry 294: 35–45.
Famulok, M. and G. Mayer. 2014. Aptamers and SELEX in Chemistry & Biology. Chemistry and Biology 21(9): 1055–1058.
FAO. 2020. http://www.fao.org/ag/againfo/themes/fr/meat/quality_trace.html (Accessed 23.12.2020)
Fayek, N.M., M.A. Farag and F.R. Saber. 2021. Metabolome classification via GC/MS and UHPLC/MS of olive fruit varieties grown in Egypt reveal pickling process impact on their composition. Food Chemistry 339(127861): 1–14.
FernándezPierna, J.A., P. Vermeulen, O. Amand, A. Tossens, P. Dardenne and V. Baeten. 2012. NIR hyperspectral imaging spectroscopy and chemometrics for the detection of undesirable substances in food and feed. Chemometrics and Intelligent Laboratory Systems 117: 233–239.
Fortunato, G., K. Mumic, S. Wunderli, L. Pillonel, J.O. Bosset and G. Gremaud. 2004. Application of strontium isotope abundance ratios measured by MC-ICP-MS for food authentication. Journal of Analytical Atomic Spectrometry 19: 227–234.
Gilbert, A., V. Silvestre, N. Segebarth, G. Tcherkez, C. Guillou, R.J. Robins, S. Akoka and G.S. Remaud. 2011. The intramolecular ^{13}C-distribution in ethanol reveals the influence of the CO2-fixation pathway and environmental conditions on the site-specific ^{13}C variation in glucose. Plant, Cell & Environment 34(7): 1104–1112.
Guyon, F., L. Gaillard, A. Brault, N. Gaultier, M.H. Salagoity and B. Medina. 2013. Potential of ion chromatography coupled to isotope ratio mass spectrometry via a liquid interface for beverages authentication. Journal of Chromatography A 1322: 62–68.
Guyon, F., P. Auberger, L. Gaillard, C. Loublanches, M. Viateau, N. Sabathie, M.H. Salagoity and B. Medina. 2014. ^{13}C/^{12}C isotope ratios of organic acids, glucose and fructose determined by HPLC-co-IRMS for lemon juices authenticity. Food Chemistry 146: 36–40.

Guyon, F., L. Gaillard, N. Sabathié and M.H. Salagoity. 2015a. Direct determination of the oxygen 18 stable isotope ratio of *in situ* water contained in pasty matrices. Analytical Methods 7: 3211–3214.

Guyon, F., K. van Leeuwen, L. Gaillard, M. Grand, S. Akoka, G.S. Remaud, N. Sabathie and M.H. Salagoity. 2015b. Comparative study of ^{13}C composition in ethanol and bulk dry wine using isotope ratio monitoring by mass spectrometry and by nuclear magnetic resonance as an indicator of vine water status. Analytical and Bioanalytical Chemistry 407: 9053–9060.

Guyon, F., F. Vaillant and D. Montet. 2020. Traceability of fruits and vegetables. Phytochemistry 173(112291): 1–7.

Jamin, E., R. Guerin, M. Retif, M. Lees and G.J. Martin. 2003. Improved detection of added water in orange juice by simultaneous determination of the oxygen-18/oxygen-16 isotope ratios of water and ethanol derived from sugars. Journal of Agricultural and Food Chemistry 51: 5202–5206.

Jia, M., J. Sha, Z. Li, W. Wang and H. Zhang. 2020. High affinity truncated aptamers for ultra-sensitive colorimetric detection of bisphenol A with label-free aptasensor. Food Chemistry 317: 1–9.

Kelly, S.D. 2003. Using stable isotope ratio mass spectrometry (IRMS) in food authentication and traceability. pp. 156–183. *In*: Lees, M. [ed.]. Food Authenticity and Traceability. Woodhead Publishing Limited, Cambridge, England.

Lebrun, M. 2002. Study of mango maturity through volatiles compounds: use of an electronic nose. Mémoire de master Natura, ENSIA Montpellier.

Lynn, W.C., R.J. Ahrens and A.L. Smith. 2002. Soil minerals, their geographic distribution, and soil taxonomy. pp. 691–709. *In*: Dixon, J.B. and D.G. Schulze [eds.]. Soil Mineralogy with Environmental Applications, volume 7. Soil Science Society of America Inc. Madison, Wisconsin, USA.

Madesis, P., I. Ganopoulos, I. Sakaridis, A. Argiriou and A. Tsaftaris. 2014. Advances of DNA-based methods for tracing the botanical origin of food products. Food Research International 60: 163–172.

Magdas, A., A. Dehelean and P. Romulus. 2012. Isotopic and elemental determination in some Romanian apple fruit juices. The Scientific World Journal 2012: 1–7.

Magdas, D.A., I. Feher, A. Dehelean, G. Cristea, T.M. Magdas, R. Puscas and O. Marincas. 2018. Isotopic and elemental markers for geographical origin and organically grown carrots discrimination. Food Chemistry 267: 231–239.

Médina, B., S. Augagneur, M. Barbaste, F.E. Grousset and P. Buat-Ménard. 2000. Influence of atmospheric pollution on the lead content of wines. Food Additives and Contaminants 17(6): 435–445.

Mohammadian, M.A., M.H. Rezadoost and S.F. Fallah. 2018. Comparative analysis and innovation of a simple and rapid method for high-quality RNA and DNA extraction of kiwifruit. MethodsX 5: 352–361.

Montet, D. and A.F. El Sheikha. 2011. Universal biological bar-code for the determination of geographical origin of foodstuffs. Mansoura Journal of Biology 37(2): 15–34. https://agritrop.cirad.fr/554363/.

Montet, D. and R.C. Ray. 2018. Editors: Food Traceability and Emerging technologies. In Food Biology series, Science Publishers/CRC Press.

Nordey, T., F. Davrieux and M. Lechaudel. 2019. Predictions of fruit shelf life and quality after ripening: Are quality traits measured at harvest reliable indicators? Postharvest Biology and Technology 153: 52–60.

Pacifico, D., L. Casciani, M. Ritota, G. Mandolino, C. Onofri, A. Moschella, B. Parisi, C. Cafiero, and M. Valentini. 2013. NMR-based metabolomics for organic farming traceability of early potatoes. Journal of Agricultural and Food Chemistry 61(46): 11201–11211.

Peres, B., N. Barlet, G. Loiseau and D. Montet. 2007. Review of the current methods of analytical traceability allowing determination of the origin of foodstuffs. Food Control 18: 228–235.

Pissard, A., V. Baeten, P. Dardenne, P. Dupont and M. Lateur. 2018. Use of NIR spectroscopy on fresh apples to determine the phenolic compounds and dry matter content in peel and flesh. Biotechnology, Agronomy, Society and Environment 22(1): 3–12.

Rummel, S., S. Hoelzl, P. Horns, A. Rossmann and C. Schlicht. 2010. The combination of stable isotope abundance ratios of H, C, N, and S with $^{87}Sr/^{86}Sr$ for geographical origin assignment of orange juices. Food Chemistry 118: 890–900.

Shujing, L., H. Yaxi, L. Wei, C. Ying, W. Fei, L. Xiaonan and Z. Wenjie. 2020. Untargeted volatile metabolomics using comprehensive two-dimensional gas chromatography-mass spectrometry—A solution for orange juice authentication, Talanta 217(121038): 1–10.

Stoltenburg, R., C. Reinemann and B. Strehlitz. 2007. SELEX—a (r)evolutionary method to generate high-affinity nucleic acid ligands. Biomolecular Engineering 24(4): 381–403.

Tan, J. and J. Xu. 2020. Applications of electronic nose (e-nose) and electronic tongue (e-tongue) in food quality-related properties determination: A review. Artificial Intelligence in Agriculture 4: 104–115.

Uhao, L., L. Lu, X. Chunhui, Y. Tianming and W. Yuanxing. 2020. UPLC-Q-TOF/MS-based untargeted metabolomics for discrimination of navel oranges from different geographical origins of China. LWT 137: 1–10.

Verbeke, W. and J. Roosen. 2009. Market differentiation potential of country-of-origin, quality and traceability labeling. Estey Centre Journal of International Law and Trade Policy 10(1): 20–35.

Vermeulen, P., J.A. Fernandez Pierna, O. Abbas, H. Rogez, F. Davrieux and V. Baeten. 2018. Authentication and traceability of agricultural and food products using vibrational spectroscopy. pp. 298–331. *In*: Montet, D. and R.C. Ray [eds.]. Food Traceability and Authenticity: Analytical Techniques. CRC Press, Boca Raton, USA.

Vidic, J. and E. Pajot-Augy. 2014. Techniques de l'ingénieur. https://www.techniques-ingenieur.fr/actualite/articles/nanobiosenseurs-les-nez-electroniques-de-derniere-generation-25582/(Accessed 23.12.2020).

Wei, Q.Z., W.Y. Fu, Y.Z. Wang, X.D. Qin, J. Wang, J. Li, Q.F. Lou and J.F. Chen. 2016. Rapid identification of fruit length loci in cucumber (*Cucumis sativus* L.) using next-generation sequencing (NGS)-based QTL analysis. Scientific Reports 6(27496): 2–11.

Wilson, A.D. and M. Baietto. 2009. Applications and Advances in Electronic-Nose Technologies. Sensors 9: 5099–5148.

Xu, L., Z. Xu and I. Strashnov. 2020. Use of information dependent acquisition mass spectra and sequential window acquisition of all theoretical fragment-ion mass spectra for fruit juices metabolomics and authentication. Metabolomics 16: 1–10.

Zalacain, A., S.A. Ordoudi, E.M. Díaz-Plaza, M. Carmona, I. Blázquez, M.Z. Tsimidou and G.L. Alonso. 2005. Near-Infrared spectroscopy in saffron quality control: determination of chemical composition and geographical origin. Journal of Agricultural and Food Chemistry 53(24): 9337–9341.

Zareef, M., Q. Chen, M.M. Hassan, M. Arslan, M.M. Hashim, W. Ahmad, F.Y.H. Kutsanedzie, and A. Agyekum. 2020. An overview on the applications of typical non-linear algorithms coupled with NIR spectroscopy in food analysis. Food Engineering Reviews 12: 173–190.

Zhang, C. and Y.J. Hao. 2020. Advances in genomic, transcriptomic, and metabolomic analyses of fruit quality in fruit crops. Horticultural Plant Journal 6(6): 361–371.

Zhang, J., Q. Yu, H. Cheng, Y. Ge, H. Liu, X. Ye and Y. Chen. 2018. Metabolomic approach for the authentication of berry fruit juice by liquid chromatography quadrupole time-of-flight mass spectrometry coupled to chemometrics. Journal of Agricultural and Food Chemistry 66(30): 8199–8208.

Index

N

O

P

Q

R

S

T

W